# Traditional Machining Technology

# Traditional Machining Technology

## Technology

### Second Edition

Helmi Youssef and Hassan El-Hofy

**CRC Press**
Taylor & Francis Group
Boca Raton London New York

CRC Press is an imprint of the
Taylor & Francis Group, an **informa** business

by CRC Press
6000 Broken Sound Parkway NW, Suite 300, Boca Raton, FL 33487-2742

and by CRC Press
2 Park Square, Milton Park, Abingdon, Oxon, OX14 4RN

© 2021 Taylor & Francis Group, LLC

CRC Press is an imprint of Taylor & Francis Group, LLC

**Library of Congress Cataloging-in-Publication Data**

Names: Youssef, Helmi A., author. | El-Hofy, Hassan, author.
Title: Traditional machining technology / Helmi Youssef and Hassan El-Hofy.
Description: Second edition. | Boca Raton, FL : CRC Press, [2020] |
Includes bibliographical references and index.
Identifiers: LCCN 2020011473 (print) | LCCN 2020011474 (ebook) | ISBN
9780367431334 (hardback) | ISBN 9781003055303 (ebook)
Subjects: LCSH: Machining. | Machine-tools.
Classification: LCC TJ1185 .Y685 2020 (print) | LCC TJ1185 (ebook) | DDC
671.3/5--dc23
LC record available at https://lccn.loc.gov/2020011473
LC ebook record available at https://lccn.loc.gov/2020011474

ISBN: 978-0-367-43133-4 (hbk)
ISBN: 978-1-003-05530-3 (ebk)

Typeset in Times
by Deanta Global Publishing Services, Chennai, India

# Dedication

---

*To our grandsons and granddaughters,*

*Helmi Youssef: Youssef, Nour, Anourine, Fayrouz, and Yousra*

*Hassan El-Hofy: Omar, Zainah, Youssef,*
*Hassan, Hana, Ali, and Hala*

# Contents

# Preface

*Traditional Machining Technology* consists of 11 chapters. Every chapter has been updated, emphasizing information on the relevant topics and providing a comprehensive description of machining technologies related to metal shaping by material removal techniques, from the basic to the most advanced, in today's industrial applications. This book is, therefore, a fundamental textbook for undergraduate students enrolled in production, materials and manufacturing, industrial, and mechanical engineering programs. Students from other disciplines can also use this book while taking courses in the area of manufacturing and materials engineering. This book covers the technologies, machine tools, and operations of several traditional machining processes. The treatment of the different subjects has been developed from the basic principles of traditional machining processes, machine-tool elements, and control systems. Solved examples, problems, and review questions are also provided.

The book describes the fundamentals, basic elements, and operations of general-purpose machine tools used for the production of cylindrical and flat surfaces by turning, drilling and reaming, shaping and planing, and milling processes. Special-purpose machines and operations used for thread cutting, gear cutting, and broaching processes are also dealt with. Semiautomatic, automatic, NC, and CNC machine tools, operations, tooling, mechanisms, accessories, tool jigs, and work fixation are discussed. Abrasion and abrasive finishing machine tools and operations such as grinding, honing, superfinishing, and lapping are described. Modern machine tools and operations, dynamometers, and hexapod machine tools and processes are described. The topics covered throughout the chapters reflect the rapid and significant advances that have occurred in various areas in machining technologies, and they are organized and described in such a manner as to attract the interest of students. The treatments throughout the chapters of the book are aimed at motivating and challenging students to explore technically and economically viable solutions to a variety of important questions regarding the optimum selection of machining operation for a given task.

In Chapter 1, the history and progress of machining, aspects of machining technology, and the basic motions in machine tools are introduced. A classification of machine tools and operations in addition to the basic motions of machining operations is also given.

Chapter 2 introduces the design considerations and requirements of machine tools, and basic elements such as beds, structures, frames, guideways, spindles and shafts, stepped and stepless drives, planetary transmission, machine-tool motors, couplings, and brakes. Material selection and heat treatment of machine-tool elements, and the testing and maintenance of machine tools, are dealt with.

Chapter 3 covers general-purpose metal-cutting machine tools, including lathes, drilling, reaming, jig-boring machines, milling machines, and machine tools of a reciprocating nature, such as shapers, planers, and slotters. Machine-tool elements, mechanisms, tooling, accessories, and operations are also explained.

Chapter 4 presents abrasion machine tools, including grinding and surface finishing machines and processes.

Chapter 5 describes the different types and applications of commonly used screw threads. Thread machining methods by cutting and grinding are described together with thread cutting machines and cutting tools.

In Chapter 6, common types of gears are given, and their applications are described. Gear productions by machining methods that include cutting, grinding, and lapping are described together with their corresponding machine tools and operations.

Chapter 7 describes capstan and turret lathes. Machine components, features, and applications are described. Tool layouts for bar-type capstan lathes and chucking-type turret lathes are described, and solved examples are given.

Semiautomatic and automatic lathes are covered in Chapter 8. Machine-tool features, components, operation, tooling, and industrial applications are described. Solved examples for typical products showing process layout and cam design are given for turret-type and long part automatics.

Chapter 9 presents computer numerical machine tools, their merits, and their industrial applications. The basic features of such machines, tooling arrangements, and programming principles and examples are illustrated in the case of machining and turning centers. An introduction to computer-assisted and CAD/CAM applications in part programming is also given.

Chapter 10 covers the fundamentals and applications of computer-integrated manufacturing, lean production, adaptive control, just-in-time manufacturing systems, smart manufacturing, artificial intelligence, and the factory of the future.

Chapter 11 describes the fundamentals, instrumentation, and operation of machine-tool dynamometers used for cutting force measurements. Examples of turning, drilling, milling, and grinding dynamometers are explained.

# Acknowledgments

Many individuals have contributed to the development of the second edition of this book. It is a pleasure to express our deep gratitude to Professor Dr. Ing. A. Visser, Bremen University, Germany, for supplying valuable materials during the preparation of this new edition. We would like to appreciate the efforts of Dr. Khaled Youssef for his continual assistance in tackling software problems during the preparation of the manuscript. Special thanks are offered to Saied Teileb of Lord Alexandria Razor Company for his fine Auto-CAD drawings.

Heartfelt thanks are due to our families for their great patience, support, encouragement, enthusiasm, and interest during the preparation of the manuscript. We extend our heartfelt gratitude to the editorial and production staff at Taylor & Francis Group for their efforts to ensure that this book is as accurate and as well-designed as possible.

We very much appreciate the permissions from all publishers to reproduce many illustrations from a number of authors as well as the courtesy of many industrial companies that provided photographs and drawings of their products to be included in this new edition of the book. Their generous cooperation is a mark of sincere interest in enhancing the level of engineering education. The credits for all this great help are given in the captions under the corresponding illustrations, photographs, and tables. It is with great pleasure that we, therefore, acknowledge the help of the following organizations:

Alexandria Engineering Journal, Alexandria, Egypt.
American Society of Mechanical Engineers.
ASM International, Materials Park, OH.
Cassell and Co. Ltd., London, UK.
Chapman and Hall, London, UK.
CIRP, Paris, France.
Dar Al-Maaref Publishing Co., Alexandria, Egypt.
El-Fath Press, Alexandria, Egypt.
Elsevier Ltd, Oxford, UK.
Hodder and Stoughton Educational, London, UK.
Industrial Press Inc., New York, NY.
Industrie-Anzeiger, Aachen, Germany.
John Wiley & Sons, Inc., New York, NY.
Khanna Publisher, Delhi, India.
Leuze Verlag, Bad Saulgau, Germany.
Machinability Data Center, Cincinnati, OH.
Marcel Dekker Inc., New York, NY.
McGraw Hill Co., NY, NY.
Mir Publishers, Moscow.
Oxford University Press, UK.

Pearson Education, Inc., NJ.
Peter Peregrines Ltd, Stevenage, UK.
Prentice Hall Publishing Co., New York, NY.
SME, Dearborn, MI.
Springer Verlag, Berlin, Germany.
Tata McGraw Hill Co., New Delhi, India.
TH-Aachen, Germany.
TH-Braunschweig, Germany.
VDI Verlag, Düsseldorf, Germany.
VEB-Verlag Technik, Berlin, Germany.
Alfred Herbert Ltd., Coventry, UK.
All Metals & Forge Group, USA.
American Gear Manufacturing Association (AGMA), USA.
British Stainless Steel Association, Sheffield, UK.
Carpenter Technology Corporation, Philadelphia, PA.
Charmilles Technologies, Geneva, Switzerland.
Cincinnati Machines, OH.
DeVlieg Machine Co., MI.
Encyclopedia Britannica, Inc., UK.
Falcon Metals Group, Waldwick, NJ.
Geodetic Inc., Melbourne, FL.
Hardinge Inc., Berwyn PA.
Heald Machine Company, Worcester, MA.
Heinemann Machine Tool Works-Schwarzwald, Germany.
Herbert Machine Tools Ltd, UK.
High Performance Alloys, Inc., USA.
Hoffman Co., Carlisle, PA.
Hottinger-Baldwin Meß-technik, Darmstadt, Germany.
Index-Werke AG, Esslingen/Neckar, Germany.
Indian Institute of Technology, Kanpur, India.
Ingersoll Waldrich Siegen Werkzeugmachinen, GmbH, Germany.
Kennametal Incorporation, Pittsburgh, PA.
Kistler Instrumente AG, Switzerland.
Krupp, Widia, GmbH, Essen, Germany.
Lehfeld Works, Heppenheim, Germany.
Liebherr Verzahntechnik, Kempten, Germany.
MAZAK Corporation, Florence, KY.
MG Industries/Steigerwald, Berlin, Germany.
Mitsubishi EDSCAN, Toyohashi, Japan.
Nassovia-Krupp, Werkzeugmaschinenfabrik, Langen/Frankfurt, Germany.
ONSRUD tool company, USA.
PittlerMaschinenfabrik AG, Langenbei Frankfurt/M, Germany.
Sandvik Coromant, Sweden.
Seco Tools, UAE.
Standard Tool Co., Athol, MA.
Thermal-Dynamic Corp., Chesterfield, MO.
VEB-Drehmaschinenwerk/Leipzig, Germany.
WMW Machinery Co, New York, NY.

# Author Biographies

 **Helmi Youssef**, born in August, 1938 in Alexandria, Egypt, acquired his BSc degree with honors in Production Engineering from Alexandria University in 1960. He then consolidated his scientific experience in the Carolo-Welhelmina, TH Braunschweig, in Germany during the period 1961–1967. In June 1964, he acquired his Dipl.-Ing. degree, and in December 1967, he completed his Dr.-Ing. degree in the domain of Nontraditional Machining. In 1968, he returned to Alexandria University, Production Engineering Department, as an assistant professor. In 1973, he was promoted to associate, and in 1978, to full professor. In the period 1995–1998, Professor Youssef was the chairman of the Production Engineering Department, Alexandria University. Since 1989, he has been a member of the scientific committee for the promotion of professors in Egyptian universities.

Based on several research and educational laboratories, which he had built, Professor Youssef founded his own scientific school in both Traditional and Nontraditional Machining Technologies. In the early 1970s, he established the first Nontraditional Machining Technologies research laboratory in Alexandria University, and maybe in the whole region. Since that time, he has carried out intensive research in his fields of specialization and has supervised many PhD and MSc theses.

Between 1975 and 1998, Professor Youssef was a visiting professor in Arabic universities, such as El-Fateh University in Tripoli, the Technical University in Baghdad, King Saud University in Riyadh, and Beirut Arab University in Beirut. Beside his teaching activities in these universities, he established laboratories and supervised many MSc theses. Moreover, he was a visiting professor in different academic institutions in Egypt and abroad. In 1982, he was a visiting professor in the University of Rostock, Germany, and during the years 1997–1998, he was a visiting professor in the University of Bremen, Germany.

Professor Youssef has organized and participated in many international conferences. He has published many scientific papers in specialized journals. He has authored many books in his fields of specialization, two of which are single authored. The first is in Arabic, titled *Nontraditional Machining Processes, Theory and Practice*, published in 2005, and the other is titled *Machining of Stainless Steels and Superalloys, Traditional and Nontraditional Techniques*, published by Wiley in 2016. Another two coauthored books were published by CRC in 2008 and 2011, respectively. The first is on *Machining Technology*, while the second deals with *Manufacturing Technology*.

Currently, Professor Youssef is an emeritus professor in the Production Engineering Department, Alexandria University. His work currently involves developing courses and conducting research in the areas of metal cutting and nontraditional machining.

 **Hassan El-Hofy** received his BSc in Production Engineering from Alexandria University, Egypt in 1976 and his MSc in 1979. Following his MSc, he worked as an assistant lecturer in the same department. In October 1980, he left for Aberdeen University in Scotland, UK and began his PhD work with Professor J. McGeough in electrochemical discharge machining. He won the Overseas Research Student award during the course of his doctoral degree, which he duly completed in 1985. He then returned to Alexandria University and resumed his work as an assistant professor. In 1990, he was promoted to the rank of associate professor. He was on sabbatical as a visiting professor at Al-Fateh University in Tripoli between 1989 and 1994.

In July 1994, he returned to Alexandria University and was promoted to the rank of full professor i November 1997. From September 2000, he worked as a professor for Qatar University. He chaired the accreditation committee for the mechanical engineering program toward ABET Substantial Equivalency Recognition, which was granted to the College of Engineering programs, Qatar University in 2005. He received the Qatar University Award and a certificate of appreciation for his role in that event.

Professor El-Hofy's first book, entitled *Advanced Machining Processes: Nontraditional and Hybrid Processes*, was published by McGraw Hill Co. in 2005. The third edition of his second book, entitled *Fundamentals of Machining Processes—Conventional and Nonconventional Processes*, was published in November 2018 by Taylor & Francis Group, CRC Press. He also coauthored the book entitled *Machining Technology—Machine Tools and Operations*, which was published by Taylor & Francis Group, CRC Press in 2008. In 2011, he released his fourth book, entitled *Manufacturing Technology—Materials, Processes, and Equipment*, which again was published by Taylor & Francis Group, CRC Press. Professor El-Hofy has published over 80 scientific and technical papers and has supervised many graduate students in the area of advanced machining. He serves as a consulting editor to many international journals and is a regular participant in many international conferences.

Between August 2007, and August 2010, he became the chairman of the Department of Production Engineering, Alexandria University. In October 2011, he was nominated as the vice dean for Education and Student's affairs, College of Engineering, Alexandria University. Between December 2012 and February 2018, he was the dean of the Innovative Design Engineering School at the Egypt-Japan University of Science and Technology in Alexandria, Egypt. He worked as acting Vice President of Research from December 2014 to April 2017 at the Egypt-Japan University of Science and Technology. Currently, he is the Professor of Machining Technology at the Department of Industrial and Manufacturing Engineering at Egypt-Japan University of Science and Technology.

# List of Symbols

| Symbol | Definition | Unit |
|--------|------------|------|
| $A$ | Included thread angle | Degree |
| $A_c$ | Uncut chip cross-sectional area | mm$^2$ |
| $a_c$ | Acme thread crest width | mm |
| $a_r$ | Acme thread root width | mm |
| $b$ | Chip-tool contact length | mm |
| $D$ | Diameter | mm |
| $D$ | Diameter of grinding wheel | mm |
| $d_{max}$ | Minimum diameter | mm |
| $d_{min}$ | Maximum diameter | mm |
| $d_a$ | Addendum diameter of gear | mm |
| $D_a$ | Burnishing gear addendum diameter | mm |
| $d_c$ | Fixation hole diameter | mm |
| $d_d$ | Dedendum diameter | mm |
| $D_0$ | Depth of thread | mm |
| $d_p$ | Pitch circle diameter | mm |
| $d_R$ | Diameter of regulating wheel | mm |
| $E$ | Young's modulus | MPa |
| $EI$ | Flexural rigidity | N mm$^2$ |
| $e_m$ | Hydraulic motor eccentricity | mm |
| $e_p$ | Hydraulic pump eccentricity | mm |
| $f$ | Feed rate | mm/rev |
| $F$ | Force | N |
| $F_a$ | Axial force | N |
| $F_b$ | Bending strength | N/mm$^2$ |
| $F_c$ | Main cutting force | N |
| $f_e$ | Frequency of exciting vibration | s$^{-1}$ |
| $F_f$ | Feed force | N |
| $f_n$ | Natural frequency | s$^{-1}$ |
| $F_r$ | Radial force | N |
| $F_r$ | Resultant cutting force | N |
| $f_r$ | Frequency | s$^{-1}$ |
| $f_{max}$ | Maximum feed rate | mm/rev |
| $F_x$ | Horizontal (passive) force | N |
| $F_y$ | Vertical force | N |
| $F_z$ | Feed force in drilling | N |
| $h$ | Chip thickness | mm |
| $h_a$ | Addendum | mm |
| $h_d$ | Dedendum | mm |
| $h_o$ | Height of thread fundamental triangle | mm |
| $h_t$ | Tooth height | mm |

| Symbol | Definition | Unit |
|---|---|---|
| $h_w$ | Working depth | mm |
| $i$ | Ratio | — |
| $i_{cg}$ | Gearing ratio | — |
| $I$ | Moment of inertia | mm$^4$ |
| $i_f$ | Transmission ratio of feed gear | — |
| $i_r$ | Transmission ratio of speed change gear | — |
| $i_x$ | Transmission ratio of indexing gear | — |
| $I_X$ | Depth of cut in the $X$-axis | mm |
| $i_y$ | Transmission ratio of differential gear | — |
| $J$ | Radius of hole to be milled minus cutter radius | mm |
| $k$ | Static stiffness | N/mm$^2$ |
| $K$ | Spring constant | M/Nm |
| $k_g$ | Gauge factor | — |
| $K_r$ | Coefficient of magneto-mechanical coupling | — |
| $K_t$ | Thread height | mm |
| $L$ | Stroke length in shaper | mm |
| $Lw$ | Lead of gear | mm |
| $l_d$ | Cantilever length | mm |
| $l_g$ | Position of displacement gauge | mm |
| $l_o$ | Maximum stock available for sharpening | mm |
| $l_r$ | Length of ring transducer | mm |
| $L_t$ | Tool travel | mm |
| $M$ | Mobility | — |
| $m$ | Mass | kg |
| $m_g$ | Module of gear | mm |
| $m_n$ | Normal module of gear | mm |
| $M_s$ | Moment at position $s$ | Nmm |
| $M_x$ | Bending moment distribution due to horizontal force | Nmm |
| $M_y$ | Bending moment distribution due to vertical force | Nmm |
| $M_z$ | Drilling torque | Nmm |
| $n$ | Rotational/reciprocating speed | rpm or stroke/min |
| $ng$ | Grinding wheel rotational speed | rpm |
| $n_R$ | Regulating wheel rotational speed | rpm |
| $N$ | Number of threads per inch | — |
| $n_{aux}$ | Auxiliary shaft rotational speed | rpm |
| $n_{cam}$ | Camshaft rotational speed | rpm |
| $n_m$ | Motor speed | rpm |
| $n_{max}$ | Maximum rotational speed | rpm |
| $n_{min}$ | Minimum rotational speed | rpm |
| $n_r$ | Reverse spindle speed | rpm |
| $n_s$ | Spindle speed | rpm |
| $p$ | Pitch | mm |
| $P_h$ | Honing stone pressure | kgf/cm$^2$ |
| $Q$ | Cutting/return speed ratio in shaping | — |
| $R$ | Resistance | $\Omega$ |

| Symbol | Definition | Unit |
|---|---|---|
| $\Delta R$ | Change in resistance | $\Omega$ |
| $R_d$ | Diameter range | — |
| $r_d$ | Displacement ratio | — |
| $R_g$ | Speed ratio | — |
| $R_n$ | Rotational speed range | — |
| $RPT$ | Rise per tooth (superelevation) | mm |
| $R_v$ | Cutting speed range | — |
| $S$ | Tooth thickness | mm |
| $t$ | Depth of cut (time) | mm (s) |
| $T_1$ | Input torque | Nmm |
| $T_2$ | Output torque | Nmm |
| $t_a$ | Auxiliary (idle or nonproductive) time | min |
| $T_{cyc}$ | Cycle time | min |
| $t_f$ | Floor-to-floor time | min |
| $t_{hel}$ | Lead of helical groove | mm |
| $t_{ls}$ | Pitch of lead screw of the lathe | mm |
| $t_m$ | Machining (production) time | min |
| $t_{mh}$ | Machine handling time | min |
| $t_o$ | Thickness | mm |
| $T_s$ | Total depth of material removed in one stroke in broaching | mm |
| $t_s$ | Setup time | min |
| $t_{th}$ | Pitch of thread to be cut on the lathe | mm |
| $t_{wh}$ | Work handling time | min |
| $U$ | Allowance for finishing in $X$-axis | mm |
| $u$ | Feed in milling | mm/min |
| $v$ | Cutting speed | m/min |
| $V_c$ | Cutting speed in shaper | m/min |
| $V_{cm}$ | Mean cutting speed of cutting stroke in shaper | m/min |
| $v_g$ | Peripheral speed of grinding wheel | m/s–m/min |
| $V_1$ | Lower speed | m/min |
| $V_{max}$ | Maximum cutting speed | m/min |
| $V_{min}$ | Minimum cutting speed | m/min |
| $v_p$ | Peripheral speed of regulating wheel | m/s |
| $V_r$ | Return speed in shaper | m/min |
| $v_r$ | Reverse speed | m/min |
| $v_{rc}$ | Reciprocation speed in honing | m/min |
| $v_{rm}$ | Mean return speed of return stroke in shaper | m/min |
| $v_{tr}$ | Workpiece traverse rate | mm/min |
| $v_{rt}$ | Surface rotation speed in honing | m/min |
| $v_t$ | Traverse speed | m/min |
| $V_u$ | Economical speed | m/min |
| $v_w$ | Peripheral speed of workpiece | m/s |
| $X$ | Chip space number | — |
| $Z$ | Number of speeds | — |
| $Z$ | Number of teeth | — |

| Symbol | Definition | Unit |
|---|---|---|
| $Z'$ | Modified number of teeth | |
| $Z_l$ | End position of the groove/thread | mm |
| $z_g$ | Number of speed steps | |

| Symbol (Greek Letters) | Definition | Unit |
|---|---|---|
| $\alpha$ | Angle of cutting stroke of shaper | Degree |
| $\beta$ | Angle of return (noncutting) stroke of shaper | Degree |
| $\delta$ | Deflection | mm |
| $\sigma_e$ | Elastic limit | $N/mm^2$ |
| $\delta_5$ | Elongation | mm |
| $\delta_n$ | Increase in speed | % |
| $\omega_o$ | Natural frequency | Hz |
| $\varphi_p$ | Progression ratio of pole change motor | — |
| $\varphi$ | Progression ratio | — |
| $\sigma_u$ | Ultimate tensile strength | $N/mm^2$ |
| $\alpha_h$ | Half cross-hatch angle (honing) | Degree |
| $\omega_h$ | Helix angle of spiral groove | Degree |
| $\alpha_l$ | Inclination angle of regulating wheel | Degree |
| $\chi$ | Setting (approach) angle | Degree |
| $\varphi$ | Diameter notation | mm |
| $\alpha_t$ | Thread helix angle | Degree |
| $\varphi_c$ | Threading tap chamfer angle | Degree |
| $\beta_g$ | Helix angle (gear) | Degree |
| $\alpha_h$ | Helix angle of the hob | Degree |
| $Y$ | Setting angle (hob) | Degree |
| $\mu$ | Coefficient of friction | — |
| $\varepsilon_s$ | Elastic strain | |

# List of Acronyms

| Abbreviation | Description |
|---|---|
| ac | Alternating current |
| AC | Adaptive control |
| ACC | Adaptive control with constraints |
| ACO | Adaptive control with optimization |
| AI | Artificial intelligence |
| AJM | Abrasive jet machining |
| ANN | Artificial neural network |
| APT | Automatically programmed tools |
| ASA | American Standards Association |
| ASCII | American Standard Code for Information Interchange |
| ASM | American Society of Metals |
| BA | British Association |
| BCD | Binary coded decimal |
| BHN | Brinell hardness number |
| BSW | British Standard Whitworth |
| BUE | Built-up edge |
| CAD | Computer-aided design |
| CAE | Computer-aided engineering |
| CAI | Computer-aided inspection |
| CAM | Computer-aided manufacturing |
| CAPP | Computer-aided process planning |
| CAR | Computer-aided reporting |
| CBN | Cubic boron nitride |
| CCP | Conventional computer program |
| ccw | Counterclockwise |
| CFG | Creep feed grinding |
| CI | Cast iron |
| CIM | Computer-integrated manufacturing |
| CLDATA | Cutter location data |
| CNC | Computer numerical control |
| CPC | Computerized part changer |
| CRT | Cathode ray tube |
| cw | Clockwise |
| DB | Database |
| dc | Direct current |
| DNC | Direct numerical control |
| DXF | Drawing exchange file |
| EBM | Electron beam machining |
| ECH | Electrochemical honing |
| ECM | Electrochemical machining |
| ED | Electrodischarge |

| | |
|---|---|
| **EIA** | Electronics Industry Alliance |
| **ES** | Expert system |
| **EXAPT** | Extended subset of APT |
| **FFT** | Floor-to-floor time |
| **FL** | Fuzzy logic |
| **FMC** | Flexible manufacturing cell |
| **FMS** | Flexible manufacturing system |
| **FOF** | Factory of the future |
| **GAC** | Geometric adaptive control |
| **GT** | Group technology |
| **GW** | Grinding wheel |
| **HAZ** | Heat-affected zone |
| **HB** | Hardness Brinell |
| **HRC** | Hardness Rockwell |
| **HSS** | High-speed steel |
| **ICE** | Internal combustion engine |
| **IMPS** | Integrated manufacturing production system |
| **ISO** | International Organization for Standardization |
| **JIC** | Just-in-case |
| **JIT** | Just-in-time |
| **KB** | Knowledge base |
| **KBS** | Knowledge base system |
| **LBM** | Laser beam machining |
| **LISP** | List processing |
| **LSG** | Low-stress grinding |
| **MCD** | Machine control data |
| **MCU** | Machine control unit |
| **MDI** | Manual data input |
| **MIT** | Massachusetts Institute of Technology |
| **MQL** | Minimum quantity lubrication |
| **MRP** | Material requirements planning |
| **MRR** | Material removal rate |
| **MS** | Manufacturing system |
| **NC** | Numerical control |
| **PROLOG** | Programming logic |
| **PTFE** | Polytetrafluoroethylene |
| **PTP** | Point-to-point |
| **PVD** | Physical vapor deposition |
| **QRM** | Quick-return motion |
| **RETAD** | Rapid exchange of tooling and dies |
| **RPT** | Rise per tooth |
| **RW** | Regulating wheel |
| **SI** | International System of Units |
| **SM** | Smart manufacturing |
| **SMED** | Single-minute exchange of die |
| **SP** | Special precision |

| **UNC** | Unified coarse |
| **UNF** | Unified fine |
| **UP** | Ultraprecision |
| **USM** | Ultrasonic machining |
| **VDU** | Visual display unit |
| **WC** | Tungsten carbide |
| **WIP** | Work in progress |
| **WP** | Workpiece |

# 1 Machining Technology

## 1.1 INTRODUCTION

Manufacturing is the industrial activity that changes the form of raw materials to create products. The derivation of the word *manufacture* reflects its original meaning: to make by hand. As the power of the hand tool is limited, manufacturing is done largely by machinery today. Manufacturing technology constitutes all methods used for shaping the raw metal materials into a final product. As shown in Figure 1.1, manufacturing technology includes plastic forming, casting, welding, and machining technologies. Methods of plastic forming are used extensively to force metal into the required shape. The processes are diverse in scale, varying from the forging and rolling of ingots weighing several tons to the drawing of wires less than 0.025 mm in diameter. Most large-scale deformation processes are performed at high temperatures, so that a minimum of force is needed, and the consequent recrystallization refines the metallic structure. Cold forming is used when smoother surface finish and high-dimensional accuracy are required. Metals are produced in the form of bars or plates. On the other hand, casting produces a large variety of components in a single operation by pouring liquid metals into molds and allowing them to solidify. Parts manufactured by plastic forming, casting, sintering, and molding are often finished by subsequent machining operations, as shown in Figure 1.2.

Machining is the removal of the unwanted material (machining allowance) from the workpiece (WP) so as to obtain a finished product of the desired size, shape, and surface quality. The practice of removal of machining allowance through cutting techniques was first adopted using simple handheld tools made from bone, stick, or stone, which were replaced by bronze or iron tools. Water, steam, and later electricity were used to drive such tools in power-driven metal cutting machines (machine tools). The development of new tool materials opened a new era for the machining industry in which machine tool development took place. Nontraditional machining techniques offered alternative methods for machining parts of complex shapes in hard, stronger, and tougher materials that are difficult to cut by traditional methods. Figure 1.3 shows the general classification of machining methods based on the material removal mechanism.

Compared with plastic-forming technology, machining technology is usually adopted whenever part accuracy and surface quality are of prime importance. The technology of material removal in machining is carried out on machine tools, which are responsible for generating the motions required for producing a given part geometry. Machine tools form around 70% of operating production machines and are characterized by their high production accuracy compared with metal-forming machine tools. Machining activities constitute approximately 20% of the manufacturing activities in the United States.

This book covers the different technologies used for material removal processes in which traditional machine tools and operations are employed. Machine tool

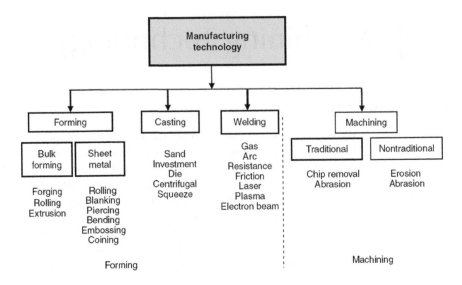

**FIGURE 1.1**   Classification of manufacturing processes.

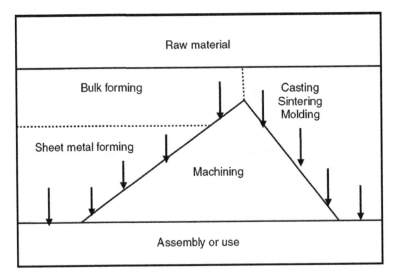

**FIGURE 1.2**   Definition of manufacturing.

elements, drives, and accessories are introduced for proper selection and understanding of their functional characteristics and technological requirements.

## 1.2   HISTORY OF MACHINE TOOLS

The development of metal-cutting machines (once briefly called machine tools) started with the invention of the cylinder, which was changed to a roller guided by a journal bearing. The ancient Egyptians used these rollers for transporting the

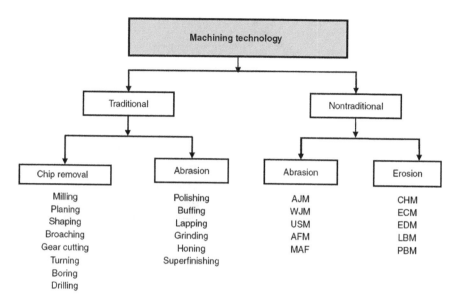

**FIGURE 1.3** Classification of machining processes. AJM, abrasive jet machining; WJM, water jet machining; USM, ultrasonic machining; AFM, abrasive flow machining; MAF, magnetic abrasive finishing; CHM, chemical machining; ECM, electrochemical machining; EDM, electrodischarge machining; LBM, laser beam machining; PBM, plasma beam machining.

required stones from a quarry to the building site. The use of rollers initiated the introduction of the first wooden drilling machine, which dates back to 4000 BC. In such a machine, a pointed flint stone tip acted as a tool. The first deep-hole drilling machine was built by Leonardo da Vinci (1452–1519). In 1840, the first engine lathe was introduced. Maudslay (1771–1831) added the lead screw, back gears, and the tool post to the previous design. Later, slide ways for the tailstock and automatic tool feeding systems were incorporated. Planers and shapers have evolved and were modified by Sellers (1824–1905). Fitch designed the first turret lathe in 1845. That machine carried eight cutting tools on a horizontally mounted turret for producing screws. A completely automatic turret lathe was invented by Spencer in 1896. He was also credited with the development of the multispindle automatic lathe. In 1818, Whitney built the first milling machine; the cylindrical grinding machine was built for the first time by Brown and Sharpe in 1874. The first gear shaper was introduced by Fellows in 1896. In 1879, Pfauter invented the gear hobber, and the gear planers of Sunderland were developed in 1908. Figures 1.4 and 1.5 show the first wooden lathe and planer machine tools.

Further developments for these conventional machines came via the introduction of copying techniques, cams, attachments, and automatic mechanisms that reduced manual labor and consequently increased product accuracy. Machine tool dynamometers are used with machine tools to measure, monitor, and control forces generated during machining operations. Such forces determine the method of holding the tool or WP and are closely related to product accuracy and surface integrity.

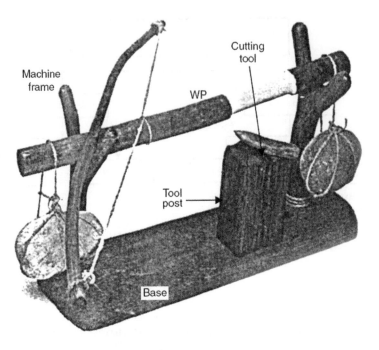

**FIGURE 1.4**  First wooden lathe machine.

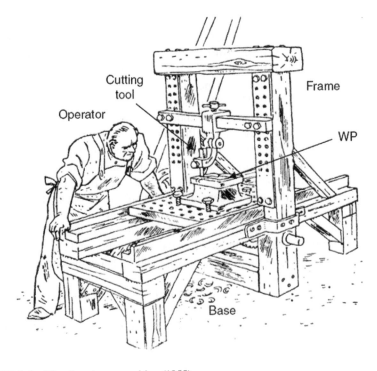

**FIGURE 1.5**  Wooden planer machine (1855).

In 1953, the introduction of numerical control (NC) technology opened doors to computer numerical control (CNC) and direct numerical control (DNC) machining centers that enhanced product accuracy and uniformity. Machine tools have undergone major technological changes through various developments in microelectronics. The availability of computers and microprocessors brought in flexibility that was not possible through conventional mechanisms.

The introduction of hard-to-machine materials has led to the use of nontraditional machining technology for the production of complex shapes in superalloys. Nontraditional machining removes material using mechanical, chemical, or thermal machining effects. Electrochemical machining (ECM) removes material by electrolytic dissolution of the anodic WP. The first patent in ECM was filed by Gussef in 1929. However, the first significant development occurred in the 1950s. Currently, ECM machines are used in automobile, die, mold, and medical engineering industries. Metal erosion by spark discharges was first noted by Sir Joseph Priestly in 1768. In 1943, B. R. Lazerenko and N. I. Lazerenko introduced their first electro-discharge machining (EDM) machine, shown in Figure 1.6. EDM machine tools

**FIGURE 1.6** First industrial EDM machine in the world. Presentation of the Eleroda D1 at the EMO exhibition in Milan, Italy, 1955. (Courtesy of Charmilles, 560 Bond St., Lincolnshire, IL.)

continued to develop through the use of novel power supplies together with computer control of process parameters, which made EDM machines widespread in the manufacturing industries. The use of high-frequency sound waves in machining was noted in 1927 by Wood and Loomis. The first patent for ultrasonic machining (USM), by Balamuth, appeared in 1945. The benefits of USM were realized in the 1950s by the production of related machines. USM machines tackle a wide range of materials, including glass, ceramic, and diamond. The earliest work on using electron beam machining (EBM) was attributed to Steigerwald, who designed the first prototype machine in 1947. Modern EBM machines are now available for drilling, perforation of sheets, and pattern generation associated with integrated circuit fabrication. The laser phenomenon was first predicted by Schawlaw and Townes. Drilling, cutting, scribing, and trimming of electronic components are typical applications of modern laser machine tools. The use of NC, CNC, computer-aided design or computer-aided manufacturing (CAD/CAM), and computer-integrated manufacturing (CIM) technologies provided robust solutions to many machining problems and made nontraditional machine tools widespread in industry. Table 1.1 summarizes the historical background of machine tools.

## 1.3 BASIC MOTIONS IN MACHINE TOOLS

In conventional machine tools, a large number of product features are generated or formed via the variety of motions given to the tool or the WP. The shape of the tool plays a considerable role in the final surface obtained. Basically, there are two types of motions in a machine tool. The primary motion, generally given to the tool or WP, constitutes the cutting speed, while the secondary motion feeds the tool relative to the WP. In some instances, combined primary motion is given either to the tool or to the WP. A classification of machine tool movements used for traditional machining is given in Table 1.2. Table 1.3 gives a classification for nontraditional machining technology. It may be concluded that movements of nontraditional machine tools are simple and mainly in the Z direction, while traditional machine tools have a minimum of two axes, that is, X and Y directions in addition to rotational movements.

## 1.4 ASPECTS OF MACHINING TECHNOLOGY

The machinability is the main aspect on which the machining technology relies. As shown in Table 1.1, the term *machinability* was first suggested by Taylor in the 1920s to describe the machining properties of workpiece materials. Since that time, it has been frequently used but seldom fully explained, as it has a variety of interpretations depending upon the view point of the person using it. Machinability is the relative susceptibility of a material to the machining process. There are various criteria used to evaluate machinability, the most important of which are:

1. Tool life
2. Forces and power
3. Surface finish
4. Ease of chip disposal

## TABLE 1.1
## Developments of Machine Tools

| | |
|---|---|
| 1200–1299 | Horizontal bench lathe appears, using foot treadle to rotate object |
| 1770 | Screw-cutting lathe invented: first to get satisfactory results (Ramsden, Britain) |
| 1810 | Lead screw adapted to lathe, leading to large-quantity machine-tool construction (Maudslay, Britain) |
| 1817 | Metal planing machine (Roberts, Britain) |
| 1818 | Milling machine invented (Whitney, United States) |
| 1820–1849 | Lathes, drilling, boring machines, and planers (most primary machine tools) refined |
| 1830 | Gear-cutting machine with involute cutters and geared indexing improved (Whitworth, Britain) |
| 1830–1859 | Milling machines, shapers, and grinding machines (United States) |
| 1831 | Surface-grinding machine patented (J. W. Stone, United States) |
| 1834 | Grinding machine developed: perhaps first (Wheaton, United States) |
| 1836 | Shaping machine invented; Whitworth soon added crank mechanism (Nasmyth, Britain) |
| 1840 ca. | Vertical pillar drill with power drive and feed in use (originated in 1750) |
| 1842 | Gear-generating machine for cutting cycloidal teeth developed (Saxton, United States) |
| 1850 | Commercially successful universal milling machine designed (Robbins and Lawrence, Howe, and Windsor, United States) |
| 1853 | Surface grinder patented (Darling, United States) |
| 1854 ca. | Commercial vertical turret lathe built for Robbins and Lawrence by Howe and Stone (Stone, Howe, Lawrence, United States) |
| 1857 | Whitney gauge lathe built (Whitney, United States) |
| 1860–1869 | First cylindrical grinder made; replaces single-point tool of engine lathe (United States) |
| 1860–1879 | Universal milling (1861–1865) and universal grinding machines (1876) produced (Brown and Sharpe, United States) |
| 1873 | Automatic screw machine invented (1893, produced finished screws from coiled wire—A2) (Spencer, United States) |
| 1887 | Spur-gear hobbing machine patented (Grant, United States) |
| 1895 | Multispindle automatic lathe introduced for small pieces (United States) |
| 1896–1940 | Heavy-duty precision, high–production rate grinding machine introduced at Brown and Sharpe (Norton, United States) |
| 1920 | The term *machinability* suggested for the first time by Taylor to describe the machining properties of workpiece materials |
| 1921 | First industrial jig borer made for precision machining: based on 1912 single-point tool (Société Genevoise, Switzerland) |
| 1943 | Electrodischarge machining (spark erosion) developed for machine tool manufacturing |
| 1944–1947 | Centerless thread-grinding machine patented (Scrivener, Britain; United States) |
| 1945 | USM patented by Balamuth |
| 1947 | First prototype of EBM designed by Steigerwald |
| 1950 | Electrochemical machines introduced into industry |
| 1952 | Alfred Herbert Ltd.'s first NC machine tool operating |
| 1958 | Laser phenomenon first predicted by Schawlaw and Townes |

From ASM International, 3 Park Ave., New York. With permission.

**TABLE 1.2**

**Tool and WP Motions for Machine Tools Used for Traditional Machining**

| Machining Process | Tool and WP Movements v — ∩ → | | f — → | | Remarks |
|---|---|---|---|---|---|
| Chip removal | | | | | |
| Turning | WP | ∩ | Tool | → | WP stationary |
| Drilling | Tool | ∩ | Tool | → | |
| Milling | Tool | ∩ | WP | → | |
| Shaping | Tool | → | WP | ⋯→ | Intermittent feed |
| Planing | WP | → | Tool | ⋯→ | |
| Slotting | Tool | → | WP | ⋯→ | |
| Broaching | Tool | ● | WP | ● | Feed motion is built into the tool |
| | WP | → | Tool | ● | |
| Gear hobbing | Tool | ∩ | WP | ∩ | |
| | | | Tool | → | |
| Abrasion | | | | | |
|   Surface grinding | Tool | ∩ | WP | → | |
|   Cylindrical grinding | Tool | ∩ | WP | ∩ | |
| | | | Tool or WP | → | |
| Honing | Tool | ∩ → | | ● | WP stationary |
| Superfinishing | WP | ∩ | Tool | → | |

∩, rotation; ●, stationary; →, linear motion; ⋯→, intermittent.

Although machinability generally refers to the work material, it should be recognized that machining performance depends on more than just material. The type of machining operation, tooling, and cutting conditions are also important factors. In addition, the machinability criterion is a source of variation. One material may yield a longer tool life, whereas another material provides a better surface finish. All these factors make the evaluation of machinability difficult. Chip disposal, an additional machinability criterion, should sometimes be seriously considered. Long thin ribbon chips, unless they are broken up with chip breakers, can interfere with the operation, leading to a hazardous cutting area. This criterion is of vital importance in automatic machine tool operation. Chip formation, friction

**TABLE 1.3**

**Tool and WP Motions for Nontraditional Machine Tools**

| Machining Process | WP | | Tool | | Remarks |
|---|---|---|---|---|---|
| | Stationary ● | Feed Movement ↓ | Stationary ● | | |
| Chemical (erosion) | | | | | In the slitting processes (plate cutting), a relative motion between tool and WP (traverse speed $v_t$) is imparted in horizontal directions $(X, Y)$. |
| CHM | ● | | ● | | |
| ECM (sinking) | ● | ↓ | | | |
| Thermal (erosion) | | | | | |
| EDM (sinking) | ● | ↓ | | | |
| EBM (drilling) | ● | | ● | | |
| LBM (drilling) | ● | | ● | | |
| PBM (drilling) | ● | | ● | | |
| Mechanical (abrasion) | | | | | |
| USM | ● | ↓ | | | |
| AJM | ● | | ● | | |
| WJM | ● | | ● | | |
| Abrasive water jet machining (AWJM) | ● | | ● | | |

at the tool/chip interface, and the built-up edge (BUE) phenomenon are determinants of machinability. A ductile material that has a tendency to adhere to the tool face or to form BUE is likely to produce a poor finish. This has been observed to be true with such materials as low-carbon steel, pure aluminum, Cu, and stainless steels.

Machining technology covers a wide range of aspects that should be understood for the proper understanding and selection of a given machining technology. Tooling, accessories, and the machine tool itself determine the nature of the machining operation used for a particular material. As shown on the right-hand side of Figure 1.7, the main objective of the technology adopted is to utilize the selected machining resources to produce the component economically and at high rates of production. Parts should be machined at levels of accuracy, surface texture, and surface integrity that satisfy the product designer and avoid the need for postmachining treatment, which, in turn, maintains acceptable machining costs. The general aspects of machining technology include the following.

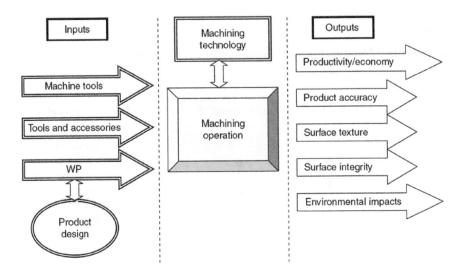

**FIGURE 1.7**   General aspects of machining technology.

### 1.4.1  MACHINE TOOL

Each machine tool is capable of performing several machining operations to produce the part required at the specified accuracy and surface integrity. Machining is performed on a variety of general-purpose machine tools, which in turn perform many operations, including chip removal and abrasion techniques, by which cylindrical and flat surfaces are produced. Additionally, special-purpose machine tools are used to machine gears, threads, and other irregular shapes. Finishing technology for different geometries includes grinding, honing, lapping, and superfinishing techniques.

Figure 1.8 shows general-purpose machine tools used for traditional machining in chip removal and abrasion techniques. Typical examples of general-purpose machine tools include turning, drilling, shaping, milling, grinding, broaching, jig-boring, and lapping machines intended for specific tasks. Gear cutting and thread cutting are examples of special-purpose machine tools. During the use of general- or special-purpose manual machine tools, product accuracy and productivity depend on the operator's participation during operation. Capstan and turret lathes are typical machines that somewhat reduce the operator's role during the machining of bar-type or chucking-type WPs at higher rates and better accuracy. Semiautomatic machine tools perform automatically controlled movements, while the WP has to be loaded and unloaded by hand. Fully automatic machine tools are those machines in which WP handling and cutting and other auxiliary activities are performed automatically. Semiautomatic and automatic machine tools are best suited for large production lots where the operator's interference is minimized or completely eliminated, and parts are machined more accurately and economically.

NC machine tools utilize a form of programmable automation by numbers, letters, and symbols using a control unit and tape reader, while CNC machine tools utilize a stored computer program to perform all the basic NC functions. NC and CNC

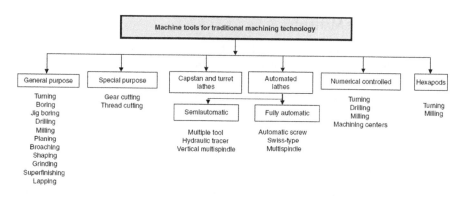

**FIGURE 1.8** Classification of machine tools for traditional machining technology.

have added many benefits to machining technology, since small and large numbers of parts can now be produced. Part geometry can be changed through the flexible control of the part programs. The integration of CAD/CAM systems to machining technology has created new industrial areas in the die, mold, aerospace, and automobile industries.

Hexapods have added a new area to machining technology in which complicated parts can be machined using a single tool that is capable of reaching the WP from many sides. The hexapod has six degrees of movement and is very dexterous, like a robot, but also offers the machine tool rigidity and accuracy that are generally beyond a robot's capability. The hexapods are used to help develop machining processes for WPs that need the dexterity offered by the hexapod design. For general-purpose machining, the hexapod is an ideal machine tool for mold- and die-machining applications. Its ability to keep a cutting tool normal to the surface being machined promotes the use of larger-radii ball nose end mills, which can cut more material with very small stepovers. In some applications, a flat nose end mill can be used very effectively for smooth surface finishes with little or no cusp. Nontraditional machining uses a wide range of machine tools, such as ECM, USM, EDM, and laser beam machining (LBM). Each machine tool is capable of performing a variety of operations, as shown in Figure 1.9. Nontraditional machining technology tackles materials including glass, ceramics, hard alloys, heat-resistant alloys, and other materials that are difficult to machine by traditional machining technology.

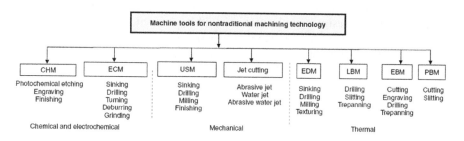

**FIGURE 1.9** Classification of machine tools for nontraditional machining technology.

### 1.4.2   WORKPIECE MATERIAL

The WP material specified for the part influences the selection of the adopted machining method. Most materials can be machined by a range of processes, some by a very limited range. In any particular case, however, the choice of the material depends on the desired shape and size, the dimensional tolerances, the surface finish, and the required quantity. It must not depend only on technical suitability but also on economic and environmental considerations.

### 1.4.3   MACHINING PRODUCTIVITY

The choice of any machining method should take into consideration a rate of production that is inversely proportional to machining time. Methods of raising productivity include the use of the following:

- High machining speeds
- High feed rates
- Multiple cutting tools
- Stacking of multiple parts
- Minimization of the secondary (noncutting) time
- Automatic feeding and tool-changing mechanisms
- High power densities

### 1.4.4   ACCURACY AND SURFACE INTEGRITY

The selection of a machining technology depends on inherent accuracy and surface quality. Below the machined surface, some alterations occur as a result of the material removal mechanisms employed. Careful examination of such a layer is essential. It affects the technological performance of the machined parts in terms of fatigue strength, corrosion, and wear resistance. In some cases, a postfinishing technology may be adopted to solve such problems, which in turn increases the production cost.

### 1.4.5   PRODUCT DESIGN FOR ECONOMICAL MACHINING

This concept is very important to produce parts accurately and economically. Product design recommendations for each operation should be strictly followed by the part designer. Design complications should be avoided so that the machining time is reduced, and consequently, the production rate is increased. Machine tool and operation capability in terms of possible accuracy and surface integrity should also be considered, so that the best technology, machine tool, and operation are selected.

### 1.4.6 ENVIRONMENTAL IMPACTS OF MACHINING

The possible hazards of the selected machining technology may affect the operator's health, the machine tool, and the surrounding environment. Reduction of such hazards requires careful monitoring, analysis, understanding, and control of environmentally clean machining technology. The hazards generated by the cutting fluids have led to the introduction of the minimum quantity lubrication (MQL), cryogenic machining, and dry machining techniques. Strict precautions are followed during LBM and abrasive jet machining (AJM), and these processes are covered in Chapter 10 of *Non-Traditional and Advanced Machining Technologies.*

## 1.5 REVIEW QUESTIONS

1.5.1 Explain what is meant by *manufacturing.*
1.5.2 What are the different manufacturing methods used for metal shaping?
1.5.3 Define the machinability of a material.
1.5.4 Explain the different mechanisms of material removal in machining technology.
1.5.5 List the main categories of machine tools used for traditional machining.
1.5.6 Classify the different nontraditional machine tools based on the material removal process.
1.5.7 Show basic motions of machine tools used for traditional and nontraditional processes.
1.5.8 Explain the different aspects of machining technology.
1.5.9 Explain what is meant by product design for economic machining.
1.5.10 Explain the importance of adopting an environmentally friendly machining technology.
1.5.11 What are the main objectives behind selecting a machining technology?

## REFERENCES

ASM International, 3 Park Ave., New York.
Charmilles, 560 Bond St., Lincolnshire, IL.

# 2 Basic Elements and Mechanisms of Machine Tools

## 2.1 INTRODUCTION

Metal-cutting machines (machine tools) are characterized by higher production accuracy compared with metal-forming machines. They are used for the production of relatively smaller numbers of pieces; conversely, metal-forming machines are economical for producing larger lots. Machine tools constitute about 70% of the total operating production machines in industry. The percentages of the different types of operating machine tools are shown in Table 2.1.

The successful design of machine tools requires the following fundamental knowledge:

1. Mechanics of the machining processes to evaluate the magnitude and direction and to control the cutting forces
2. The machinability of the different materials to be processed
3. The properties of the materials used to manufacture the different parts of the machine tools
4. The manufacturing techniques that are used to produce each machine tool part economically
5. The durability and capability of the different tool materials
6. The principles of engineering economy

The productivity of a machine tool is measured either by the number of parts produced in a unit of time, by the volumetric removal rate, or by the specific removal rate per unit of power consumed. Productivity levels can be enhanced using the following methods:

1. Increasing the machine speeds and feed rates
2. Increasing the machine tool available power
3. Using several tools or several workpieces (WPs) machined simultaneously
4. Increasing the traverse speed of the operative units during the nonmachining parts of the production time
5. Increasing the level of automation for the machine tool operative units and their switching elements

**TABLE 2.1**

**Percentages of Different Types of Operating Machine Tool**

| Type of Machine Tool | Percentage |
|---|---|
| Lathes, including automatics | 34 |
| Grinding | 30 |
| Milling | 15 |
| Drilling and boring | 10 |
| Planers and shapers | 4 |
| Others | 7 |

6. Adopting modern control techniques such as numerical control and computer numerical control
7. Selecting the machining processes properly based on the machined part material, shape complexity, accuracy, and surface integrity
8. Introducing jigs and fixtures that locate and clamp the work parts in the minimum possible time

Machine tools are designed to achieve the maximum possible productivity and to maintain the prescribed accuracy and the degree of surface finish over their entire service life. To satisfy these requirements, each machine tool element must be separately designed to be as rigid as possible and then checked for resonance and strength. Furthermore, the machine tool, as whole, must have an adequate stability and should possess the following general requirements:

1. High static stiffness of the different machine tool elements such as structure, joints, and spindles
2. Avoidance of unacceptable natural frequencies that cause resonance of the machine tool
3. Acceptable level of vibration
4. Adequate damping capacity
5. High speeds and feeds
6. Low rates of wear in the sliding parts
7. Low thermal distortion of the different machine tool elements
8. Low design, development, maintenance, repair, and manufacturing cost

Machine tools are divided according to their specialization into the following categories:

- General-purpose (universal) machines, which are used to machine a wide range of products
- Special-purpose machines, which are used for machining articles similar in shape but different in size
- Limited-purpose machines, which perform a narrow range of operations on a wide variety of products

Machine tools are divided according to their level of accuracy into the following categories:

1. Normal-accuracy machine tools, which include the majority of general-purpose machines
2. Higher-accuracy machine tools, which are capable of producing finer tolerances and have more accurate assembly and adjustments
3. Machine tools of super-high accuracy, which are capable of producing very accurate parts

The main functions of a machine tool are holding the WPs to be machined, holding the tool, and achieving the required relative motion to generate the part geometry required.

Machine tools include the following elements:

1. A structure that is composed of bed, column, or frame
2. Slides and tool attachments
3. Spindles and spindle bearings
4. A drive system (power unit)
5. Work-holding and tool-holding elements
6. Control systems
7. A transmission linkage

Stresses produced during machining, which tend to deform the machine tool or a WP, are usually caused by one of the following factors:

1. Static loads that include the weight of the machine and its various parts
2. Dynamic loads that are induced by the rotating or reciprocating parts
3. Cutting forces generated by the material removal process

Both the static and the dynamic loads affect the machining performance in the finishing stage, while the final degree of accuracy is also affected by the deflection caused by the cutting forces.

## 2.2 MACHINE TOOL STRUCTURES

The machine tool structure includes a body, which carries and accommodates all other machine parts. Figure 2.1 shows a typical machine tool bed for a lathe and a frame for a drilling machine. The main functions of the machine structure include the following:

1. Ability of the structure or the bed to resist distortion caused by static and dynamic loads
2. Stability and accuracy of the moving parts
3. Wear resistance of the guideway
4. Freedom from residual stresses
5. Damping of vibration

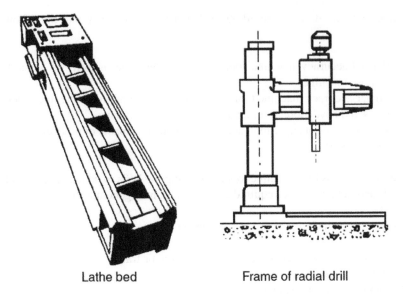

Lathe bed                           Frame of radial drill

**FIGURE 2.1**   Typical bed of center lathe and frame of a drilling machine.

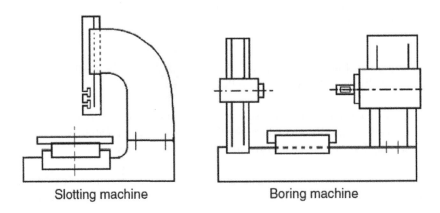

Slotting machine                    Boring machine

**FIGURE 2.2**   Examples of open frames (C-frames).

Machine tool structures are classified by layouts into open (C-frames) and closed frames. Open frames provide excellent accessibility to the tool and the WP. Typical examples of open frames are found in turning, drilling, milling, shaping, grinding, slotting, and boring machines (Figure 2.2). Closed frames find application in planers, jig-boring, and double-spindle milling machines (Figure 2.3). A machine tool structure mounts and guides the tool and the WP and maintains their specified relative position during the machining process. Machine tool structures must therefore be designed to withstand and transmit, without deflection, the cutting forces and weights of the moving parts of the machine onto the foundation. For a multiunit structure, the units must be designed to locate and guide each other in accordance with the required position between the tool and the WP.

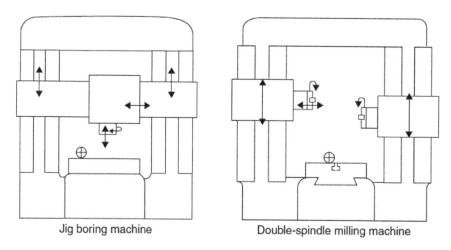

| Jig boring machine | Double-spindle milling machine |

**FIGURE 2.3**   Examples of closed frames.

The configuration of machine tool structure is governed by the arrangement of the necessary cutting and feed movements and their stroke lengths as well as the size and capacity of the machine. In this regard, chip disposal, transport, erection, and maintenance are also considered. The rate of material removal determines the power capacity of the machine tool and hence the magnitude of the cutting forces. The grade of production accuracy is affected by the deformation and deflections of the structure, which should be kept within specified limits. The assessment of the behavior of machine tool structure is obtained by evaluating its static and dynamic characteristics.

*Static characteristics.* These characteristics concern the steady deflection under steady operational cutting forces, the weight of the moving components, and the friction and inertia forces. They affect the accuracy of the machined parts and are usually measured by the static stiffness.

*Dynamic characteristics.* The dynamic characteristics are determined mainly by the dynamic deflection and natural frequencies. They affect the machine tool chatter and hence the stability of the machining operation.

The static and dynamic deflections of a machine tool structure depend on the manner by which the operational forces are transmitted and distributed and the behavior of each structural unit under operating conditions. A beam-like element, having a cross-section in the form of a hollow rectangle, is the preferred element. A typical application of this concept is given in the lathe bed shown in Figure 2.4; the adverse effect of cast holes on the stiffness of the closed-box cross-section is minimized by reducing their number and size. As can be seen in Figure 2.5, closed-frame structures, although deformed under load, keep the alignment of their centerline axes unchanged. This, in turn, results in an axial (not lateral) displacement of the tool relative to the WP, which does not affect the accuracy of machined parts. An open frame can, therefore, be supplemented with a supporting element to close its frame during the machining operation, as shown in the radial drilling machine in Figure 2.6.

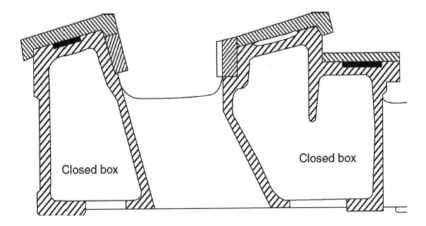

**FIGURE 2.4**   Hollow box sections of the lathe bed.

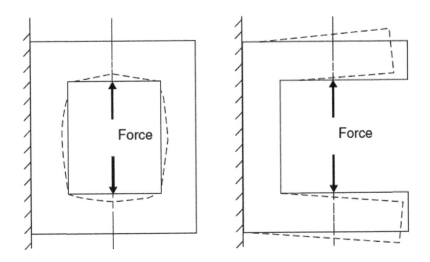

**FIGURE 2.5**   Deformation in open and closed frames.

Machine-tool stiffness and damping of the machine-tool structure depend on the number and type of joints used to connect the different units of the structure. As a rule, the fewer the joints, the greater the stiffness of the structure and the smaller its damping capability. The ribbing system is an effective method for increasing the stiffness of the machine tool structures. In this regard, simple vertical stiffeners, seen in Figure 2.7a, increase the stiffness of the vertical bending but do not improve horizontal bending. The diagonal stiffness arrangement, shown in Figure 2.7b, gives higher stiffness in both bending and torsion. In some cases, to eliminate the tilting movement that usually acts on the tailstock of the lathe machine, raised rear guideways are introduced, as shown in Figure 2.8. Machine-tool frames can be produced using cast or welded construction. Welded structures ensure great savings of

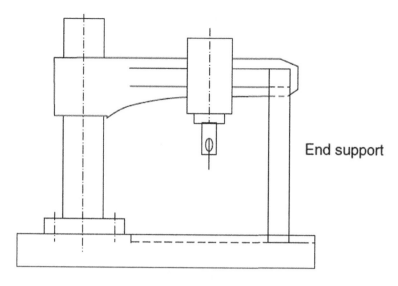

**FIGURE 2.6**  Radial drilling machine with end support.

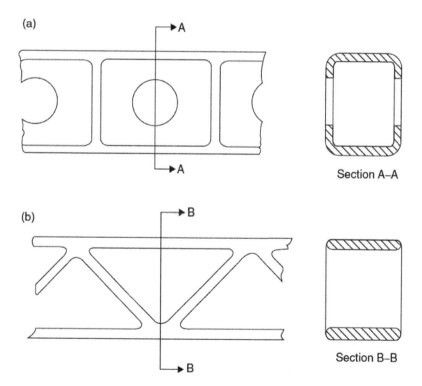

**FIGURE 2.7**  Arrangement of stiffeners in machine tool beds: (a) vertical and (b) diagonal stiffeners.

## Raised guideways

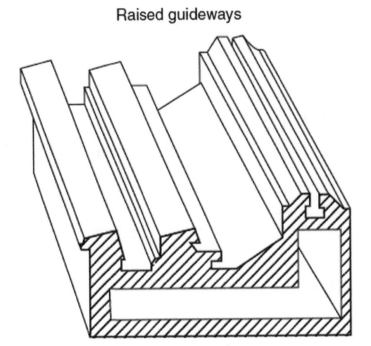

**FIGURE 2.8**   Lathe bed with raised rear guideways.

(a)

(b)

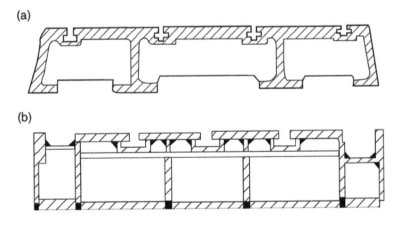

**FIGURE 2.9**   Cast and fabricated structures: (a) cast and (b) welded machine-tool bases.

material and pattern costs. Figure 2.9 shows typical cast and fabricated machine-tool structures. A cast iron (CI) structure ensures the following advantages:

- Better lubricating property (due to the presence of free graphite). Most suitable for beds in which rubbing is the main criterion
- High compressive strength

- Better damping capacity
- Easily cast and machined

### 2.2.1  LIGHT- AND HEAVY-WEIGHT CONSTRUCTIONS

Machine tool structures are classified according to their natural frequency as light- or heavy-weight construction. The natural frequency $\omega_0$ of a machine tool can be described by

$$\omega_0 = \sqrt{\frac{k}{m}} \tag{2.1}$$

where
$k$ = structure static stiffness
$m$ = mass

$$k = \frac{F}{\delta} \tag{2.2}$$

where
$F$ = force (N)
$\delta$ = deflection (mm)

To avoid resonance and thus reduce the dynamic deflection of the machine-tool structure, $\omega_0$ should be far below or far above the exciting frequencies, which are equal to multiples of the rotational speed of the machine.

If the natural frequency of the machine structure is kept far below the speed working range of the machine tool, then

$$\omega_0 < \text{exciting frequency}$$

or

$$\sqrt{\frac{k}{m}} < \text{exciting frequency}$$

This requirement is achieved by the increase of the mass $m$, which, in turn, leads to a heavyweight construction. On the other hand, lightweight constructions are made when

$$\omega_0 > \text{exciting frequency}$$

or

$$\sqrt{\frac{k}{m}} > \text{exciting frequency}$$

Chip disposal, in the case of high-production machine tools, affects the construction of the machine tool frame as shown in Figure 2.10.

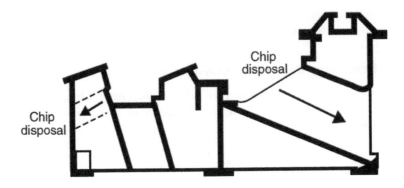

**FIGURE 2.10**   Chip disposal in a lathe bed.

## 2.3   MACHINE-TOOL GUIDEWAYS

Machining occurs as a result of a relative motion between the tool and the WP. Such a motion is a rotary, linear, or rectilinear one. Guideways are required to perform the necessary machine-tool motion at a high level of accuracy under severe machining conditions. Generally, guideways control the movement of the different parts of the machine tool in all positions during machining and nonmachining times. Besides the accuracy requirements, ease of assembly, and economy in manufacturing guideways, the following features should be provided:

- Accessibility for effective lubrication
- Wear resistance, durability, and rigidity
- Possibility of wear compensation
- Restriction of motion to the required directions
- Proper contact all over the sliding area

Guideways are classified as sliding friction, rolling friction, and externally pressurized (Figure 2.11).

### 2.3.1   SLIDING FRICTION GUIDEWAYS

Sliding friction guideways consist of any one of or a combination of the flat, vee, dovetail, and cylindrical guideway elements. Flat circular guideways are used for guiding the rotating table of the vertical turning and boring machines. Figure 2.12 shows the different types of guideways that are normally used to guide sliding parts in the longitudinal directions. Holding strips may be provided to prevent the moving part from lifting or tilling by the operational forces. Scraping and the introduction of thin shims are used for readjustments that may be required to compensate wear of the sliding parts.

Vee-shaped guideways are either male or female type, which are self-adjusting under the weight of the guided parts. Practically, a vee guideway is usually combined with a flat one, as in the case of the carriage guides of the center lathe to ensure proper contact all over the sliding surfaces. The combination of two vee guideways has an unfavorable effect on the machining accuracy and is limited to guideways

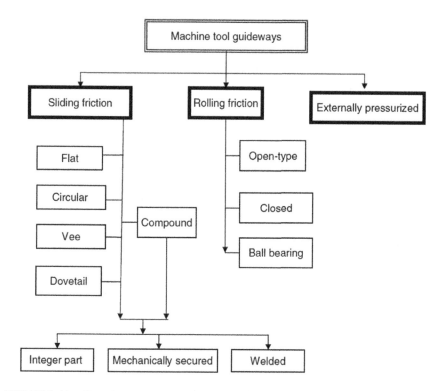

**FIGURE 2.11** Classification of machine-tool guideways.

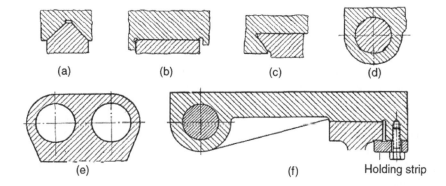

**FIGURE 2.12** Types of guideways: (a) vee, (b) flat, (c) dovetail, (d) cylinder, (e) cylindrical–cylindrical, and (f) cylindrical–flat.

of relatively small distance between the two vees. Circular vee guideways carry the operational loads and provide self-location for the rotating table. Dovetail guideways, shown in Figure 2.12c, are used separately or in a combination of half-dovetail and flat guideways. Cylindrical guides, shown in Figure 2.12d, are either male or female type and must be accurately manufactured. They require special devices to adjust their working clearances. The column of the drilling machine is a typical example of the male type, while the sleeve of the drilling-machine spindle is a female type. The combinations of cylindrical guideways are shown in Figure 2.12e (cylindrical–cylindrical) and f (cylindrical–flat).

For the sliding surfaces, the bulk of the load is carried on the metal-to-metal contact. The load carried by the lubricating oil film is very small. The localized pressures cause elastic or plastic deformation to the supporting asperities of the surface, which in turn results in an instability of the sliding motion usually known as the *stick–slip effect*. This phenomenon can be reduced or eliminated by the use of proper lubricants or through the introduction of externally pressurized guideways. Friction condition and, consequently, the wear of the guideways are affected by

1. Material properties of the fixed and moving element
2. Surface dimensions of the guideways
3. Acting pressure
4. Accumulation of dirt, chip, and wear debris

When the machine parts rub together, loss of material from one or both surfaces occurs, which in turn results in a change of the designed dimensions and geometry of the guideway system. Wear of guideways may be caused by the cutting action of the hard particles (adhesive wear), which is often accompanied by oxidation of the wear debris, leading to additional abrasive wear. Wear of guideways can be minimized by

1. Minimizing the sliding surface roughness
2. Increasing the hardness of the sliding surfaces
3. Removing the abrasive wear particles from the guideways system
4. Reducing the pressure acting on the guiding surfaces

Guideways are equipped with devices for initially adjusting and periodically compensating the working clearance. Clearance adjustment is accomplished by using suitable metallic strips, as shown in Figure 2.13. Guideways may be an integral part of the machine tool or mechanically secured to the bed by fastening or welding. In the first arrangement, the bed and the guideway are made from the same material, and flame or induction hardening is employed upon the guiding surfaces. In the mechanically secured guideways, separate steel guideways are secured to the CI beds, as shown in Figure 2.14a.

In plastic guideways, plates of phenolic resin bonded fiber are inserted into one of the sliding surfaces. These guideways reduce friction and stick–slip effect. They also reduce the danger of seizure when lubricant is inadequate and minimize vibrations. The design and arrangement of the guideways must prevent chip and dirt accumulation, which promotes the rate of wear. Methods of protecting guideways against foreign matter include:

1. Extending the length of the moving parts using cover plates that protect the guideways
2. Providing covering belts or a telescopic plate that surrounds the guideways and seals them from external materials

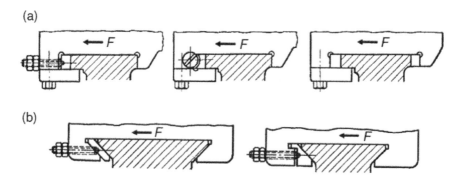

**FIGURE 2.13** Wear compensation in guideways: (a) flat and (b) dovetail guideways. *F* is the side force acting on the carriage.

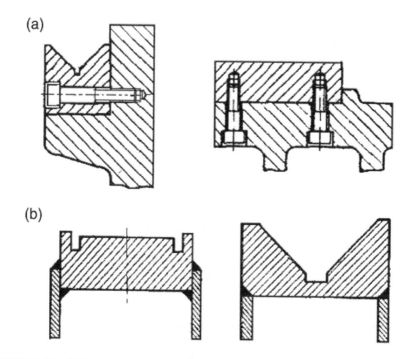

**FIGURE 2.14** (a) Mechanically secured and (b) welded guideways.

## 2.3.2 Rolling Friction Guideways

In rolling friction guideways, rollers, needles, or balls are inserted between the moving parts to minimize the frictional resistance, which is kept constant irrespective of the traveling speed. Rolling friction guideways find wide applications in numerically controlled and medium-size machine tools in which the setting accuracy is decisive. Their expensive manufacturing, their complicated construction, and the short life of the rolling elements create problems. Rolling friction guideways are either open or closed. The open type (Figure 2.15) is used when the load acts downward, which

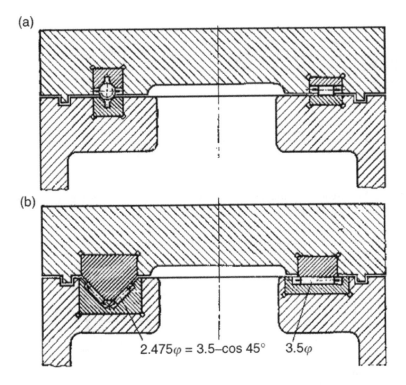

**FIGURE 2.15** Open-type rolling friction guideways: (a) flat and (b) vee–flat guideways.

makes this type self-adjusting for wear in the guideways. In the closed type, wear compensation requires adjusting elements. For long strokes, recirculating rolling elements (as shown in Figure 2.16) or ball- or roller-bearing guideways (Figure 2.17) are used to shorten the length of the slider.

Circular rolling friction guideways find applications in high-speed vertical lathes. The size and the distribution of the load on the rolling elements and the deformation of the guideways are affected by:

1. Magnitude, distribution, and type of loading
2. Stiffness of the rolling elements
3. Manufacturing errors of the rolling elements
4. Form error of the guideways
5. Magnitude of preloading
6. Stiffness of the table, bed, fixture, and WP

### 2.3.3 EXTERNALLY PRESSURIZED GUIDEWAYS

The load-carrying capacity and stiffness of ordinary lubricated guideways are excellent; however, their friction levels are undesirable. To overcome such a problem, externally pressurized guideways are used in which the sliding elements are separated by a thin film of pressurized fluid, as shown in Figure 2.18. Such an arrangement

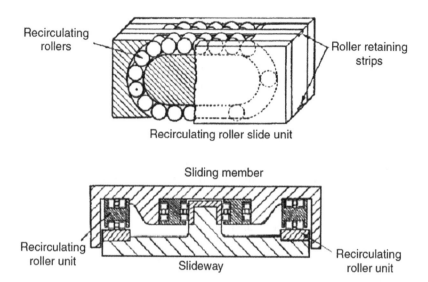

**FIGURE 2.16**    Recirculating rolling friction guideways.

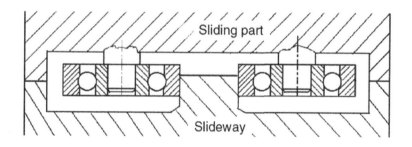

**FIGURE 2.17**    Ball-bearing guideway.

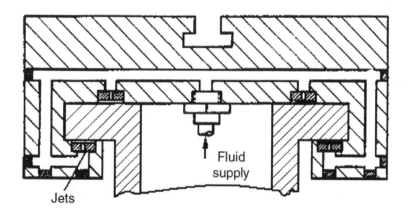

**FIGURE 2.18**    Externally pressurized guideways.

prevents contact between the sliding surfaces and hence avoids the occurrence of wear. The load-carrying capacity is independent of the sliding speed, and the reaction forces are distributed over the full bearing area. Externally pressurized guideways are ideal guideways in terms of stiffness, uniformity of travel, low friction, large damping, and better heat dissipation capacity. Generally, the service properties of machine-tool guideways can be improved by

1. Providing favorable frictional conditions, which can be achieved by using
    a. combined sliding and rolling guides
    b. proper lubricants and materials for guideways
    c. hydrostatic ways with high-rigidity oil film and automatic control systems
2. Providing adequate protection of guideways
3. Using optimal cross-section of slideways
4. Using optimal surface finishes

## 2.4 MACHINE-TOOL SPINDLES

Machine-tool spindles are used to locate, hold, and drive the tool or the WP. These spindles possess a high degree of rigidity, rotational accuracy, and wear resistance. Spindles of general-purpose machine tools are subjected to heavier loads compared with precision ones. In the former class of spindles, rigidity is the main requirement; in the second, the manufacturing accuracy is the prime consideration. Spindles are normally made hollow and provided with an internal taper at the nose end to accommodate the center or the shank of the cutting tool (Figure 2.19). A thread can be added at the nose end to fix a chuck or a face plate. Medium-carbon steel containing 0.5% C is used for making spindles in which hardening is followed by tempering to produce a surface hardness of about 40 Rockwell (HRC). Low-carbon steel containing 0.2% C can also be carburized, quenched, and tempered to produce a surface hardness of 50–60 HRC. Spindles for high-precision machine tools are hardened by nitriding, which provides sufficient hardness with the minimum possible deformation. Manganese steel is used for heavy-duty machine tool spindles.

### 2.4.1 SPINDLE BEARINGS

Machine-tool spindles are supported inside housings by means of ball, roller, or antifriction bearings. Precision bearings are used for a precision machine tool. The

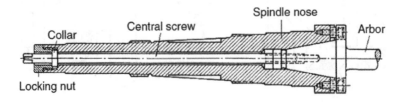

**FIGURE 2.19** Typical milling machine spindle.

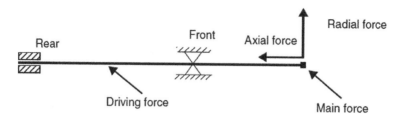

**FIGURE 2.20**   Forces acting on machine-tool spindles.

geometrical accuracy and surface finish of the machined components depend on the quality of the spindle bearings. The considerable attention paid to the spindle design, the selection and proper mounting of its bearings, and the construction of the housing of bearings makes the spindle system one of the most expensive parts of machine tools. Drive shafts, which are subjected to bending and tensional stresses, are designed on the basis of strength, while spindles are designed on the basis of stiffness. Generally, machine-tool spindle bearings must provide the following requirements:

1. Minimum deflection under varying loads
2. Accurate running under loads of varying magnitudes and directions
3. Adjustability to obtain minimum axial and radial clearances
4. Simple and convenient assembly
5. Sufficiently long service
6. Maximum temperature variation throughout the speed ranges
7. Sufficient wear resistance

The forces acting on a machine-tool spindle are the cutting force, which acts at the spindle nose, and the driving force, which acts in between the spindle bearings (Figure 2.20). The cutting force can be resolved into two components with respect to the spindle. The spindle bearings have to take radial and axial components of the cutting and driving forces. In this manner, when the machine tool spindle is mounted at two points, the bearing at one point takes the axial component besides the reaction of the radial component, while the other takes only the reaction of the radial component. The bearings that carry the axial component should prevent the axial movement of the spindle under the effect of the cutting and driving forces (fixed bearing). The other bearing (floating bearing) provides only a radial support and provides axial displacement due to differential thermal expansion of the spindle shaft and the housing.

The arrangement shown in Figure 2.21a is used in most high-speed machine tools, because the free length of the spindle (from nose to the fixed bearing) is limited, which minimizes nose deflection. Additionally, the effect of differential thermal expansion of the spindle and spindle housing acts toward the floating (rear) end, which in turn reduces the axial displacements of the spindle nose. Figure 2.22 shows typical spindle-bearing mounting arrangements. Figure 2.23 presents a machine-tool spindle with the fixed front bearing, while the rear end axially slides

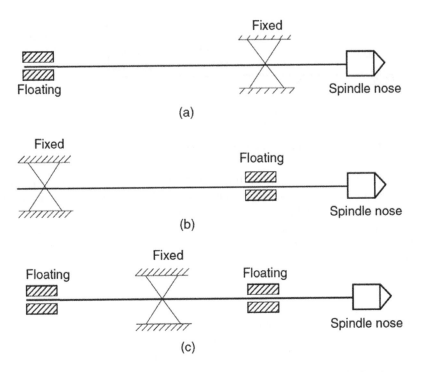

**FIGURE 2.21** Fixed and floating bearing arrangements: (a) fixed front, (b) fixed rear, and (c) fixed middle.

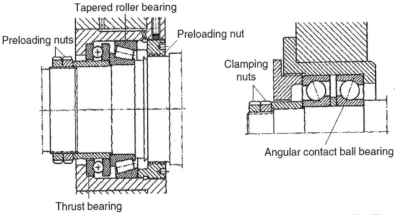

**FIGURE 2.22** Typical spindle-bearing arrangements. (From Browne, J. W., *The Theory of Machine Tools, Book-1*, Cassell and Co. Ltd., London, 1965.)

at the outer race of the roller bearing. The various considerations in the selection of bearings are:

1. Direction of load relative to the bearing axial
2. Intensity of load

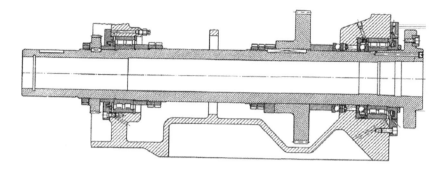

**FIGURE 2.23** Typical machine-tool spindle. (From Koenigsberger, F., *Berechnungen, Konstruktionsgrundlagen und Bauelemente spanender Werkzeugmaschinen*, Springer, Berlin, 1961. With permission.)

3. Speed of rotation
4. Thermal stability
5. Stiffness of the spindle shaft
6. Class of accuracy of the machine

Ball bearings sustain considerable loads; roller bearings are preferred for severe conditions and shock loads. Tapered roller bearings are suitable for high axial and radial forces (combined loads). To increase the accuracy of ball and roller bearings, these are fitted with very high-interference fits, which eliminate the radial play between the bearing and the spindle. Angular contact ball bearings or roller bearings, installed in pairs, are preloaded by the adjustments made during their assembly.

## 2.4.2 SELECTION OF SPINDLE-BEARING FIT

The high accuracy requirements of a machine tool have direct implications for the method of bearing mounting and the type of fit in the spindle assembly. To prevent creep, roll, or excessive interference fitting of bearing on either the spindle or the housing, it is important to select the correct fit between the bearing and the seats. A bearing fit (inner race on the spindle or outer race in the housing) is either interference, transition, or clearance fit. A correct interference fit provides proper support around the whole circumference and hence provides a correct load distribution; moreover, the load-carrying ability of the bearing is fully utilized. In the case of floating bearings, which are made free to move axially, interference fit is unacceptable. Figure 2.24 shows the recommended types of fit for machine-tool spindle bearings. Apart from thrust types of bearings, the fixed bearing on the spindle has j or k types of fit, while the housing has M, K, or J to ensure sufficient stiffness. In the case of floating bearing, the h fit is used for the spindle, and the H fit is used for the housing. Because the stiffness of the thrust types of bearings is not affected by the type of fit, the spindle has a transition fit, while the housing has either a clearance or a transition fit.

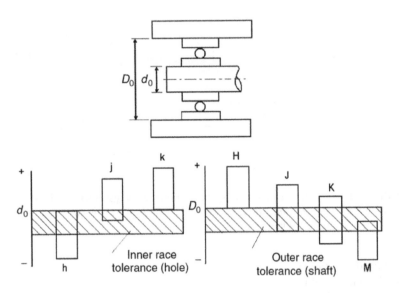

**FIGURE 2.24**   Recommended bearing fits.

**TABLE 2.2**
**Recommended Fits for Machine-Tool Spindle Bearings**

| Bearing Type | Spindle | | | Working Conditions | Housing | | |
|---|---|---|---|---|---|---|---|
| | **P6** | **P5-SP** | **P4-UP** | | **P6** | **P5-SP** | **P4-UP** |
| Deep groove ball bearing | j5 | j4 | j3 | Point load | J5 | J4 | J5 |
| Angular contact ball bearing | | | | Rotating load | M6 | M5 | M4 |
| Cylindrical roller bearing | k5 | k4 | – | Point load | K6 | K5 | K4 |
| | | | | Rotating load | M6 | M5 | M4 |
| Tapered roller bearing | k5 | k4 | – | Loose adjustable | J6 | J5 | – |
| | | | | Tight adjustable | K6 | K5 | – |
| | | | | Rotating load | M6 | M5 | – |
| Angular contact ball thrust bearing | – | h5 | h4 | | K6 | K5 | K4 |
| Ball thrust bearing | h6 | h5 | – | | H8 | H8 | – |

Table 2.2 shows the recommended types of fit applied to machine-tool bearings. According to International Organization for Standardization (ISO) recommendations, rolling bearings are manufactured in normal tolerance grade, close tolerance grades (P6, P5), the special precision (SP) grade, and the ultraprecision (UP) grade (P4). Bearings of normal tolerance grades are of general use, while SP and UP grades are used in spindles of high-precision machine tools. For comparison of the fits for machine-tool spindles, see Table 2.3.

**TABLE 2.3**

**Bearing Tolerances (Microns)**

| Tolerance Grade | Bore (80 mm) | | | Outer Diameter (80 mm) | | |
|---|---|---|---|---|---|---|
| | Bore | Width | Radial Runout | Diameter | Width | Radial Runout |
| Normal | 5–20 | 25 | 25 | 5–20 | 25 | 25 |
| P4 | 0–8 | 4 | 5 | 0–8 | 5 | 6 |
| SP | 0–10 | 7 | 5 | 0–10 | 7 | 6 |
| UP | 0–8 | 3 | 3 | 0–8 | 3 | 3 |

### 2.4.3 SLIDING FRICTION SPINDLE BEARING

Rolling bearings are used at a speed and diameter range of $n \cdot d \leq 2 \times 10^5$, where $n$ is the spindle rotational speed in revolutions per minute and $d$ is the diameter of the spindle in millimeters. At higher running speeds, the bearing life is reduced due to the gyratory action, especially in bearings that take combined loads. At high spindle speeds, as in case of grinding, sliding friction (journal) bearings that have high damping capacity compared with rolling bearings are used. Their load-carrying capacity increases as the spindle speed increases due to the hydrodynamic action created within the bearing. For optimum performance, the radial clearance between journal and bearing should be properly maintained, as it affects bearing friction, load-carrying capacity, and the efficiency of heat dissipation of the bearing. The main types of sliding friction bearings include the following:

1. Sliding bearing with radial play adjustment using segments that can be adjusted radically to control the clearance.
2. Bearing with axial play adjustment, in which a bush with a cylindrical bore and external taper has a slot along its length and is made to fit in a taper hole in the housing. When the bush slides axially, through two opposing nuts, on the two ends of the bush, radial play can be finely adjusted and controlled (Figure 2.25).
3. Mackensen bearing is used in highly accurate machine tool spindles, running at extremely high speeds, under limited applied load. As shown in Figure 2.26, an elastic bearing bush is supported at three points in the housing. This bush has nine equally spaced axial slots along its circumference. When the shaft is running, the bush deforms into a triangular shape, and three wedge-shaped oil pockets are formed, which constitute the load-carrying parts of the bearing.
4. Hydrodynamic multipad spindle bearing of high radial and axial thrust capacity, high stiffness, and practically no clearance during operation.

Sliding bearing materials should have high compressive strength to withstand the bearing pressure, low coefficient of friction, and high thermal conductivity. They

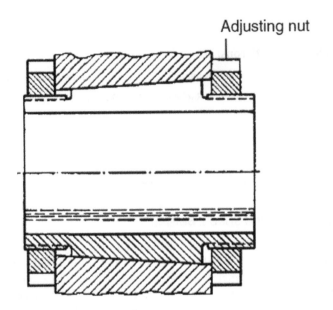

**FIGURE 2.25**  Sliding friction bearing.

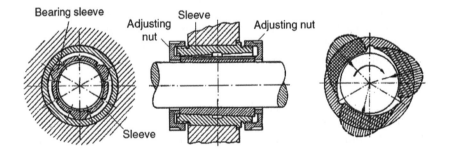

**FIGURE 2.26**  Mackensen bearing.

should possess high wear resistance and maintain a continuous oil film. The various sliding bearing metals include:

1. Copper-base bearing metals (85% Cu, 10% Sn, 5% Zn), which are used for heavy loads
2. Tin-base bearing (babbit) metals (85% Sn, 10% Sb, 5% Cu), which are used for higher loads
3. Lead-base bearing metals (10–30% Pb, 10–15% Sb, and the rest is copper), which are used for light loads
4. Cadmium-base bearing metals (95% Cd and a very small amount of iridium), which have higher compressive strength and more favorable properties at higher temperatures

## 2.5 MACHINE TOOL DRIVES

To obtain a machined part by a machine tool, coordinated motions must be imparted to its working members. These motions are either primary (cutting and feed) movements, which remove the chips from the WP, or auxiliary motions that are required to prepare for machining and ensure the successive machining of several surfaces of one WP or a similar surface of different WPs. Principal motions may be either rotating or straight reciprocating. In some machine tools, this motion is a combination of rotating and reciprocating motions. Feed movement may be continuous (lathes, milling machine, drilling machine) or intermittent (shapers, planers). As shown in Figure 2.27, stepped motions are obtained using belting or gearing. Stepless speeds are achieved by mechanical, hydraulic, and electrical methods.

### 2.5.1 STEPPED SPEED DRIVES

#### 2.5.1.1 Belting

The belting system, shown in Figure 2.28, is used to produce four running rotational speeds $n_1$, $n_2$, $n_3$, and $n_4$. It is cheap and absorbs vibrations. It has the limitations of

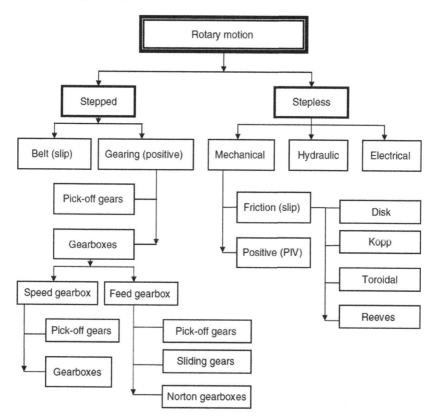

**FIGURE 2.27** Classification of transmission of rotary motion.

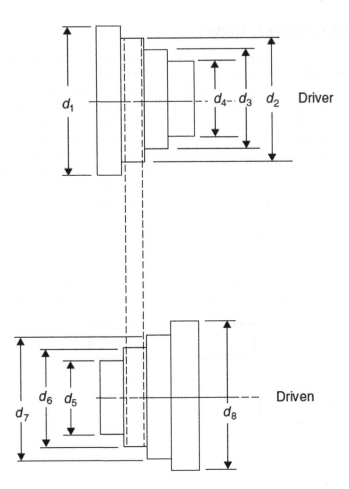

**FIGURE 2.28**  Belting transmission.

low-speed changing, slip, and the need for more space. Based on the driver speed $n_1$, the following speeds can be obtained in a decreasing order:

$$n_1 = n\frac{d_1}{d_5} \tag{2.3}$$

$$n_2 = n\frac{d_2}{d_6} \tag{2.4}$$

$$n_3 = n\frac{d_3}{d_7} \tag{2.5}$$

This type is commonly used for grinding and bench-type drilling machines.

### 2.5.1.2  Pick-Off Gears

Pick-off gears are used for machine tools of mass and batch production (automatic and semiautomatic machines, special-purpose machines, and so on) when the

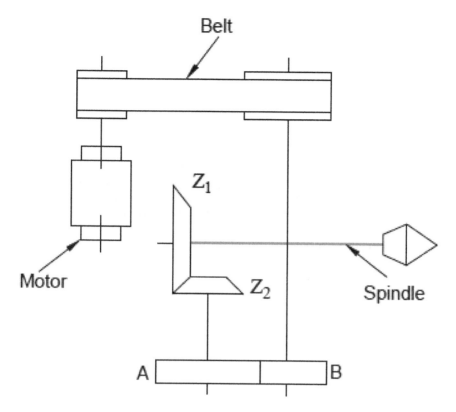

**FIGURE 2.29**  Pick-off gears.

changeover from job to job is comparatively rare. Pick-off gears may be used in speed or feed gearboxes. As shown in Figure 2.29, the change of speed is achieved by setting gears A and B on the adjacent shafts. As the center distance is constant, correct gear meshing occurs if the sum of teeth of gears A and B is constant.

### 2.5.1.3  Gearboxes

Machine tools are characterized by their large number of spindle speeds and feeds to cope with the requirements of machining parts of different materials and dimensions using different types of cutting-tool materials and geometries. The cutting speed is determined on the bases of the cutting ability of the tool used, surface finish required, and economic considerations.

A wide variety of gearboxes utilize sliding gears or friction or jaw coupling. The selection of a particular mechanism depends on the purpose of the machine tool, the frequency of speed change, and the duration of the working movement. The advantage of a sliding gear transmission is that it is capable of transmitting higher torque and is small in radial dimensions. Among the disadvantages of these gearboxes is the impossibility of changing speeds during running. Clutch-type gearboxes require small axial displacement for speed changing and less engagement force compared with sliding gear mechanisms and therefore can employ helical gears. The extreme spindle speeds of a machine-tool main gearbox, $n_{max}$ and $n_{min}$, can be determined by

$$n_{max} = \frac{1000\,V_{max}}{\pi d_{max}} \qquad (2.6)$$

$$n_{min} = \frac{1000\,V_{min}}{\pi d_{max}} \qquad (2.7)$$

where

$V_{max}$ = maximum cutting speed (m/min) used for machining the softest and most machinable material with a cutting tool of the best cutting property

$V_{min}$ = minimum cutting speed (m/min) used for machining the hardest material using a cutting tool of the lowest cutting property or the necessary speed for thread cutting

$d_{max}$, $d_{min}$ = maximum and minimum diameters (mm) of WP to be machined

The speed range $R_n$ becomes

$$R_n = \frac{n_{max}}{n_{min}} = \frac{V_{max}}{V_{min}} \cdot \frac{d_{max}}{d_{min}} = R_v.R_d \qquad (2.8)$$

where

$R_v$ = cutting speed range
$R_d$ = diameter range

In the case of machine tools having rectilinear main motion (planers and shapers), the speed range $R_n$ is dependent only on $R_v$. For other machine tools, $R_n$ is a function of $R_v$ and $R_d$, and large cutting speeds and diameter ranges are required. Generally, when selecting a machine tool, the speed range $R_n$ is increased by 25% for future developments in the cutting-tool materials. Table 2.4 shows the maximum speed ranges in modern machine tools.

### 2.5.1.4 Stepping of Speeds According to Arithmetic Progression

Let $n_1, n_2, \ldots, n_z$ be arranged according to arithmetic progression. Then

$$n_1 - n_2 = n_3 - n_2 = \text{constant} \qquad (2.9)$$

**TABLE 2.4**
**Speed Range for Different Machine Tools**

| Machine | Range |
|---|---|
| Numerically controlled lathes | 250 |
| Boring | 100 |
| Milling | 50 |
| Drilling | 10 |
| Surface grinding | 4 |

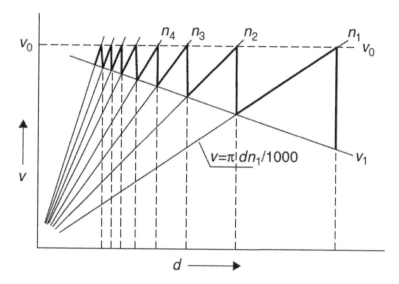

**FIGURE 2.30** Speed stepping according to arithmetic progression.

The sawtooth diagram in such a case is shown in Figure 2.30. Accordingly, for an economical cutting speed $v_0$, the lowest speed $v_1$ is not constant; it decreases with increasing diameter. Therefore, the arithmetic progression does not permit economical machining at large diameter ranges. The main disadvantage of such an arrangement is that the percentage drop from step to step, $\delta_n$, decreases as the speed increases. Thus, the speeds are not evenly distributed but are more concentrated and closely stepped in the small diameter range than in the large one. Stepping speeds according to arithmetic progression are used in Norton gearboxes or gearboxes with a sliding key when the number of shafts is only two.

### 2.5.1.5 Stepping of Speeds According to Geometric Progression

As shown in Figure 2.31, the percentage drop from one step to the other is constant, and the absolute loss of economically expedient cutting speed $\Delta v$ is constant all over the whole diameter range. The relative loss of cutting speed $\Delta v_{max}/v_0$ is also constant. Geometric progression, therefore, allows machining to take place between limits $v_0$ and $v_u$ independently of the WP diameter, where $v_0$ is the economical cutting speed and $v_u$ is the allowable minimum cutting speed. Now suppose that $n_1, n_2, n_3, \ldots, n_z$ are the spindle speeds. According to the geometric progression,

$$\frac{n_2}{n_1} = \frac{n_3}{n_2} = \varphi \tag{2.10}$$

where $\varphi$ is the progression ratio. The spindle speeds can be expressed in terms of the minimal speed $n_1$ and the progression ratio $\varphi$.

| $n_1$ | $n_2$ | $n_3$ | $n_4$ | $n_z$ |
|---|---|---|---|---|
| $n_1$ | $n_1\varphi$ | $n_1\varphi^2$ | $n_1\varphi^3$ | $n_1\varphi^{z-1}$ |

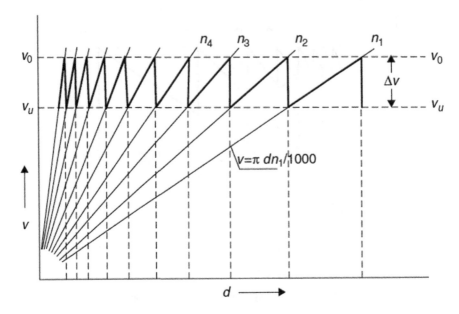

**FIGURE 2.31**   Speed stepping according to geometric progression.

Hence, the maximum spindle speed $n_z$ is given by

$$n_z = n_1 \varphi^{z-1} \tag{2.11}$$

where $z$ is the number of spindle speeds; therefore,

$$\varphi = \sqrt[z-1]{\frac{n_z}{n_1}} = \sqrt[z-1]{R_n} = \left(R_n\right)^{1/(z-1)} \tag{2.12}$$

from which

$$z = \frac{\log R_n}{\log \varphi} + 1 \tag{2.13}$$

Progression ratios are standardized according to ISO standards in such a way as to allow standard speeds and feeds, including full load induction motor speeds of 2800, 1400, and 710 rpm, to be used. Table 2.5 shows the standard values of $\varphi$ according to ISO/R229. Similarly, machine-tool speeds are standardized according to ISO/R229. Such speeds enable the direct drive of machine-tool spindles using induction motors with changing poles. The full load speeds of induction motors are 236, 280, 322, 472, 200, 710, 920, 1400, and 2800 rpm. Tables 2.6 and 2.7 show the standard speeds and feeds according to ISO/R229.

## TABLE 2.5
## Standard Values of Progression Ratio φ According to ISO/R229 and Deutsches Institüt für Normung (DIN) 323

| Basic and Derived Series | Standard Value | Accurate Value | Percentage Drop | Application |
|---|---|---|---|---|
| R20 | $20\sqrt{10} = 1.12$ | 1.1221 | 10 | Seldom used |
| R20/2 | $\left(20\sqrt{10}\right)^2 = 1.26$ | 1.258 | 20 | Machines of large $z$ |
| R20/3 | $\left(20\sqrt{10}\right)^3 = 1.4$ | 1.4125 | 30 | Machines of large $R_n$ and small $z$ |
| R20/4 | $\left(20\sqrt{10}\right)^4 = 1.6$ | 1.5849 | 40 | |
| R20/6 | $\left(20\sqrt{10}\right)^6 = 2.0$ | 1.9953 | 50 | Drilling machines |

$z$, Number of speeds; $R_n$, speed range.

## Illustrative Example 1

The following speeds form a geometric progression. Find the progression ratio and the percentage increase in the speed series.

### SOLUTION

| $n_1$ (rpm) | $n_2$ (rpm) | $n_3$ (rpm) | $n_4$ (rpm) | $n_5$ (rpm) | $n_6$ (rpm) |
|---|---|---|---|---|---|
| 14 | 18 | 22.4 | 28 | 35.2 | 45 |

$$\varphi = \frac{n_2}{n_1} = \frac{18}{14} = 1.25$$

or

$$\varphi = \sqrt[5]{\frac{45}{15}} = 1.25$$

The percentage increase in speed $\delta_n$

$$\delta_n = \frac{n_2 - n_1}{n_1} = \frac{\varphi n_1 - n_1}{n_1} = \left(\varphi - 1\right) \times 100$$

hence, $\delta_n = (1.25 - 1) \times 100 = 25\%$.

**TABLE 2.6**
**Standard Speeds According to ISO/R229 and DIN 804**

| Accurate Value (rpm) | Basic Series | | Derived Series | | | | | Limiting Values | Considering 2% Mechanical Tolerance |
|---|---|---|---|---|---|---|---|---|---|
| | R20 (φ = 1.12) | R20/2 (φ = 1.25) | R20/3 (φ = 1.4) | R20/4 1400–800 (φ = 1.6) | φ = 1.6 | R20/6 2800 (φ = 2.0) | φ = 2.0 | −2% | +2% |
| 100 | 100 | | | | | | | 98 | 102 |
| 112.2 | 112 | 112 | 11.2 | | 112 | | 11.2 | 110 | 114 |
| 162.89 | 125 | | 125 | | | | | 123 | 128 |
| 141.25 | 140 | 140 | | 1400 | | 1400 | | 138 | 144 |
| 158.49 | 160 | | 16 | | | | | 155 | 162 |
| 177.83 | 180 | 180 | 180 | | 180 | 180 | | 174 | 181 |
| 199.52 | 200 | | | 2000 | | | | 193 | 204 |
| 223.87 | 224 | 224 | 22.4 | | | | 22.4 | 219 | 228 |
| 251.19 | 250 | | 250 | | | | | 246 | 256 |
| 281.84 | 280 | 280 | | 2800 | | 2800 | | 276 | 287 |
| 316.23 | 315 | | 31.5 | | | | | 310 | 323 |
| 354.81 | 355 | 355 | 355 | | | 355 | | 348 | 368 |
| 398.11 | 400 | | | 4000 | | | | 390 | 406 |
| 446.68 | 450 | 450 | 45 | | 450 | | 45 | 448 | 456 |
| 501.19 | 500 | | 500 | | | | | 491 | 511 |
| 562.34 | 560 | 560 | | 5600 | | 5600 | | 551 | 574 |
| 630.96 | 630 | | 63 | | | | | 618 | 643 |
| 707.95 | 710 | 710 | 710 | | 710 | 710 | | 694 | 722 |
| 794.33 | 800 | | | 8000 | | | | 778 | 810 |
| 891.25 | 900 | 900 | 90 | | | | 90 | 873 | 909 |
| 1000 | 1000 | | 1000 | | | | | 980 | 1020 |

## TABLE 2.7
## Standard Feeds According to ISO/R229 and DIN 803

| | | Nominal Values | | | | | | |
|---|---|---|---|---|---|---|---|---|
| | | | R20/3 | | | | R20/6 | |
| R20 | R20/2 | | ...1... | | R20/4 | | ...1... | |
| φ = 1.12 | φ = 1.25 | | φ = 1.4 | | φ = 1.6 | | φ = 2.0 | |
| 1.00 | 1.0 | | 1.0 | | 1.0 | | 1.0 | |
| 1.12 | | | | 11.2 | | | | |
| 1.25 | 1.25 | 0.125 | | | | 0.125 | | |
| 1.40 | | | 1.4 | | | | | |
| 1.60 | 1.6 | | | 16 | 1.6 | | | 16 |
| 1.80 | | 0.18 | | | | | | |
| 2.00 | 2.0 | | 2.0 | | | | 2.0 | |
| 2.24 | | | | 20 | | | | |
| 2.50 | 2.5 | 0.25 | | | 2.5 | 0.25 | | |
| 2.80 | | | 2.8 | | | | | |
| 3.15 | 3.15 | | | 31.5 | | | | 31.5 |
| 3.55 | | 0.355 | | | | | | |
| 4.00 | 4.0 | | 4.0 | | 4 | | 4.0 | |
| 4.50 | | | | 45 | | | | |
| 5.00 | 5.0 | 0.5 | | | | 0.5 | | |
| 5.60 | | | 5.6 | | | | | |
| 6.30 | 6.3 | | | 63 | 6.3 | | | 63 |
| 7.10 | | 0.71 | | | | | | |
| 8.00 | 8.0 | | 8.0 | | | | 8.0 | |
| 9.00 | | | | 90 | | | | |
| 10.00 | 10.0 | | 1000 | | 10 | | | |

## Illustrative Example 2

Given $n_1 = 2.8$ rpm, $n_z = 31.50$ rpm, and $\varphi = 1.41$, calculate the speed range $R_n$ and the number of speeds $z$.

### SOLUTION

$$R_n = \frac{n_z}{n_1} = \frac{31.50}{2.8} = 11.2$$

since

$$\varphi = \left(R_n\right)^{1/(z-1)}$$

or

$$z = \frac{\log R_n}{\log \varphi} + 1$$

hence,

$$z = \frac{\log 11.2}{\log 1.41} + 1 = 8$$

### 2.5.1.6 Kinetic Calculations of Speed Gearboxes

Consider the six-speed gearbox shown in Figure 2.32. There are two possibilities of the kinematic diagrams for this gearbox, $z = 6 = 3 \times 2$ or $z = 6 = 2 \times 3$. The structural diagram for the first arrangement is shown in Figure 2.33, and the speed chart for the structural diagram $z = 3 \times 2$ is shown in Figure 2.34. In the first group, the motor speed $n_m$ is taken as 1000 rpm, thus allowing a speed reduction of $1 : \varphi^4 (1 : 2.5)$. In the second group, the ratio of speed $R_g$ is $\varphi^3 = 2$, which is lower than the permissible speed reduction in machine tools (1:4). Based on the transmission ratios shown in Figure 2.34, the number of gear teeth $z_1$ through $z_{10}$ can be calculated.

### 2.5.1.7 Application of Pole-Changing Induction Motors

The use of multispeed induction motors through pole changing in machine tools simplifies machine-tool gearboxes. The possibility of changing speeds while the machine is running is an advantage of pole-changing motors. It reduces the auxiliary time and enables the automatic change of spindle speeds and feeds during operation in automatic machine tools. The pole-changing motor with its standard speeds replaces one of the transmission groups depending on its speed ratio $\varphi_p$, where $p$ is the progression ratio of the gearbox to be constructed. For example, if a two-speed pole-changing motor of 1500 and 3000 rpm (full load speeds 1400 and 2800 rpm) is

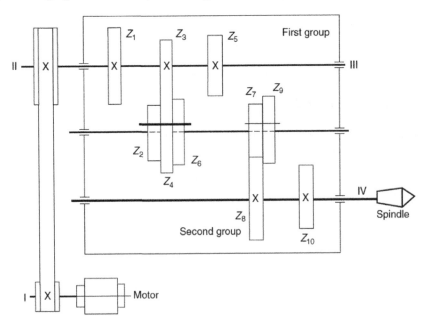

**FIGURE 2.32** Six-speed gearbox.

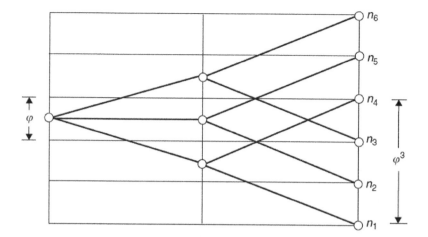

**FIGURE 2.33** Six-speed gearbox structural diagram.

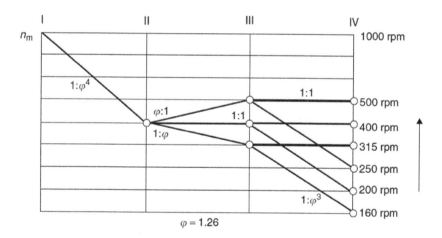

**FIGURE 2.34** Speed chart for six speed gearbox.

used, it can be used as the main group of a number of steps 2 and a progression ratio $\varphi_p = 2$. If the gearbox to be designed has a progression ratio $\varphi = 1.25$, then this motor is used as the first extension group of $\varphi_p = \varphi^3 = 2$. The number of speed steps of the main group $z_g$ following the electrical group is given by

$$z_g = \frac{\log \varphi_p}{\log \varphi} = 3 \qquad (2.14)$$

Figure 2.35 shows the kinematic diagram and the speed chart for a six-speed gearbox driven by a two-speed pole-changing motor. Accordingly, it is clear that the gearbox has been simplified by using the two-speed induction motor, in which the number of shafts and gears has been reduced (two shafts instead of three and six gears instead of 10).

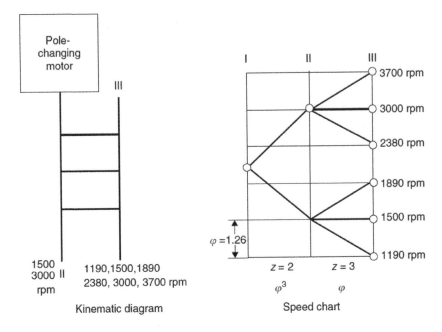

**FIGURE 2.35** Kinematic diagram and speed chart for six-speed gearbox driven by pole-changing induction motor of two speeds.

### 2.5.1.8 Feed Gearboxes

Feed gearboxes are designed to provide the feed rates required for the machining operation. The values of feed rates are determined by the specified surface finish, the tool life, and the rate of material removal. The classification of feed gearboxes according to the type of mechanism used to change the rate of feed is as follows:

1. *Feed gearboxes with pick-off gears.* Used in batch-production machine tools with infrequent changeover from job to job, such as automatic, semi-automatic, single-purpose, and special-purpose machine tools. These gearboxes are simple in design and are similar to those used for speed changing (Figure 2.29).
2. *Feed gearboxes with sliding gears.* These gearboxes are widely used in general-purpose machine tools, transmit high torques, and operate at high speeds. Figure 2.36 shows a typical gearbox that provides four different ratios. Accordingly, gears $Z_2$, $Z_4$, $Z_6$, and $Z_8$ are keyed to the drive shaft and mesh, respectively, with gears $Z_1$, $Z_3$, $Z_5$, and $Z_7$, which are mounted freely on the driven key shaft. The sliding key engages any gear on the driven shaft. The engaged gear transmits the motion to the driven shaft, while the rest of the gears remain idle. The main drawbacks of such feed boxes are the power loss and wear occurring due to the rotation of idle gears and insufficient rigidity of the sliding key shaft. Feed boxes with

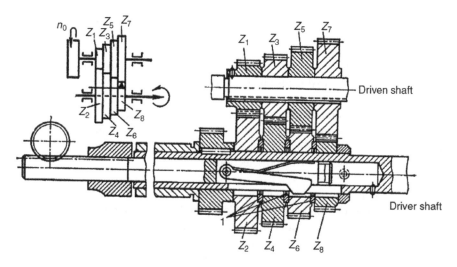

**FIGURE 2.36**   Feed gearbox with sliding gears.

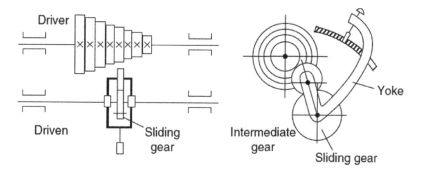

**FIGURE 2.37**   Norton gearbox.

sliding gears are used in small- and medium-size drilling machines and turret lathes.

3. *Norton gearboxes.* These gearboxes provide an arithmetic series of feed steps that is suitable for cutting threads, and so are widely used in engine lathe feed gearboxes, as shown in Figure 2.37.

### 2.5.1.9   Preselection of Feeds and Speeds

Preselection mechanisms in machine tools are used to select the speeds and feeds for the next machining operation during the machining time of the current operation. Once the current operation is finished, the selected speed and feed are automatically switched on with the press of a button. The main advantage of such a system in machine tools is to save the significant secondary time normally used for selecting the speeds and feeds at the end of each machining operation. Consequently, the total production time is reduced. The adoption of preselection mechanisms is justified whenever the speeds and feeds of the machine tool are frequently changed. Preselection has the following three steps:

1. Positioning the switching elements in the required position corresponding to the required (next) feeds and speeds without actuation, which is carried out during the cutting time of the current operation
2. Switching on directly after the current machining operation is finished by bringing the corresponding coupling and shiftable gears in mesh
3. Returning the switching elements to the original position automatically to be ready for the following preselection

Preselection may be carried out mechanically, electrically, and hydraulically. Figure 2.38 shows an example of mechanical preselection for a nine-speed gearbox. The process is carried out by adjusting the preselection dial (a) to the required speed. Hence, the preselection drum (b) is rotated to the required position. Once the current machining operation is finished, the drums (b) and (c) are shifted axially against each other by pulling lever 1 in the switching-on position. The shifting forks ($k_1$) and

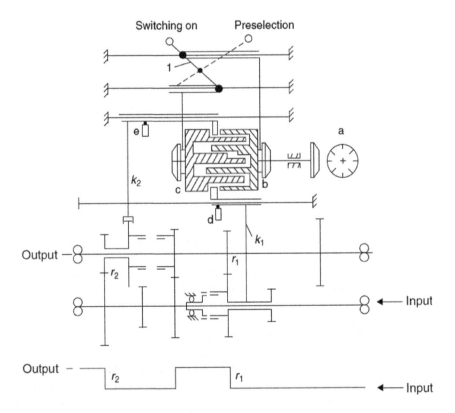

**FIGURE 2.38** Preselection of spindle speeds. (From Youssef, H. et al., *Design and Construction of Machine Tool Elements*, Dar El-Maaref Publishing Co., Alexandria, Egypt, 1976. With permission.)

$(k_2)$ are moved using fingers (d) and (e) to the required position. Consequently, the blocks $r_1$ and $r_2$ are switched to mesh, giving the required speed.

### 2.5.2 STEPLESS SPEED DRIVES

#### 2.5.2.1 Mechanical Stepless Drives

Infinitely variable speed (stepless) drives provide output speeds, forming infinitely variable ratios to the input ones. Such units are used for main as well as feed drives to provide the most suitable speed or feed for each job, thereby reducing the machining time. They also enable machining to be achieved at a constant cutting speed, which leads to an increased tool life and ensures uniform surface finish. The easy and smooth changing of the speed or feed, without stopping the machine, results in an appreciable reduction in the production time, which increases the productivity of the machine tool. Stepless speed drives may be mechanical, hydraulic, or electric. The selection of the suitable drive depends on the purpose of the machine tool, power requirements, speed range ratio, mechanical characteristics of the machining operation, and cost of the variable speed unit. In most stepless drives, the torque transmission is not positive. Their operation involves friction and slip losses. However, they are more compact, less expensive, and quieter in operation than the stepped speed control elements. Mechanical stepless drives include the following types.

*Friction Stepless Drive*

Figure 2.39 shows the disk-type friction stepless mechanism. Accordingly, the drive shaft rotates at a constant speed $n_1$ as well as the friction roller of diameter $d$. The output speed of the driven shaft rotates at a variable speed $n_2$ that depends on the instantaneous diameter $D$.

Because

$$n_1 d = n_2 D \tag{2.15}$$

hence

$$n_2 = n_1 \frac{d}{D} \tag{2.16}$$

The diameter ratio $d/D$ can be varied in infinitely small steps by the axial displacement of the friction roller. If the friction force between the friction roller and the disk is $F$,

$$F = \frac{\text{input torque}\,(T_1)}{\text{input radius}\,(d/2)} = \frac{\text{output torque}\,(T_2)}{\text{output radius}\,(D/2)} \tag{2.17}$$

If the power, contact pressure, transmission force, and efficiency are constant, the output torque $T_2$ is inversely proportional to the speed of the output shaft $n_2$.

$$T_2 \alpha T_1 \frac{n_1}{n_2} \tag{2.18}$$

Due to the small contact area, a certain amount of slip occurs, which makes this arrangement suitable for transmitting small torques and is limited to reduction ratios not more than 1:4.

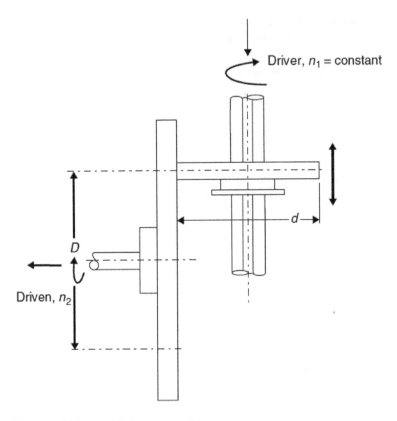

**FIGURE 2.39**   Disk-type friction stepless drive.

*Kopp Variator*

In the Kopp variator, shown in Figure 2.40, the drive balls (4) mounted on inclinable axes (3) run in contact with identical, effective radii $r_1 = r_2$, and drive cones (1 and 2) are fixed on coaxial input and output shafts. When the axes of the drive balls (3) are parallel to the drive shaft axes, the input and output speeds are the same. When they are tilted, $r_1$ and $r_2$ change, which leads to increase or decrease of the speed. Using the Kopp mechanism, a speed range of 9:1, efficiency of higher than 80%, and 0.25–12 hp capacity are obtainable.

*Toroidal and Reeves Mechanisms*

Figure 2.41 shows the principle of toroidal stepless speed transmission. Figure 2.42 shows the Reeves variable speed transmission, which consists of a pair of pulleys connected by a V-shaped belt; each pulley is made up of two conical disks. These disks slide equally and simultaneously along the shaft and rotate with it. To adjust the diameter of the pulley, the two disks on the shaft are made to approach each other so that the diameter is increased or decreased. The ratio of the driving diameter to the driven one can be easily changed, and therefore, any desired speed can be obtained without stopping the machine. Drives of this type are available with up to 8:1 speed range and 10 hp capacity.

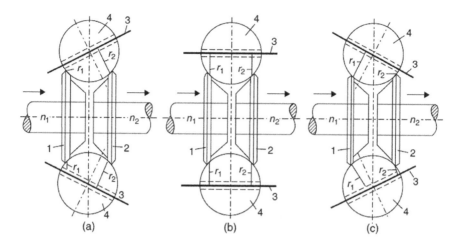

**FIGURE 2.40**　Kopp stepless speed mechanism: (a) $n_2 < n_1$, (b) $n_2 = n_1$, and (c) $n_2 > n_1$.

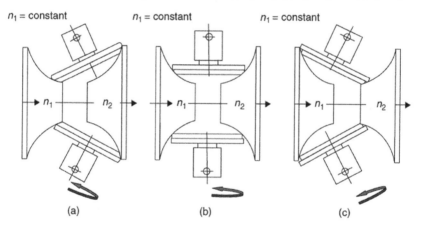

**FIGURE 2.41**　Toroidal stepless speed transmission: (a) $n_2 < n_1$, (b) $n_2 = n_1$, and (c) $n_2 > n_1$.

*Positive Infinitely Variable Drive*

Figure 2.43 shows a positive torque transmission arrangement that consists of two chain wheels, each of which consists of a pair of cones that are movable along the shafts in the axial direction.

The teeth of the chain wheels are connected by a special chain. Rotating the screw causes the levers to move, thus changing the location of the chain pulleys and hence the speed of rotation. This drive provides a speed ratio of up to 6 and is available with power rating up to 50 hp. The use of infinitely variable speed units in machine tool drives and feed units is limited by their higher cost and lower efficiency or speed range.

## 2.5.2.2　Electrical Stepless Speed Drive

Figure 2.44 shows the Leonard set, which consists of an induction motor, which drives the direct current (dc) generator, and an exciter (E). The dc generator provides

Movement of cone pulleys

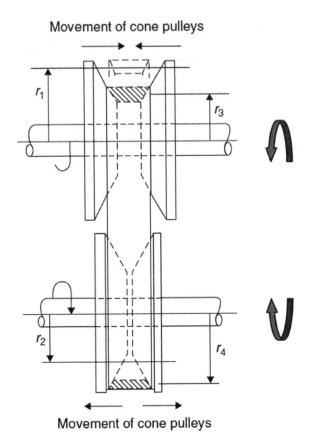

FIGURE 2.42 Reeves variable speed transmission.

the armature current for the dc motor, and the exciter provides the field current; both are necessary for the dc motors that drive the machine tool. The speed control of the dc motor takes place by adjusting both the armature and the field voltages by means of the variable resistances A and F, respectively. By varying the resistance A, the terminal voltage of the dc generator and hence the rotor voltage of the dc motor can be adjusted between zero and a maximum value. The Leonard set has a limited efficiency: it is large, expensive, and noisy. Nowadays, dc motors and thyristors that permit direct supply to the dc motors from an alternating current (ac) mains are available, and therefore, the Leonard set can be completely eliminated. Thyristor feed drives can be regulated such that the system offers infinitely variable speed control.

### 2.5.2.3 Hydraulic Stepless Speed Drive

The speeds of machine tools can be hydraulically regulated by controlling the oil discharge circulated in a hydraulic system consisting of a pump and hydraulic motor, both of the vane type, as shown in Figure 2.45. This is achieved by changing either the eccentricity of the pump $e_p$ or the eccentricity of the hydraulic motor $e_m$ or both. The vane pump, running approximately at a constant speed, delivers the pressurized

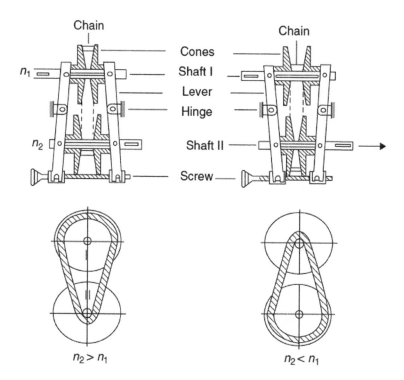

**FIGURE 2.43** Positive infinitely variable drive.

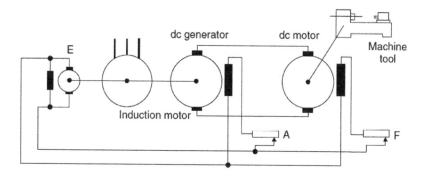

**FIGURE 2.44** Leonard set (electrical stepless speed drive).

oil to the vane-type hydraulic motor, which is coupled to the machine-tool spindle. To change the direction of rotation of the hydraulic motor, the reversal of the pump eccentricity is preferred. Speed control in hydraulic circuits can be accomplished by throttling the quantity of fluid flowing into or out of the hydrocylinders or hydromotor. The advantages of a hydraulic system are as follows:

1. Has a wide range of speed variation
2. Changes in the magnitude and direction of speed can be easily performed

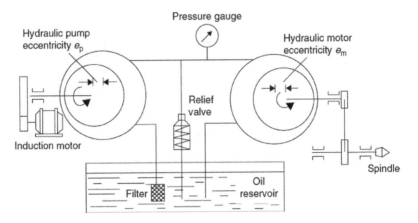

**FIGURE 2.45**  Hydraulic stepless speed drive.

3. Provides smooth and quiet operation
4. Ensures self-lubrication
5. Has automatic protection against overloads

The major drawback of a hydraulic system is that the operation of the hydraulic drive becomes unstable at low speeds. Additionally, the oil viscosity varies with temperature and may cause fluctuations in feed and speed rates.

## 2.6  PLANETARY TRANSMISSION

Figure 2.46 shows a planetary transmission with bevel gears that is widely used in machine tools. Accordingly, any two members may be the driving members, while the third one is the driven member. The differential contains central gears $Z_1$ and $Z_4$

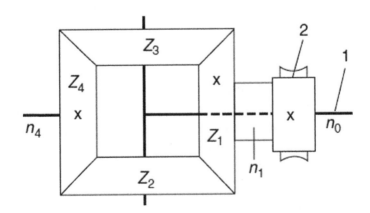

**FIGURE 2.46**  Planetary transmission.

and satellites $Z_2$ and $Z_3$ (an additional wheel) rotated by worm gear 2. The differential can operate as follows (Chernov, 1975):

1. $Z_4$ is a driving member, the carrier is a driven member, and worm gear 2 is stationary.
2. The carrier is a driving member, gear $Z_4$ is a driven member, and worm gear 2 is stationary.
3. Gear wheel $Z_1$ is a driving member (rotated by worm gear 2), gear wheel $Z_4$ is a driven member, and the carrier is fixed.
4. The carrier is a driving member, so is gear $Z_1$, and gear wheel $Z_4$ is a driven member.
5. Gear wheels $Z_1$ and $Z_4$ are driving members, and the carrier is a driven member.

The principal relationship between axis speeds is described by Willis formula, with $Z_2 = Z_3$ and $Z_1 = Z_4$, as follows:

$$i = \frac{n_4 - n_0}{n_1 - n_0} = \frac{Z_2 Z_1}{Z_4 Z_3} = -1$$

where

$i$ = conversion ratio
$n_0$ = speed of carrier rotation
$n_1, n_2$ = rotational speeds of $Z_1$ and $Z_4$, respectively

The minus sign in this equation indicates that gear wheels $Z_1$ and $Z_4$ rotate in opposite direction when the carrier is stationary. Willis also suggested the following relations:

$$n_0 = \frac{n_4}{2}, \qquad \text{i.e., } i = \frac{1}{2} \tag{2.19}$$

$$n_4 = 2n_0, \qquad \text{i.e., } i = 2 \tag{2.20}$$

$$n_4 = n_1, \qquad \text{i.e., } i = 1 \tag{2.21}$$

$$n_0 = \frac{n_1}{2} \pm \frac{n_4}{2} \tag{2.22}$$

The plus sign in Equation 2.22 indicates opposite rotational directions, and the minus sign indicates the same direction of the differential driving members.

## 2.7  MACHINE-TOOL MOTORS

Most machine-tool drives operate on standard three-phase 50 Hz, 400/440 V ac supply. The selection of motors for machine tools depends on the following:

1. Motor power
2. The power supply used (ac/dc)
3. Electrical characteristics of the motor

**TABLE 2.8**
**Machine Tool Motors**

| Machine Tool | Types of Motor |
|---|---|
| Lathe | |
|    Main drive and traverse drive | Multispeed squirrel cage |
| | Adjustable-speed dc |
|    Traverse drive | dc series |
| | High-slip squirrel cage |
| Shapers and slotters | Constant-speed squirrel cage |
| Planers | Multispeed squirrel cage |
| | dc adjustable voltage |
| Drilling machines | Constant-speed squirrel cage |
| | dc shunt motor |
| Milling machines | Squirrel cage |
| | dc shunt motor |
| Power saws | Constant-speed squirrel cage |
| Grinding machines | |
|    Wheel | Constant-speed squirrel cage |
| | Adjustable-speed dc |
|    Traverse | Constant-speed squirrel cage |

From Nagpal, G.R., in *Machine Tool Engineering*, Khanna Publishers, Delhi, India, 1999.

4. Mechanical features that include mounting, transmission of drive, noise level, and the type of cooling
5. Overload capacity

Squirrel-cage induction motors are the most popular due to their simplicity, robustness, availability with a wide range of operating characteristics, and low cost. ac motors can provide infinitely variable speed over a wide range; however, their cost is high. dc shunt motors with field and armature control are commonly used for the main drives. For traverse drives, dc series or compound wound motors are preferred. Table 2.8 shows the different machine-tool motors recommended for machine tools (Nagpal, 1996).

## 2.8 REVERSING MECHANISMS

Movements of machine-tool elements can be reversed by mechanical, electrical, and hydraulic devices. Among these are mechanisms with spur gears and bevel gears. Figure 2.47 shows reversing mechanisms with sliding spur gears (a) and those with fixed gears and clutches (b). Figure 2.47 also shows a reversing mechanism with bevel gears and a double-claw clutch (c). Hydraulic reversal of motion is effected by redirection of the oil delivered to an operative cylinder using a directional control

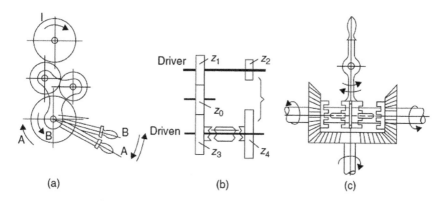

**FIGURE 2.47** Reversing mechanisms: (a) tumbler yoke gear, (b) spur gear with clutch, and (c) bevel gear with clutch.

valve, and electrical reversal is achieved by changing the direction of the drive motor rotation.

## 2.9 COUPLINGS AND BRAKES

Shaft couplings are used to fasten together the ends of two coaxial shafts. Permanent couplings cannot be disengaged while clutches engage and disengage shafts in operation. Safety clutches avoid the breakdown of the engaging mechanisms due to a sharp increase in load, while overrunning clutches transmit the motion in only one direction. Figure 2.48 shows permanent couplings. Figure 2.49 shows a typical claw clutch (a) and a toothed clutch (b). These two clutches cannot be engaged when the difference between the speeds of shafts is high. However, a friction clutch (c) can be engaged regardless of the speeds of its two members. Additionally, they can slip in the case of overloading. Other types of clutch include friction multidisk, contactless magnetic, and hydraulic clutch (Chernov, 1984). Brakes are used in machine tools to quickly slow or completely stop their moving parts. This step can be performed using mechanical, electrical, or hydraulic (or a combination of these) devices. Figure 2.50

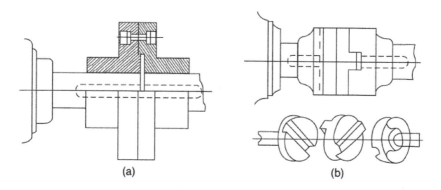

**FIGURE 2.48** (a) Flanged coupling and (b) Oldham coupling.

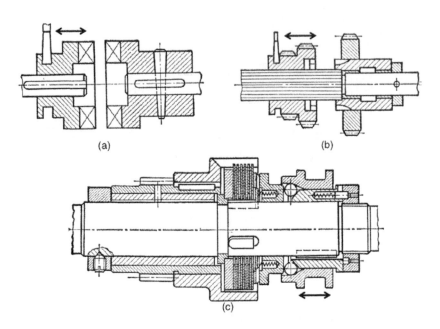

**FIGURE 2.49**   (a) Claw clutch, (b) toothed clutch, and (c) friction clutch. (From Chernov, N., *Machine Tools*, Mir Publishers, Moscow, 1975. With permission.)

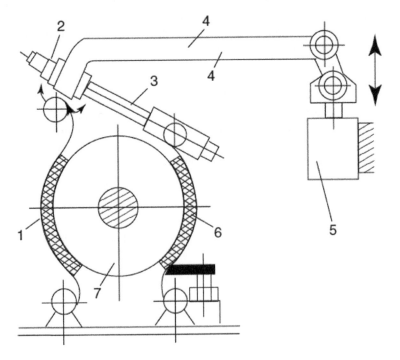

**FIGURE 2.50**   Shoe brake. (From Chernov, N., *Machine Tools*, Mir Publishers, Moscow, 1975. With permission.)

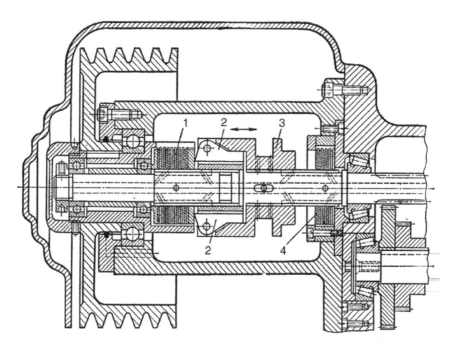

**FIGURE 2.51** Friction brake. (From Chernov, N., *Machine Tools*, Mir Publishers, Moscow, 1975. With permission.)

shows the shoe brake, in which shoes (1 and 6) are connected by a rod (3), whose length is controlled by a nut (2) that controls the clearance between the shoes and the pulley (7). Braking is achieved by pressing the shoe against the pulley by an arm (4) driven by the brake actuator (5). Band brakes operate frequently by electromagnetic or solenoid actuators. In a multiple-disk friction brake, shown in Figure 2.51, when the shaft sleeve (3) is moved to the left, it engages with its lever (2), which in turn compresses the clutch disks, thereby engaging the clutch. For braking, the sliding sleeve (3) is moved to the right, disengaging the clutch (1) and engaging the friction brake (4).

## 2.10 RECIPROCATING MECHANISMS

### 2.10.1 QUICK-RETURN MECHANISM

Ruled flat surfaces are machined on the shaping or planing machines by the combined reciprocating motion and the side feed of the tool and WP. Figure 2.52 shows the quick-return mechanism of the shaper machine. Accordingly, the length of the stroke is controlled by the radial position of the crank pin and sliders A and B. The time taken for the crank pin to move through the angle corresponding to the cutting stroke $\alpha$ is less than that of the noncutting stroke $\beta$ (the usual ratio is 2:1). Velocity curves for the cutting and reverse strokes are shown in Figure 2.52. The maximum speed occurs when the link is vertical.

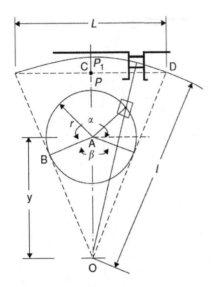
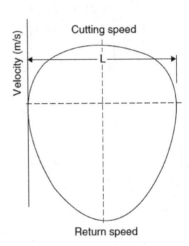

**FIGURE 2.52** Quick-return mechanism.

The speed of the link at point $P$ for a given stroke length $L$ will be that at the corresponding crank radius $r$; hence, the cutting speed $v_c$ at point $P_1$ is

$$v_c = 2\pi rn \frac{l}{y+r} \text{ m/min} \tag{2.23}$$

where
   $n$ = number of strokes per minute
   $l$ = length of crank arm (constant)

Similarly, the maximum reverse speed $v_r$ is given by the following equation:

$$v_r = 2\pi rn \frac{l}{y-r} \text{ m/min} \tag{2.24}$$

In terms of the stroke length for maximum radius using similar triangles OBA and OCD,

$$\frac{OD}{OA} = \frac{DC}{AB}$$

$$\frac{l}{y} = \frac{L}{2r} \tag{2.25}$$

hence

$$v_c = \pi n \left[ \frac{lL}{l+L/2} \right] \tag{2.26}$$

and

$$v_r = \pi n \left[ \frac{lL}{l-L/2} \right] \tag{2.27}$$

therefore, the speed ratio, $Q$

$$Q = \frac{V_r}{V_c} = \frac{2l + L}{2l - L} \tag{2.28}$$

## Illustrative Example 3

In the slotted arm quick-return mechanism of the shaping machine, the maximum quick-return ratio is 3/2, and the stroke length is 400 mm. Calculate the length of the slotted arm. Calculate the maximum quick-return ratio if the stroke length is 180 mm.

### SOLUTION

The quick-return ratio $Q$

$$Q = \frac{V_r}{V_c} = \frac{2l + L}{2l - L}$$

$$Q = \frac{3}{2} = \frac{2l + 400}{2l - 400}$$

$$l = 1200 \text{mm}$$

The quick-return ratio $Q$ for $L = 180$ mm

$$Q = \frac{2 \times 1200 + 180}{2 \times 1200 - 180} = 1.11$$

## 2.10.2 Whitworth Mechanism

This arrangement is shown in Figure 2.53; when AB rotates, it drives CE about D by means of the slider F so that G moves horizontally along MN. AB moves through an angle $(360° - \alpha)$, while CE moves through 180°, which is less than $360° - \alpha$. Also, the crank moves through $\alpha$ while CE moves through 180°, which is greater than $\alpha$. Hence, with a uniformly rotating crank, the link moves through one-half of its revolution more quickly than the other. The angle $\alpha$ is used for the return stroke. Hence,

$$\frac{\text{Time for cutting stroke}}{\text{Time for return stroke}} = \frac{360 - \alpha}{\alpha} \tag{2.29}$$

The stroke can be changed by altering the radius DE, with the angle $\alpha$ being unchanged. Provided that the fixed center D lies on the line of movement of G, the ratio of the cutting speed to the return speed lies between 1:2 and 1:2.5.

## 2.10.3 Hydraulic Reciprocating Mechanism

As shown in Figure 2.54, the electrically driven pump supplies the fluid under pressure to the operating cylinder through the solenoid-operated valve. The piston is connected to the machine table. At the end of the forward stroke, the direction control

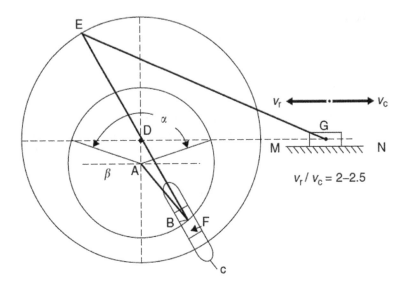

**FIGURE 2.53**   Whitworth quick-return mechanism.

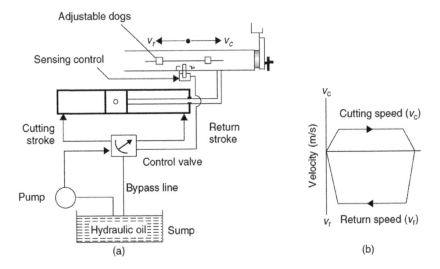

**FIGURE 2.54**   Reciprocating mechanism (a) and velocity diagram (b) of hydraulic shaper.

valve reverses the direction of the flow through limit switches set at the stroke limits, and the table moves backward.

## 2.11   MATERIAL SELECTION AND HEAT TREATMENT OF MACHINE-TOOL COMPONENTS

The operating characteristics of a machine-tool component depend on the proper choice of the material of each component. The most extensively used materials in machine-tool components include CI and steels.

## 2.11.1 Cast Iron

In the majority of cases, machine-tool beds and frames are made of gray CI (see Table 2.9) because of its good damping characteristics. If the guideways are cast as an integral part of the bed, frame, column, and so on, high–wear resistance grade CI (GG22 or A48–30B) with pearlitic matrix is recommended for medium-size machine-tool beds and frames for a wall thickness of 10–30 mm and the grade GG26 or A48–40B for a wall thickness of 20–60 mm. High-strength, wear-resistant special gray CI of the grade GG30 or A48–50B with a pearlitic structure can be used for heavy machine-tool beds with a wall thickness of more than 20 mm.

Due to the drawbacks associated with the manufacture of beds and frames by casting, beds and frames are made by welding rolled steel sheets. The elastic limit and the mechanical properties of such steel are higher than those of CI. Therefore, much less material (50–75%) is required for welded steel structures or beds than CI ones to be subjected to the same forces and torques, if the rigidity and stiffness of the two structures are made equal. CI beds are more often used in large-lot production, while welded steel beds and frames are preferable in job or small-lot production.

## 2.11.2 Steels

The majority of machine-tool components, such as spindles, guides, shafts, springs, keys, forks, and levers, are generally made of steels. Since the Young's modulus of various types of steels cannot vary by more than $\pm 3\%$, the use of the alloy steels for machine-tool components does not provide any advantages unless their application is mandated by other requirements. Tables 2.10 and 2.11 show the different

---

**TABLE 2.9**

**Grades of Gray CI According to DIN 1691, American Iron and Steel Institute (AISI), Society of Automotive Engineers/American Society for Testing and Materials (SAE/ASTM)**

| DIN 1691 | AISI, SAE/ ASTM | C (%) | Brinell Hardness Number (BHN) (kg/mm²) | Applications | Approximate Composition (%) |
|---|---|---|---|---|---|
| GG 12 | A48-20B | 3.5 | 160 | No acceptance test for parts of no special requirements | C = 3.2–3.6, Si = 1.7–3, Mn = 0.5, P = 0.5, S = 0.12 |
| GG 14 | A48-26B | 3.4 | 180 | | |
| GG 18 | A48-30B | 3.3 | 200 | | |
| GG 22 | A48-30B | 3.3 | 210 | Machine parts and frames | |
| GG 26 | A48-40B | 3.2 | 230 | To withstand high stresses | |
| GG 30 | A48-50B | 2.8 | 240 | Machine parts and frames of special quality | C = 2.8–3.0, Si = 1.5–1.7, Mn = 0.8–1.8, P = 0.3, S = 0.12 |

**TABLE 2.10**

**Structural Steel According to DIN 17100 and AISI, SAE/ASTM**

| DIN 17100 | AISI, SAE/ASTM | Mechanical Properties | | | | Hardening Temperature (°C) | Properties | Applications |
|---|---|---|---|---|---|---|---|---|
| | | C (%) | $\sigma_u$ (kg/mm²) | $\sigma_e$ (kg/mm²) | $\delta_5$ (%) | | | |
| St 34 | – | 0.17 | 34–42 | 18 | 30 | 920 | Case hardenable and weldable | Case-hardened parts |
| St 37 | – | 0.20 | 37–45 | – | 25 | 920 | Low grade, low weldability T[a] or M[a] | General machine constructions |
| St 42 | – | 0.25 | 42–50 | 23 | 25 | 880–900 | Case hardenable, hard core, machinable, not weldable | Machine elements and shafts withstanding variable loads |
| St 50 | A570Gr50 | 0.35 | 50–60 | 27 | 22 | 820–850 | Not case hardenable, not weldable, may be hardened, machinable | Machine elements and shafts withstanding heavy loads, not hardened gears |
| St 52 | – | 0.17 | 52–64 | 35 | 22 | 920 | High strength, weldable | Welded steel construction in bridges and automotives |
| St 60 | – | 0.45 | 60–70 | 30 | 17 | 800–820 | Can be hardened and toughened | Same applications as St 50 but for higher loads, keys, gears, worms |
| St 70 | – | 0.60 | 70–85 | 35 | 12 | 780–800 | Can be hardened and toughened | For parts in which wear resistance is recommended |

[a] T, Thomas; M, Martin.

**TABLE 2.11**

**Case-Hardened Steels According to DIN 17210 and AISI, SAE/ASTM**

| DIN 17210 | Quenching | AISI, SAE/ ASTM | Composition (%) | | | | Mechanical Properties | | | Applications |
|---|---|---|---|---|---|---|---|---|---|---|
| | | | C | Mn | Cr | Ni | $\sigma_u$ (kg/mm) | $\sigma_e$ (kg/mm$^2$) | $\delta_5$ (%) | |
| C 10 | Water | 1010 | 0.06–0.12 | 0.25–0.5 | – | – | 50 | 29 | – | Typewriter parts |
| C 15 | | 1015 | 0.12–0.18 | 0.25–0.5 | – | – | 55 | 35 | – | Levers, bolts, sleeves |
| CK 10* | | 1010 | 0.06–0.12 | 0.25–0.5 | – | – | 50 | 30 | 20 | Levers, bolts, pins of good surface finish |
| CK 15* | | 1015 | 0.12–0.18 | 0.25–0.5 | – | – | 55–60 | 35 | 15 | |
| 15Cr3 | | – | 0.12–0.18 | 0.4–0.6 | 0.5–0.8 | – | 70–90 | 49 | 12 | Spindles, cam shafts, piston pins, bolts, measuring tools |
| 16MnCr3 | Oil | 5115 | 0.14–0.19 | 1–1.3 | 0.8–1.1 | – | 85–110 | 60 | 20–10 | Pinions, automotive shafts, machine shafts |
| 15CrNi6 | | – | 0.12–0.17 | 0.4–0.6 | 1.4–1.7 | 1.4–1.7 | 95–120 | 70–90 | 15–6 | Highly stressed small gears |
| 20MnCr5 | | 5120 | 0.17–0.22 | 1.1–1.4 | 1.0–1.3 | – | 110–145 | 75 | 12–7 | Medium-size gears, automotive shafts, machine shafts |
| 18CrNi8 | | | 0.15–0.22 | 0.4–0.6 | 1.8–2.1 | 1.8–2.1 | 120–145 | 90–110 | 14–7 | Highly stressed gears, shafts, spindles, differential gears |
| 41Cr4 | Cy | 5140 | 0.38–0.40 | 0.5–0.8 | 0.9–1.2 | – | 160–190 | 130–140 | 12–7 | Cyanided gears |

CK 10* and CK 15* are carbon steels of quality better than C10 and C15 due to smaller contents of S and P; Cy, cyaniding.

types of structural and alloy steels frequently used in machine tools. Structural steels are used when no special requirements are needed. Case-hardening steels of carbon content <0.25% and phosphorus (P) or sulfur (S) not exceeding 0.40% are used when the surface hardness of the component needs to be very high while the core remains tough. Typical applications of case-hardening steels are in gears, shafts, and spindles. Tempered steels, shown in Table 2.12, have a higher carbon content than case-hardened steels. They are used when high strength and toughness are required. Nonalloy tempered steels are used for machine components that are not heavily loaded. For components that are heavily loaded, such as gears, spindles, and shafts, the alloy type is recommended. Nitriding steels (see Table 2.13) contain aluminum as the main alloying element. After nitriding, the components possess an extraordinary surface hardness and therefore are used for machine parts subjected to wear, such as spindles, guideways, and gears. The main advantage of the nitriding steel is minimum distortion after nitriding.

## 2.12   TESTING OF MACHINE TOOLS

After the manufacture or repair of any machine tool, a machine-tool test (usually called an *acceptance test*) should be performed according to the approved general specification. Such tests are essential, because the accuracy and surface quality of the parts produced depend on the performance of the machine tool used. Testing machine tools has the following general advantages:

1. Determines the precision class and the accuracy capabilities of the machine tool
2. Prepares plans for preventive maintenance
3. Determines the actual condition and hence the expected life of the machine tool

Machine-tool tests are classified into two categories: geometrical alignment tests and performance tests.

Geometrical tests cover the manufactured accuracy of machine tools. These tests are carried out to determine the various relationships between the various machine-tool elements when idle and unloaded (static test). They include checking parallelism of the spindle and a lathe bed, squareness of the table movement to the milling machine spindle, straightness of guideways, and so on. Static tests are inadequate to judge the machine-tool performance, because they do not reveal the machine-tool rigidity or the accuracy of machining. The normal procedure for acceptance tests is made through the following steps:

1. Checking the principal horizontal and vertical planes and axes using a spirit level
2. Checking the guiding and bearing surfaces for parallelism, flatness, and straightness, using dial gauge, test mandrel, straight edge, and squares
3. Checking the various movements in different directions using dial gauges, mandrels, straight edges, and squares

## TABLE 2.12
## Tempered Steels According to DIN 17100, AISI, SAE/ASTM

| DIN 17100 | AISI, SAE/ASTM | C | Si | Mn | Cr | Mo | Others | BHN | σ_u (kg/mm²) | σ_e (kg/mm²) | δ_5 (%) |
|---|---|---|---|---|---|---|---|---|---|---|---|
| C22 | 1020 | 0.18–0.25 | 0.15–0.36 | 0.3–0.6 | – | – | – | 155 | 50–60 | 30 | 22 |
| C35 | 1035 | 0.32–0.40 | 0.15–0.36 | 0.4–0.7 | – | – | – | 172 | 60–72 | 37 | 18 |
| C45 | 1045 | 0.42–0.50 | 0.15–0.36 | 0.5–0.8 | – | – | – | 206 | 65–80 | 40 | 16 |
| C60 | 1060 | 0.57–0.65 | 0.15–0.36 | 0.5–0.8 | – | – | – | 243 | 75–90 | 40 | 14 |
| CK22 | 1020–1023 | 0.18–0.25 | 0.15–0.36 | 0.3–0.6 | – | – | – | 155 | 50–60 | 30 | 22 |
| CK35 | 1035 | 0.32–0.40 | 0.15–0.36 | 0.4–0.7 | – | – | – | 172 | 60–72 | 37 | 18 |
| CK45 | 1045 | 0.42–0.50 | 0.15–0.36 | 0.5–0.8 | – | – | – | 206 | 65–80 | 49 | 16 |
| CK60 | 1055 | 0.57–0.65 | 0.15–0.36 | 0.5–0.8 | – | – | – | 243 | 75–90 | 40 | 14 |
| 40Mn4 | 1039 | 0.36–0.44 | 0.25–0.50 | 0.8–1.1 | – | – | – | 217 | 80–95 | 55 | 14 |
| 30Mn5 | 1330 | 0.27–0.34 | 0.15–0.35 | 1.2–1.5 | – | – | – | 217 | 88–95 | 55 | 14 |
| 37MnSi5 | – | 0.38–0.41 | 1.1–1.4 | 1.1–1.4 | – | – | – | 217 | 90–105 | 56 | 12 |
| 42MnV7 | – | 0.38–0.45 | 0.15–0.35 | 1.6–1.9 | – | – | 0.07–0.12 V | 217 | 100–120 | 80 | 11 |
| 34Cr4 | – | 0.30–0.37 | 0.15–0.55 | 0.5–0.8 | 0.9–1.2 | – | – | 217 | 90–105 | 65 | 12 |
| 41Cr4, 42Cr4 | 5140 | 0.38–0.44 | 0.15–0.55 | 0.5–0.8 | 0.9–1.2 | – | – | 217 | 90–105 | 65 | 12 |
| 25CrMo4 | 4130 | 0.22–0.29 | 0.15–0.55 | 0.5–0.8 | 0.9–1.2 | 0.15–0.25 | – | 217 | 80–95 | 55 | 14 |
| 34CrMo4 | 4135—4137 | 0.30–0.37 | 0.15–0.55 | 0.5–0.8 | 0.5–0.15 | – | – | 217 | 90–105 | 65 | 12 |
| 42CrMo4 | 4140^1142 | 0.38–0.45 | 0.15–0.55 | 0.5–0.8 | 0.9–1.2 | – | – | 217 | 100–120 | 80 | 11 |
| 50CrMo4 | 4150 | 0.46–0.54 | 0.15–0.55 | 0.5–0.8 | 0.9–1.2 | – | – | 235 | 110–130 | 90 | 10 |
| 30CrMoV9 | – | 0.26–0.34 | 0.15–0.55 | 0.4–0.7 | 2.3–2.1 | – | 0.1–0.2 V | 248 | 125–145 | 105 | 9 |
| 36CrNiMo4 | 9840 | 0.32–0.40 | 0.15–0.55 | 0.5–0.8 | 0.9–1.2 | – | 0.9–1.2 Ni | 217 | 100–120 | 80 | 11 |
| 34CrNiMo6 | 4340 | 0.30–0.38 | 0.15–0.55 | 0.4–0.7 | 1.4–1.7 | – | 1.4–1.7 Ni | 235 | 110–130 | 90 | 10 |
| 30CrNiMo8 | – | 0.26–0.34 | 0.15–0.55 | 0.3–0.6 | 1.8–2.1 | – | 1.8–2.1 Ni | 248 | 125–145 | 105 | 9 |
| 27NiCrV4 | – | 0.24–0.30 | 0.15–0.55 | 1.0–1.3 | 0.6–0.9 | – | 0.07–0.12 V | 217 | 80–95 | 55 | 14 |
| 36Cr6 | – | 0.32–0.40 | 0.15–0.55 | 0.3–0.6 | 1.4–1.7 | – | – | 217 | 100–105 | 65 | 12 |
| 42CrV6 | – | 0.38–0.46 | 0.15–0.55 | 0.5–0.8 | 1.4–1.7 | – | 0.07–0.12 V | 217 | 100–120 | 80 | 11 |
| 50CrV4 | 6150 | 0.47–0.56 | 0.15–0.55 | 0.8–1.1 | 0.9–1.12 | – | 0.07–0.12 V | 235 | 110–130 | 90 | 10 |

**TABLE 2.13**
**Nitriding Steels**

| Not Specified in DIN | AISI, SAE/ ASTM | Composition (%) | | | | | Mechanical Properties | | | Applications |
| --- | --- | --- | --- | --- | --- | --- | --- | --- | --- | --- |
| | | C | Cr | Al | Mn | Others | $\sigma_u$ (kg/ mm$^2$) | $\sigma_e$ (kg/ mm$^2$) | $\delta_5$ (%) | |
| 27CrAl6 | – | 0.27 | 1.5 | 1.1 | 0.6 | – | 85–80 | 45 | 16 | Valve stems |
| 34CrAl6 | A355Cl.D | 0.34 | 1.5 | 1.1 | 0.6 | – | 80–100 | 60 | 12 | Shafts, measuring instruments |
| 32AlCrMo4 | – | 0.32 | 1.1 | 1.1 | 0.6 | 0.2 Mo | 80–95 | 60 | 12 | Steam machinery shafts |
| 32AlNi7 | – | 0.33 | 0.7 | 1.7 | 0.5 | 1.0 Ni | 88–100 | 60 | 14 | Piston rods, shafts |
| 31CrMoV9 | – | 0.31 | 2.3 | – | 0.6 | 0.15Mo/0.1Ni | 90–115 | 75 | 12 | Cam- and crankshafts |
| 30CrAlNi7 | – | 0.30 | 0.3 | 0.9 | 0.5 | 0.5 Ni | 65–80 | 45 | 14 | Spindles and shafts |

4. Testing the spindle concentricity, axial slip, and accuracy of axis
5. Conducting working tests to check whether the accuracy of machined parts is within the specified limits
6. Preparing acceptance charts for the machine tool that specify the type of test and the range of allowable limits of deformation, deflection, error in squareness, flatness eccentricity, parallelism, and amplitude of vibrations

In contrast, dynamic tests are used to check the working accuracy of machine tools through the following steps:

1. Performing an idle run test and operation check mechanisms
2. Checking for geometrical accuracy and surface roughness of the machined parts
3. Performing rigidity and vibration tests

Standards for testing machine tools are covered by Schlesinger (1961).

## 2.13 MAINTENANCE OF MACHINE TOOLS

Machine tools cannot produce accurate parts throughout their working life if there is excessive wear in their moving parts. Machine-tool maintenance delays the possible deterioration in machine tools and avoids the machine stoppage time that leads to lower productivity and higher production cost. Maintenance is classified under the following schemes.

### 2.13.1 PREVENTIVE MAINTENANCE

Preventive maintenance is mainly carried out to reduce wear and prevent disruption of the production program. Lubrication of all the moving parts that are subjected to sliding or rolling friction is essential. A regular planned preventive maintenance consists of minor and medium repairs as well as major overhaul. The features of a well-conceived preventive maintenance scheme include

1. Adequate records covering the volume of work
2. Inspection frequency schedule
3. Identification of all items to be included in the maintenance program
4. Well-qualified personnel

Preventive maintenance of machine tools ensures reliability, safety, and the availability of the right machine at the right time. Figure 2.55 shows preventive maintenance of a machine tool.

### 2.13.2 CORRECTIVE MAINTENANCE

When a machine tool is in use, it should be regularly checked to determine whether wear has reached the level when corrective maintenance should be carried out to

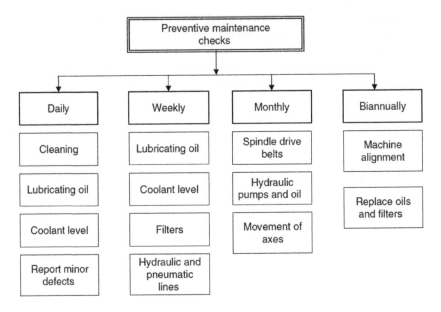

**FIGURE 2.55** Preventive maintenance scheme.

avoid machine-tool failure. A record of all previous repairs shows those elements of the machine tool that need frequent inspection. Additionally, such records are used for decisions regarding the need for machine-tool reconditioning and replacement.

### 2.13.3 RECONDITIONING

The need for machine-tool recondition is determined by the frequency of the corrective maintenance repairs. Every machine-tool component has a certain life span beyond which it becomes unserviceable despite the best preventive maintenance. A major overhaul or reconditioning is required.

Inspection reports of the machine indicate the components to be replaced, labor time, and the cost estimate. As a general rule, it is undesirable to recondition the machine if the cost exceeds 50% of the cost of buying new equipment.

## 2.14   REVIEW QUESTIONS

2.14.1   State the main requirements of a machine tool.

2.14.2   Give examples of open and closed machine-tool structures.

2.14.3   Explain why closed-box elements are best suited for machine-tool structures.

2.14.4   Sketch the different types of ribbing systems used in machine-tool frames.

2.14.5   Explain what is meant by light- and heavyweight construction in machine tools.

2.14.6   Sketch the different types of machine-tool guideways.

2.14.7   Show how wear is compensated for in machine-tool guideways.

2.14.8   Differentiate between cast and welded structures.

2.14.9   Distinguish among the kinematic, structural, and speed diagrams of gearboxes.

2.14.10   Show an example of externally pressurized and rolling friction guideways.

2.14.11   Show the different schemes of spindle mounting in machine tools.

2.14.12   What are the main applications of pick-off gears, feed gearboxes with a sliding gear, and Norton gearboxes?

2.14.13   Compare toroidal and disk-type stepless speed mechanisms.

2.14.14   Give examples of speed-reversing mechanisms in machine tools.

2.14.15   Derive the relationship between the cutting and the reverse speeds of the quick-return mechanism used in the mechanical shaper.

2.14.16   State the main objectives behind machine-tool testing.

2.14.17   Compare corrective and preventive maintenance of machine tools.

## REFERENCES

Browne, JW 1965, *The theory of machine tools, book-1*, Cassell and Co. Ltd., London.

Chernov, N. 1975, *Machine tools*, Mir Publishers, Moscow.

DIN 1691—Grades of gray cast iron.

DIN 17100—Tempered and structural steels.

DIN 17210—Case hardened steels.

DIN 323—Standard values of progression ratio.

DIN 803—Standard feeds.

DIN 804—Standard speeds.

ISO/R229—Standard feeds and speeds.

ISO/R229—Standard values of progression ratio.

Koenigsberger, F 1961, *Berechnungen, Konstruktionsgrundlagen und Bauelemente spanender Werkzeugmaschinen*, Springer, Berlin.

Nagpal, GR 1996, *Machine tool engineering*, Khanna Publishers, Delhi, India.

Schlessinger, G 1961, *Testing machine tools*, The Machine Publishing Company, London.

Youssef, H, Ragab, H & Issa, S 1976, *Design and construction of machine tool elements*, Dar Al-Maaref Publishing Company, Alexandria.

# 3 General-Purpose Metal-Cutting Machine Tools

## 3.1 INTRODUCTION

Machine tools are factory equipment used for producing machines, instruments, tools, and all kinds of spare parts. Therefore, the size of a country's stock of machine tools, and their technical quality and condition, characterize its industrial and technical potential fairly well. Metal-cutting machine tools are mainly grouped into the following categories:

- General-purpose machine tools. These are multipurpose machines used for a wide range of work.
- Special-purpose machine tools. These are machines used for making one type of part of a special configuration, such as screw thread and gear-cutting machines.
- Capstan, turret, and automated lathes.
- Numerical and computer numerical controlled machine tools.

In this chapter, the general-purpose machine tools are characterized and dealt with in brief. This group of machine tools comprises lathes, drilling machines, milling machines, shapers, planers, slotters, boring machines, jig boring machines, broaching machines, and microfinishing machines.

## 3.2 LATHE MACHINES AND OPERATIONS

Lathes are generally considered to be the oldest machine tools still used in industry. About one-third of the machine tools operating in engineering plants are lathe machines. Lathes are employed for turning external cylindrical, tapered, and contour surfaces and for boring cylindrical and tapered holes, machining face surfaces, cutting external and internal threads, knurling, centering, drilling, counterboring, countersinking, spot facing and reaming of holes, cutting off, and other operations. Lathes are used in both job and mass production.

### 3.2.1 TURNING OPERATIONS

In operations performed on lathes (turning operations), the primary cutting motion $v$ (rotary) is imparted to the workpiece (WP), and the feed motion $f$ (in most cases straight along the axis of the WP) is imparted to a single-point tool. The tool feed

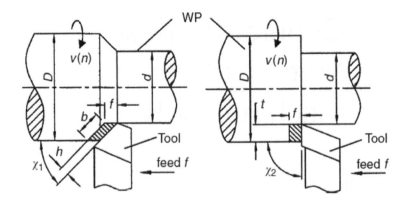

**FIGURE 3.1**   Basic machining parameters in turning.

rate *f* is usually very much smaller than the surface speed *v* of the WP. Figure 3.1 visualizes the basic machining parameters in turning, which include:

1. Cutting speed *v*

$$v = \frac{\pi D n}{1000} \text{ m/min} \tag{3.1}$$

   where
   *D* = initial diameter of the WP (mm)
   *n* = rotational speed of the WP (rpm)
2. Rotational speed *n*

$$n = \frac{1000v}{\pi D} \text{ rpm} \tag{3.2}$$

3. Feed rate *f*, which is the movement of the tool cutting edge in millimeters per revolution of the WP (mm/rev)
4. Depth of cut *t*, which is measured in a direction perpendicular to the WP axis, for one turning pass

$$t = \frac{D - d}{2} \text{ mm} \tag{3.3}$$

   where *d* is the diameter of the machined surface
5. Undeformed chip cross-section area $A_c$

$$A_c = f \cdot t = h \cdot b \text{ mm}^2 \tag{3.4}$$

   where
   *h* = chip thickness in millimeters ($h = f \sin\chi$ mm)
   *b* = contact length in millimeters
   $\chi$ = cutting edge angle (setting angle)

Different types of turning operations using different tools together with cutting motions *v, f* are illustrated in Table 3.1.

## TABLE 3.1
## Lathe Operations and Relevant Tools

| Lathe Operation and Relevant Tool | Sketch and Directions of Cutting Movements |
|---|---|
| 1. Cylindrical turning with a straight-shank turning tool | 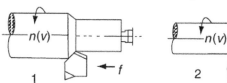 |
| 2. Taper turning with a straight-shank turning tool | |
| 3. Facing of a WP with:<br>  a. Facing tool while the WP is clamped by a half center<br>  b. Facing tool while the WP is mounted in a chuck | |
| 4. Finish turning with:<br>  a. Broad-nose finishing tool<br>  b. Straight finishing tool with a nose radius | |
| 5. Necking or recessing with:<br>  a. Recessing tool<br>  b. Wide recessing tool<br>  c. Wide recessing using narrow recessing tool | |
| 6. Parting off with parting-off tool | 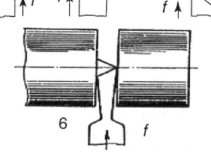 |

*(Continued)*

## TABLE 3.1 (CONTINUED)
## Lathe Operations and Relevant Tools

| Lathe Operation and Relevant Tool | Sketch and Directions of Cutting Movements |
|---|---|
| 7. Boring of cylindrical hole with:<br>  a. Bent rough-boring tool<br>  b. Bent finish-boring tool | |
| 8. Threading with:<br>  a. External threading tool<br>  b. Internal threading tool | |
| 9. Drilling and core drilling with a twist drill:<br>  a. Originating with a twist drill<br>  b. Enlarging with a twist drill<br>  c. Enlarging with a core drill | |
| 10. Forming with:<br>  a. Straight forming tool<br>  b. Flat dovetailed tool<br>  c. Circular form tool | |

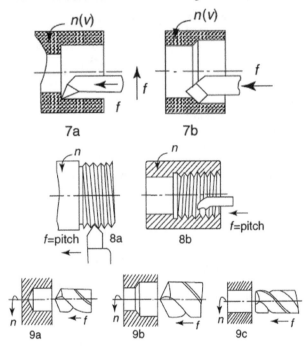

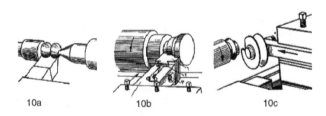

### 3.2.2 METAL-CUTTING LATHES

Every engine lathe provides a means for traversing the cutting tool along the axis of revolution of the WP and at right angles to it. Beyond this similarity, the lathe may embody other characteristics common to several classifications according to fields of application, which range from manual to fully automatic machining. Metal-cutting lathes may differ in size and construction. Among these are the general-purpose machines, which include universal engine lathes, plain turning lathes, facing lathes, and vertical turning and boring mills.

### 3.2.2.1 Universal Engine Lathes

Universal engine lathes are widely employed in job and lot production as well as for repair work. Parts of very versatile forms may be machined by these lathes. Their size varies from small bench lathes to heavy-duty lathes for machining parts weighing many tons. Figure 3.2 illustrates a typical universal engine lathe. The bed (2) carries the headstock (1), which contains the speed gearbox. The bed also mounts the tailstock (6), whose spindle usually carries the dead center. The work may be held between centers, clamped in a chuck, or held in a fixture mounted on a faceplate. If a long shaft (5) is to be machined, it will be insufficient to clamp one end in a chuck; therefore, it is necessary to support the other end by the tailstock center. In many cases, when the length of the shaft exceeds 10 times its diameter ($\ell > 10D$), a steady rest or follower rest is used to support these long shafts.

Single-point tools are clamped in a square turret (4) mounted on the carriage (3). Tools such as drills, core drills, and reamers are inserted in the tailstock spindle after removing the center. The carriage (3), to which the apron (10) is secured, may traverse along the guideways either manually or using power. The cross slide can also be either manually or power traversed in the cross direction.

Surfaces of revolution are turned by longitudinal traverse of the carriage. The cross slide feeds the tool in the cross direction to perform facing, recessing, forming, and knurling operations. Power traverse of the carriage or cross slide is obtained through the feed mechanism. Rotation is transmitted from the spindle through change gears and the quick-change feed gearbox (11) to either the lead screw (8) or the feed rod (9). From either of these, motion is transmitted to the carriage. Powered motion of the lead screw is used only for cutting threads using a threading tool. In all other cases, the carriage is traversed by hand or powered from the feed rod. Carriage feed is obtained by a pinion and rack (7) fastened to the bed. The pinion may be actuated manually or powered from the feed rod. The cross slide is powered by the feed rod through a gearing system in the apron (10). Figure 3.3 shows an isometric

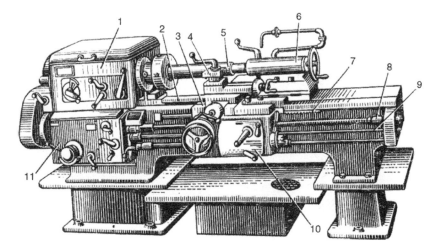

**FIGURE 3.2** Typical engine lathe.

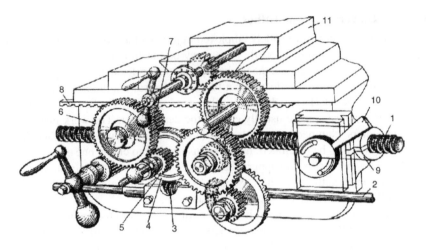

**FIGURE 3.3**  Lathe apron mechanism.

view of the apron mechanism. During thread cutting, the half nuts (9) are closed by the lever (10) over the lead screw (1).

### 3.2.2.1.1  Specifications of an Engine Lathe

Figure 3.4 shows the main dimensions that indicate the capacity of an engine lathe. These are:

- Maximum diameter $D$ of work accommodated over the bed (swing over bed). According to most national standards, $D$ varies from 100 to 6300 mm, arranged in geometric progression $\varphi = 1.26$.
- Maximum diameter $D_1$ of work accommodated over the carriage.

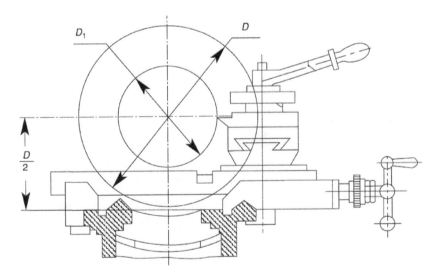

**FIGURE 3.4**  Main dimensions of an engine lathe.

- Distance between centers, which determines the maximum work length. It is measured with the tailstock shifted to its extreme right-hand position without overhanging.
- Maximum bore diameter of spindle, which determines the bar capacity (maximum bar stock).

In addition to these dimensions, other important specifications are:

- Number of spindle speeds and speed range
- Number of feeds and feed range
- Motor power and speed
- Overall dimensions and net weight

### 3.2.2.1.2 Setting up the Engine Lathe for Taper Turning

Tapered surfaces are turned by employing one of the following methods (Figure 3.5):

a. *By swiveling the compound rest to the required angle α.* Before performing the operation, the compound rest is to be clamped in this position. The tool is fed manually by rotating handle (1). This method is used for turning short internal and external tapers with large taper angles, while the work is commonly held in a chuck, and a straight turning tool is used (Figure 3.5a).

b. *By using a straight-edge broad-nose tool.* The tool of width that exceeds the taper being turned is cross-fed. The work is held in a chuck or clamped on a faceplate (Figure 3.5b).

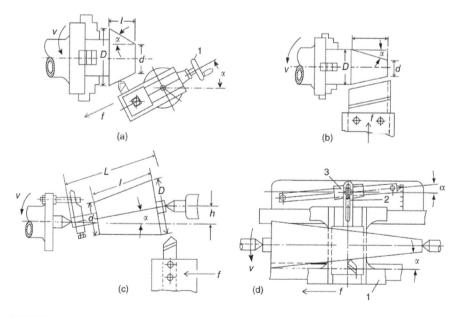

**FIGURE 3.5** Methods of taper turning: (a) swiveling the compound rest; (b) using a straight-edge broad-nose tool; (c) setting over the tailstock; (d) using a taper-turning attachment.

c. *By setting over the tailstock.* The angle of taper α should not exceed 8°. Since the turned surface is parallel to the spindle axis, the powered feed of the carriage can be used, while the work is to be mounted between centers as shown in Figure 3.5c. Before turning cylindrical surfaces, it is a good practice to check whether the tailstock is not previously set over for taper turning; otherwise, tapered surfaces are produced.

d. *By using a taper-turning attachment.* This is best suited for long tapered work. The cross slide (1) is disengaged from the cross feed screw and is linked through the tie (2) to the slide (3) (Figure 3.5d).

### 3.2.2.1.3 *Setting up the Engine Lathe for Turning Contoured Surfaces with a Tracer Device*

Longitudinal contoured surfaces are produced using a tracer device similar to the taper-turning attachment, except that the template of the required profile is substituted by the guide bar. The disadvantages of such mechanical duplication are the difficulties in making a template sufficiently accurate and strong to withstand the cutting force and the rapid wear of such templates. A mechanical tracer for turning spherical surfaces, shown in Figure 3.6, operates by similar principles. Accordingly, the template (1) is clamped in the tailstock spindle, and a roller (2) is clamped in the square turret opposite the tool (3) and in contact with the template. If the cross feed is transmitted to the cross slide, the profile of the template will be produced on the WP. When much contour turning work is to be done with longitudinal feeds, a hydraulic tracer slide is often installed on the engine lathe, whereby the stylus sliding on the template does not carry the cutting force.

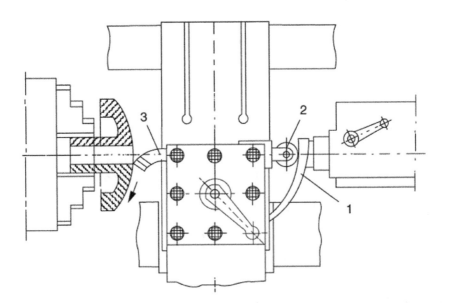

**FIGURE 3.6**  Turning of a spherical surface.

### 3.2.2.1.4 Setting up the Engine Lathe for Cutting Screw Threads

In some cases, when the machine does not have a quick-change gearbox, or when the thread pitch to be cut is nonstandard, change gears must be used and set up on the quadrant as shown in Figure 3.7. Because one revolution of the spindle provides the pitch $t_{th}$ of the screw thread to be produced, the kinematic linkage is given by the following equation:

$$t_{th} = t_{ls} \cdot i_{cg} \tag{3.5}$$

or

$$i_{cg} = \frac{t_{th}}{t_{ls}} = \frac{a}{b} \times \frac{c}{d} \tag{3.6}$$

where

$t_{ls}$ = pitch of the lead screw of the lathe
$i_{cg}$ = gearing ratio of the quadrant
$a, b, c, d$ = number of teeth of change gears

### 3.2.2.1.5 Holding the Work on an Engine Lathe

WP fixation on an engine lathe depends mainly upon the geometrical features of the WP and the precision required. The WP can be held between centers, on a mandrel, in a chuck, or on a faceplate:

1. *Holding the WP between centers.* A dog plate (1) and a lathe dog (2) are used (Figure 3.8a). It is an accurate method for clamping a long WP. The tailstock center may be a dead center (Figure 3.8b), or a live center (Figure 3.8c), when the work is rotating at high speed. In such a case, rests are used to support long WPs to prevent their deflection under the action of the cutting forces. The steady rest (Figure 3.9a) is mounted on the guideways of the bed, while the follower rest (Figure 3.9b) is mounted on the saddle of the carriage.

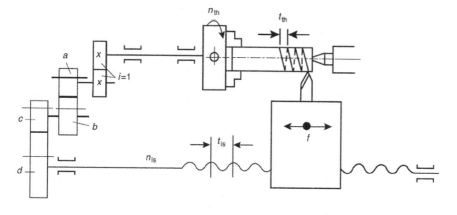

**FIGURE 3.7** Setting up the engine lathe for thread cutting.

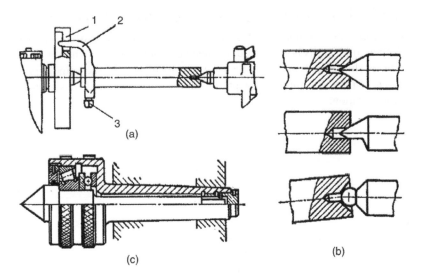

**FIGURE 3.8** Holding the work between centers: (a) dog plate, (b) dead centers, (c) live centers.

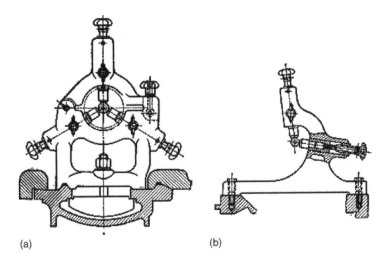

**FIGURE 3.9** Steady and follower rest of an engine lathe: (a) steady rest, (b) follower rest.

2. *Clamping hollow WPs on mandrels.* Mandrels are used to hold WPs with previously machined holes. The WP to be machined (2) is tightly fitted on a conical mandrel, tapered at 0.001, and provided with center holes to be clamped between centers using a dog plate and a lathe dog (Figure 3.10a). The expanding mandrel (Figure 3.10b) consists of a conical rod (1), a split sleeve (2), and nuts (3 and 4). The work is held by expansion of a sleeve (2) as the latter is displaced along the conical rod (1) by nut (3). Nut (4) removes the work from the mandrel. There is a flat (5) on the left of the conical rod used for the setscrew of the driving lathe dog.

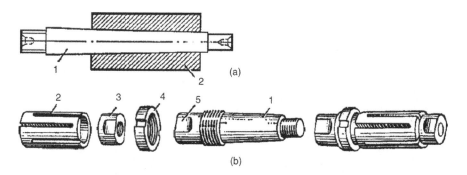

**FIGURE 3.10** Mounting WPs on a mandrel: (a) conical mandrel, (b) expanding mandrel.

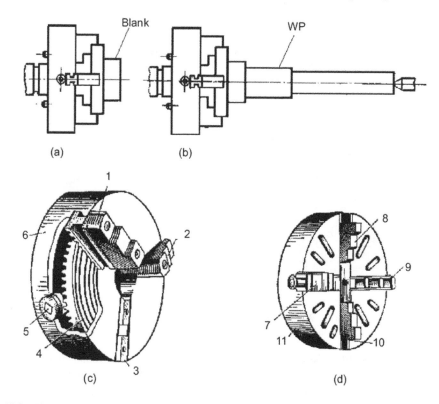

**FIGURE 3.11** Clamping WPs in chucks: (a) short WP, (b) long WP, (c) three-jaw chuck, (d) four-jaw chuck.

3. *Clamping the WP in a chuck.* The most commonly employed method of holding short work is to clamp it in a chuck (Figure 3.11a). If the work length is considerably large relative to its diameter, supporting the free end with the tailstock dead or live center (Figure 3.11b) is also used. Chucks may be universal (self-centering) with three jaws, which are expanded and drawn simultaneously (Figure 3.11c), or they may be independent with four jaws

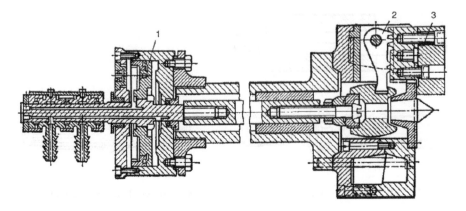

**FIGURE 3.12**   Pneumatic chuck.

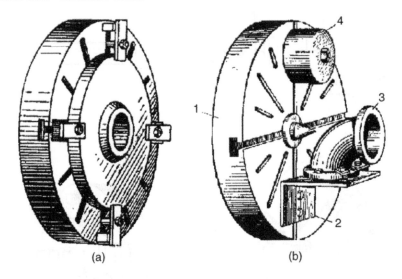

**FIGURE 3.13**   Mounting WPs on faceplates: (a) direct mounting on a faceplate, (b) mounting on a plate fixed to face plate.

(Figure 3.11d). The three-jaw chucks are used to clamp circular and hexagonal rods, whereas the independent four-jaw chucks are especially useful in clamping irregular and nonsymmetrical WPs. Air-operated (pneumatic) chucks are commonly used in batch or mass production by increasing the degree of automation (Figure 3.12). The piston (1) is attached to a rod that moves it to the right or to the left depending on which chamber of the pneumatic cylinder is fed with compressed air. The end of the rod is connected to three levers (2), which expand the jaws (3) in a radial direction to clamp or release the WP.

4. *Clamping the WP on a faceplate.* Large WPs cannot be clamped in a chuck and are, therefore, either mounted directly on a faceplate (Figure 3.13a) or mounted on a plate fixture (2) that is attached to a faceplate (1)

(Figure 3.13b). The work (3) and angle plate (2) must be counterbalanced by using the counterweight (4) mounted at the opposite position on the faceplate. The plate fixture has been proved to be highly efficient in machining asymmetrical work of complex and irregular shape.

### 3.2.2.2  Other Types of General-Purpose Metal-Cutting Lathes

These include plain turning lathes, facing lathes, and vertical turning and boring mills. Facing lathes, vertical turning and boring mills, and heavy-duty plain turning lathes are generally used for heavy work. They are characterized by low speeds, large feeds, and high cutting torques.

1. *Plain turning lathes.* Plain turning lathes differ from engine lathes in that they do not have a lead screw. They perform all types of lathe work except threading and chasing. The absence of the lead screw substantially simplifies the kinematic features and the construction of the feed gear trains. Their dimensional data are similar to those of engine lathes. Plain turning lathes are available in three different size ranges: small, medium, and heavy duty. Heavy-duty plain turning lathes have several common carriages that are powered either from a common feed rod, linked kinematically to the lathe spindle or from a variable-speed direct current (dc) motor mounted on each carriage. The tailstock traverses along the guideway by a separate drive.

2. *Facing lathes.* These are used to machine work of large diameter and short length in single-piece production and for repair jobs. These machines are generally used for turning external, internal, and taper surfaces, facing, boring, and so on. Facing lathes have relatively small length and large diameter of faceplates (up to 4 m). Sometimes, they are equipped with a tailstock. The construction of this lathe differs, to some extent, from the center lathe. It consists of the base plate (1), headstock (4) with faceplate (5), bed (2), carriage (3), and tailstock (6) (Figure 3.14). The work is clamped on the faceplate using jaws, or clamps, and T-slot bolts. It may be additionally supported by the tailstock center. The feed gear train is powered from a separate motor to provide the longitudinal and transverse feeds. Facing lathes have been almost superseded by vertical turning and boring mills; however, because of their simple construction and low cost, they are still employed.

3. *Vertical turning and boring mills.* These machines are employed in machining heavy pieces of large diameters and relatively small lengths. They are used for turning and boring cylindrical and tapered surfaces, facing, drilling, countersinking, counterboring, and reaming. In vertical turning and boring mills, the heavy work can be mounted on rotating tables more conveniently and safely as compared with facing lathes. The horizontal surface of the worktable completely excludes the overhanging load on the spindle of the facing lathes. This facilitates the application of high-velocity machining and, at the same time, enables high accuracy to be attained. These small machines are called vertical turret lathes. As their name implies, they are equipped with turret heads, which increase their productivity.

**FIGURE 3.14**   Facing lathe.

## 3.3   DRILLING MACHINES AND OPERATIONS

### 3.3.1   DRILLING AND DRILLING-ALLIED OPERATIONS

#### 3.3.1.1   Drilling Operation

Drilling is an extensively used process by which through or blind holes are origi-
nated or enlarged in a WP. This process involves feeding a rotating cutting tool
(drill) along its axis of rotation into a stationary WP (Figure 3.15). The axial feed
rate $f$ is usually very small when compared with the peripheral speed $v$. Drilling is
considered a roughing operation, and therefore, the accuracy and surface finish in
drilling are generally not of much concern. If high accuracy and good finish are
required, drilling must be followed by some other operation such as reaming, boring,
or grinding.

The most commonly employed drilling tool is the twist drill, which is available
in diameters ranging from 0.25 to 80 mm. A standard twist drill (Figure 3.16) is
characterized by a geometry in which the normal rake and the velocity of the cutting
edge are a function of their distance from the center of the drill. Referring to the
terminology of the twist drill shown in Figure 3.17, the helix angle of the twist drill
is the equivalent of the rake angle of other cutting tools. The standard helix is 30°,
which, together with a point angle of 118°, is suitable for drilling steel and cast iron
(CI) (Figure 3.17a). Drills with a helix angle of 20°, known as slow-helix drills, are
available with a point of 118° for cutting brass and bronze (Figure 3.17b) and with a
point of 90° for cutting plastics. Quick helix drills, with a helix angle of 40° and a
point of 100°, are suitable for drilling softer materials such as aluminum alloys and
copper (Figure 3.17c). Figure 3.18 visualizes the basic machining parameters in drilling and enlarging holes.

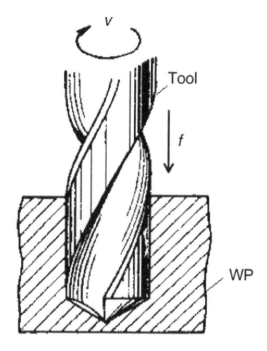

**FIGURE 3.15** Drilling operation.

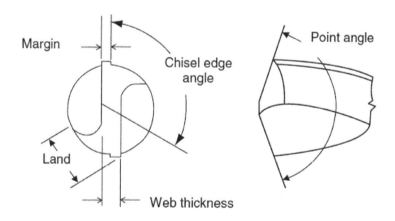

**FIGURE 3.16** Terminology of a standard point twist drill.

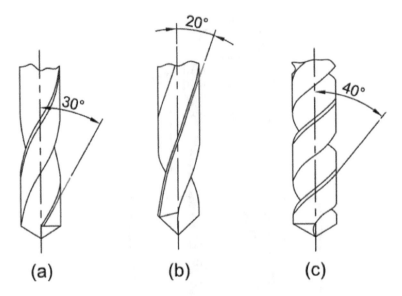

**FIGURE 3.17**   Helix drills of different helix angles: (a) standard, (b) slow, (c) quick.

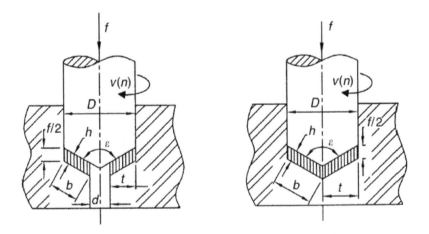

**FIGURE 3.18**   Basic machining parameters in drilling.

### 3.3.1.2   Drilling-Allied Operations

Drilling-allied or alternative operations such as core drilling, center drilling, counterboring, countersinking, spot facing, reaming, tapping, and other operations can
also be performed on drilling machines, as shown in Figure 3.19. Accordingly, the
main and feed motion are the same as in drilling; that is, the drill rotates while it is
fed into the stationary WP. In these processes, the tool shape and geometry depend
upon the machining process to be performed. The same operations can be accomplished in some other machines by holding the tool stationary and rotating the work.
The most general example is performing these processes on a center lathe, in which

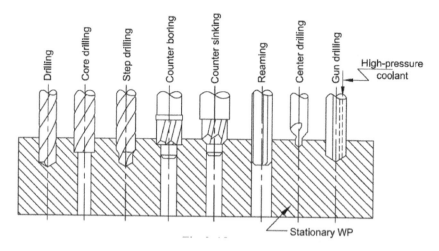

**FIGURE 3.19** Drilling and drilling-allied operations.

the tool (drill, counterbore, reamer, tap, and so on) is held in the tailstock, and the work is held and rotated by a chuck (Figure 3.20). The most important drilling-allied processes are as follows:

1. Core drilling, which is performed for the purpose of enlarging holes, as shown in Figure 3.19. Higher dimensional and form accuracy and improved surface quality can be obtained by this operation. It is usually an intermediate operation between drilling and reaming. Similar allowances should be considered for both reaming and core drilling. Core drills are of three or four flutes; they have no web or chisel edge and consequently provide better guidance into the hole than ordinary twist drills, which produces better and more accurate performance (third and fourth grades of accuracy). Core drilling is a more productive operation than drilling, since at the same cutting speeds, the feeds used may be two to three times larger. It is recommended to enlarge holes with core drills wherever possible instead of drilling with a larger drill. This process is much more efficient than boring large-diameter holes with a single drill.

2. Counterboring, countersinking, and spot facing, which are performed with various types of tools. Counterboring and countersinking (Figure 3.19) are used for machining cylindrical and tapered recesses in previously drilled holes. Such recesses are used for embedding the heads of screws and bolts, when these heads must not extend over the surface (Figure 3.21a and b). Spot facing is the process of finishing the faces of bosses for washers, thrust rings, nuts, and other pieces (Figure 3.21c). Spot facing tools cut only to a very limited depth. The tools used in these processes are made of high-speed steel (HSS) and have a guide or pilot, which is usually interchangeable. For these processes, cutting speeds and feeds are similar to those of core drilling.

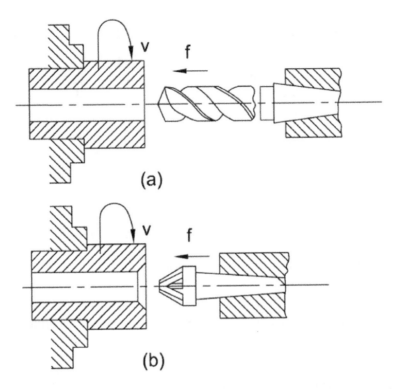

**FIGURE 3.20**  Drilling and drilling-allied operations as performed on an engine lathe: (a) drilling and (b) countersinking.

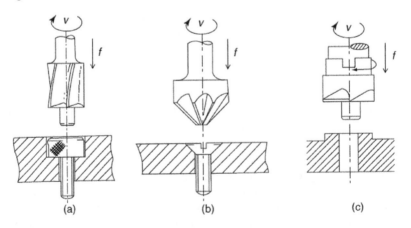

**FIGURE 3.21**  (a) Counterboring, (b) Countersinking, and (c) Spot facing operations.

3. Center drilling is a combined operation of drilling and countersinkings. Center drills are used for making center holes in blanks and shaft (Figure 3.19).
4. Reaming is a hole-finishing process intended to true up the hole to obtain high dimensional and form accuracy. Although it is recommended to

perform this process after core drilling, it may be performed after drilling. Depending upon the hole diameter, a reaming allowance of 40–400 μm should be provided. For HSS reamers, and depending on the WP material, low cutting speeds ranging from 2 to 20 m/min and small feeds ranging from 0.1 to 1 mm/rev are used. The preceding values are doubled when carbide reamers are used. The produced holes are always slightly (up to 20 μm) larger than the reamers. However, when using worn reamers or reaming holes in ductile material, the hole after reaming may have a smaller diameter than that of the reamer. Therefore, all these factors should be considered in selecting the reamer. Reamers may be hand or mechanical, cylindrical or taper, straight or helical fluted, and standard or adjustable.

5. Tapping is the process of generating internal threads in a hole using a tap, which is basically a threading tool. There are two possibilities for performing tapping on drilling machines:
   • The tapping of blind holes, where the machine should be provided with a reversing device together with a safety tap chuck.
   • The tapping of through holes, which does not necessitate a reversing device and a safety tap chuck.

6. Deep-hole drilling, where the length-to-diameter ratio of the hole is 10 or more, the work is rotated by a chuck and supported by a steady rest, and the drill is fed axially. The following special types of drills are used:
   • Gun drills for drilling holes up to 25 mm in diameter
   • Half-round drills for drilling holes over 25 mm in diameter
   • Trepanning drills for annular drilling of holes over 80 mm in diameter, leaving a core that enters the drill during operation

### 3.3.2 General-Purpose Drilling Machines

The general-purpose drilling machines are classified as:

• Bench-type sensitive drill presses
• Upright drill presses
• Radial drills
• Multispindle drilling machines
• Horizontal drilling machines for drilling deep holes

The most widely used in the general engineering industries are the upright drill presses and radial drills.

#### 3.3.2.1 Bench-Type Sensitive Drill Presses

These drill presses are used for machining small-diameter holes of 0.25–12 mm diameter. Manual feeding characterizes these machines, and this is why they are called "sensitive." High speeds are typical for bench-type sensitive drill presses.

#### 3.3.2.2 Upright Drill Presses

These machines are used for machining holes up to 50 mm diameter in relatively small-size work. Figure 3.22 shows a typical drilling machine. They have a wide

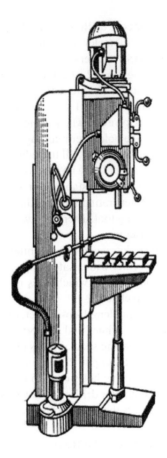

**FIGURE 3.22**  Typical upright drill press.

range of spindle speeds and feeds. Therefore, they are employed not only for drilling from solid material but also for core drilling, reaming, and tapping operations. Figure 3.23 illustrates the gearing diagram of the machine.

*Cutting movements.* As shown in the gearing diagram (Figure 3.23), the kinematic chain equations for the maximum spindle speed and feed are given by

$$n_{max} = 1420 \cdot \frac{27}{27} \cdot \frac{33}{33} \cdot \frac{52}{26} = 2840 \text{ rpm} \tag{3.7}$$

and

$$f_{max} = 1 \cdot \frac{22}{42} \cdot \frac{24}{24} \cdot \frac{32}{21} \cdot \frac{17}{44} \cdot \frac{1}{60} \times \pi \times 2.5 \times 14 = 0.56 \text{ mm/rev} \tag{3.8}$$

*Auxiliary movements.* The drill head, housing the speed and feed gearboxes, moves along the machine column through the gear train: worm gearing 1/20-rack and pinion ($z = 14$, $m = 2$). The machine table can be moved vertically by hand through bevels 18/45 and an elevating screw driven by means of a handle (Figure 3.23).

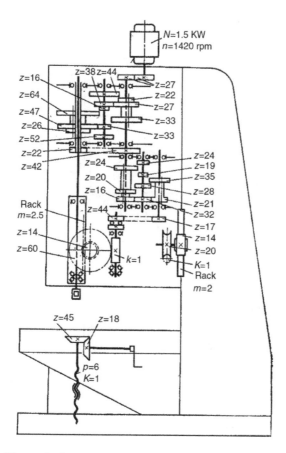

**FIGURE 3.23** Kinematic diagram of an upright drill press.

### 3.3.2.3 Radial Drilling Machines

These machines are especially designed for drilling, counterboring, countersinking, reaming, and tapping holes in heavy and bulky WPs that are inconvenient or impossible to machine on the upright drilling machines. They are suitable for multitool machining in individual and batch production. Radial drilling machines (Figure 3.24) differ from upright drill presses in that the spindle axis is made to coincide with the axis of the hole being machined by moving the spindle in a system of polar coordinates to the hole while the work is stationary. This is achieved by

1. Swinging the radial arm (4) about the rigid column (2)
2. Raising or lowering the radial arm on the column by the arm-elevating and -clamping mechanism (3) to accommodate the WP height
3. Moving the spindle head (5) along the guideways of the radial arm (4)

Accordingly, the tool is located at any required point on the stationary WP, which is set either on the detachable table (6) or directly on the base (1). After the maneuvering tasks performed by the radial arm and spindle head, they are held in position

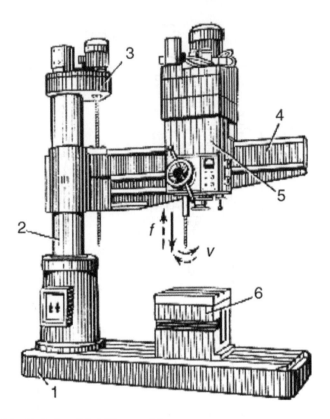

**FIGURE 3.24** Typical radial drilling machine.

using power-operated clamping devices. The spindle-head gearing diagram of the radial drilling machine is very similar to that of the upright drill press.

### 3.3.2.4 Multispindle Drilling Machines

These are mainly used in lot production for machining WPs requiring simultaneous drilling, reaming, and tapping of a large number of holes in different planes of the WP. A single-spindle drilling machine is not economical for such purposes, as not only is a considerably large number of machines and operators required but also the machining cycle is longer. There are three types of multiple-spindle drilling machines:

a. *Gang multispindle drilling machines.* The spindles (2–6) are arranged in a row, and each spindle is driven by its own motor. The gang machine is in fact several upright drilling machines having a common base and single worktable (Figure 3.25). They are used for consecutive machining of different holes in one WP or for the machining of a single hole with different cutting tools.

b. *Adjustable-center multispindle vertical drilling machines.* These differ from gang-type machines in that they have a common drive for all working spindles. The spindles are adjusted in the spindle head for drilling holes of varying diameters at random locations on the WP surface (Figure 3.26).

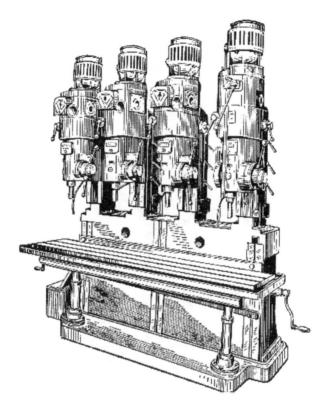

**FIGURE 3.25**   Gang, multiple-spindle drilling machine.

c. *Unit-type multispindle drilling machines.* These are widely used in mass production. They are, as a rule, chiefly built of standard units. Such machines are designed for machining a definite component held in a jig and are frequently built into an automatic transfer machine (Figure 3.27).

### 3.3.2.5  Horizontal Drilling Machines for Drilling Deep Holes

Such machines are usually equipped with powerful pumps, which deliver cutting fluid under high pressure, either through the hollow drilling tool or through the clearance between the drill stem and the machined hole. The cutting fluid washes out the chips produced by drilling.

In deep-hole drilling, the work is rotated by chuck and supported by a steady rest while the drill is fed axially. This process reduces the amount by which the drill departs from the drilled hole center. Deep-hole drilling machines (also called drill lathes) are intended for drilling holes having a length-to-diameter ratio of 10 or more.

### 3.3.3  Tool-Holding Accessories of Drilling Machines

Twist drills are either of a straight shank (for small sizes) or of a tapered shank (for medium to large sizes). A self-centering, three-jaw drill chuck (Figure 3.28) is used

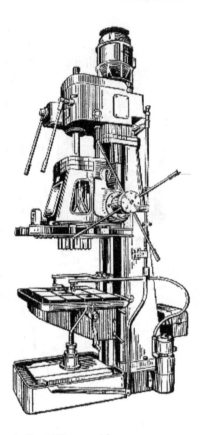

**FIGURE 3.26**  Multiple-spindle drilling machine.

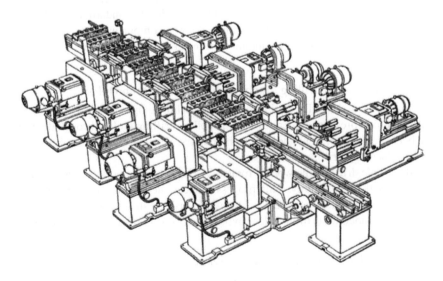

**FIGURE 3.27**  Unit-type multiple-spindle drilling machine.

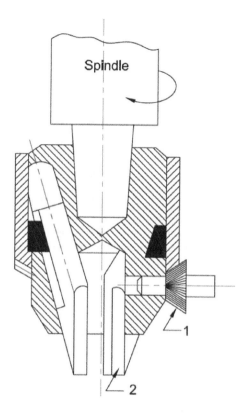

**FIGURE 3.28** Three-jaw drilling chuck.

to hold small drilling tools (up to 15 mm) with straight shanks. The rotation of the chuck wrench with the bevel pinion (1) closes or opens the jaws (2). The chuck itself is fitted with a Morse-taper shank, which fits into the spindle socket. Tapered sleeves (Figure 3.29a) are used for holding tools with taper shanks in the spindle socket (Figure 3.29b). The size of a Morse-taper shank is identified from smallest to largest by the numbers 1–6 and depends on the drill diameter (Table 3.2).

The included angle of the Morse taper is in the range of 3°. If the two mating tapered surfaces are clean and in good condition, such a small taper is sufficient to provide a frictional drive between the two surfaces. At the end of the taper shank of the tool or the taper shank of the sleeve, two flats are machined, leaving a tang. The purpose of the tang is to remove the tool from the spindle socket by a drift, as shown in Figure 3.29c. When the cutting tool has a Morse taper smaller than that of the spindle socket, the difference is made up by using one or two tapered sleeves (Figure 3.29d).

If a single hole is to be machined consecutively by several tools in a single operation, quick-change chucks are used for reducing the handling times in operating drilling machines. They enable tools to be changed rapidly without stopping the machine.

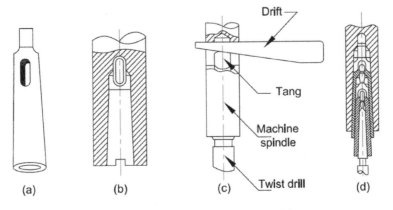

**FIGURE 3.29** Holding drills in spindle socket or sleeves and drifting out from a socket or sleeve. (a) Tapered sleeve, (b) spindle socket, (c) drifting out, and (d) holding by different sleeves.

**TABLE 3.2**
**Morse Taper Sizes**

| Morse Taper Number | 1 | 2 | 3 | 4 | 5 | 6 |
|---|---|---|---|---|---|---|
| Drill diameter (mm) | up to 14 | 14.25–23 | 23.25–31.75 | 32–50.5 | 51–76 | 77–100 |

A quick-change chuck for tapered-shank tools is shown in Figure 3.30. The body (1) has a tapered shank inserted into the machine spindle. A sliding collar (2) may be raised (for releasing) or lowered (for chucking). Interchangeable tapered sleeves (3) into which various tools have been secured are inserted into the chuck. When the collar (2) is lowered, it forces the balls (4) into the recess b and the torque is transmitted. The sleeve is rapidly released by raising the collar upward.

A safety-tap chuck (Figure 3.31) is used in tapping blind holes on machines having a reversing device. It is difficult to time the reversal at the proper moment; if a safety chuck is not used, the tap may run up against the bottom of the hole and break. The safety-tap chuck is secured in the machine spindle by the taper shank of the central shaft (4). Clutch member (2) is keyed on shaft (4), and the second clutch member (3) is mounted freely on shaft 4. Both members are held in engagement by the action of the spring (1). The compression of the spring is adjusted by a nut (6). Rotation is transmitted to the sleeve (5) through clutch member (3). When the actual torque exceeds the preset value, clutch member (2) begins to slip, the tap stops rotating, and the spindle is then reversed.

### 3.3.4  WORK-HOLDING DEVICES USED ON DRILLING MACHINES

The type of work-holding device used depends upon the shape and size of the WP, the required accuracy, and the production rate. It should be stressed that the work

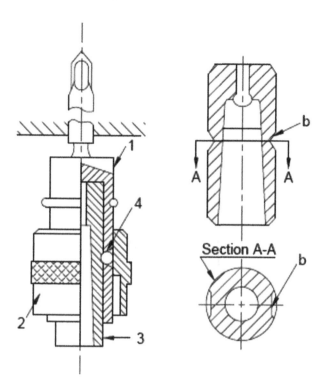

**FIGURE 3.30** Quick-change chuck.

being drilled should never be held by hand. High torque is transmitted by a revolving drill, especially when the drill is breaking through the bottom surface, which can wrench the work from the hand. The resulting injuries can vary from a small hand cut to the loss of a finger. Generally, work is held on a drilling machine by clamping to the worktable, in a vise, or in the case of mass production, in a drilling jig. Standard equipment in any workshop includes a vise and a collection of clamps, studs, bolts, nuts, and packing, which are simple and inexpensive. Vises do not accurately locate the work and provide no means for holding cutting tools in alignment. A small WP can be held in a vise, whereas larger work and sheet metal are best clamped on to the worktable surface, which is provided with standard tee slots for clamping purposes.

Drilling jigs are special devices designed to hold a particular WP and guide the cutting tool. Jigs enable work to be done without previously laying out the WP. Drilling using jigs is, therefore, accurate and quicker than standard methods. However, larger quantities of WPs must be required to justify the additional cost of the equipment. Jigs are provided with jig bushings to ensure that the hole is machined in the correct location. Jig bushings are classified as press-fit bushings for jigs used in small-lot production for machining holes using a single tool. Slip renewable bushings are used for mass production. Bushings are made of hardened steels to ensure

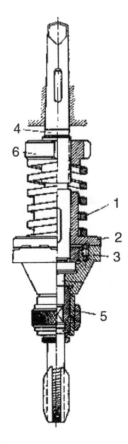

**FIGURE 3.31**   Safety tap chuck.

the required hardness to resist the wear. The drilling jigs are generally produced on jig boring-machines.

According to Figure 3.32, the plate jig (2) is mounted on the surface of the WP (1), where the holes are to be drilled. The WP is clamped under the plate jig with screws (3).

Figure 3.33 shows the jig used for drilling three holes in thin gauge components. The press-fit drill bushings are pressed into a separate top plate, which is doweled and screwed to the jig body and the base plate. A post jig used to make eight holes (up and down) in the flanges of the cylindrical component is shown in Figure 3.34. Accordingly, clamping is achieved by the finger nut. The previously drilled holes are located by the spring-loaded location pin in the jig base to enable the holes to be drilled in line. Figure 3.35 illustrates a jig design that enables a hole to be drilled at an angle to the component centerline. A special drill bushing is used to take the drill as close to the component as possible. Figure 3.36 shows an inverted post jig with four legs, and Figure 3.37 presents an indexing jig used for drilling six equally spaced holes around the periphery of the component.

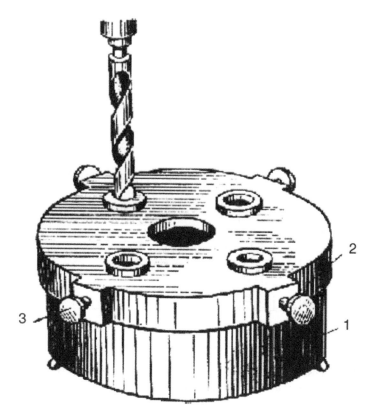

**FIGURE 3.32**  Simple plate jig.

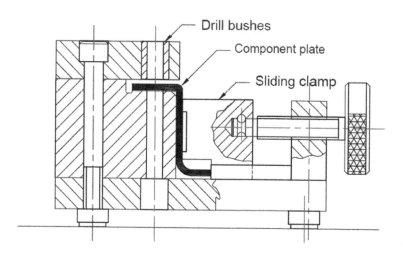

**FIGURE 3.33**  Thin plate drilling jig. (From Mott, L. C., *Engineering Drawing and Construction*, Oxford University Press, Oxford, 1976. With permission.)

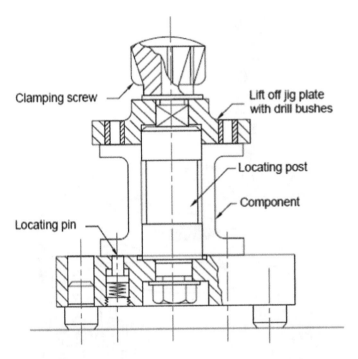

**FIGURE 3.34** Post jig to drill holes into flanged, cylindrical WP. (From Mott, L. C., *Engineering Drawing and Construction*, Oxford University Press, Oxford, 1976. With permission.)

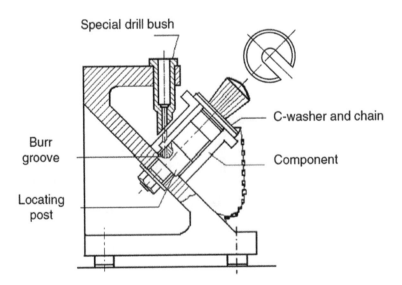

**FIGURE 3.35** Angle drilling jig. (From Mott, L. C., *Engineering Drawing and Construction*, Oxford University Press, Oxford, 1976. With permission.)

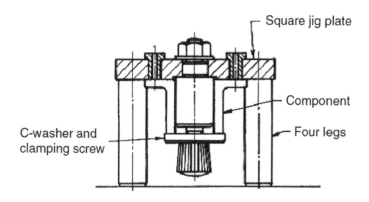

**FIGURE 3.36** Inverted post jig. (From Mott, L. C., *Engineering Drawing and Construction,* Oxford University Press, Oxford, 1976. With permission.)

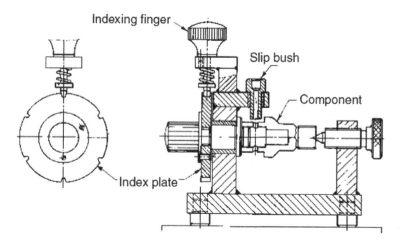

**FIGURE 3.37** Indexing drilling jig. (From Mott, L. C., *Engineering Drawing and Construction,* Oxford University Press, Oxford, 1976. With permission.)

## 3.4 MILLING MACHINES AND OPERATIONS

### 3.4.1 MILLING OPERATIONS

Milling is the removal of metal by feeding the work past a rotating multitoothed cutter. In this operation, the material removal rate (MRR) is enhanced, as the cutter rotates at a high cutting speed. The surface quality is also improved due to the multi-cutting edges of the milling cutter. The action of the milling cutter is totally differ-ent from that of a drill or a turning tool. In turning and drilling, the tools are kept continuously in contact with the material to be cut, whereas milling is an intermittent process, as each tooth produces a chip of variable thickness. Milling operations may be classified as peripheral (plain) milling or face (end) milling (Figure 3.38).

**FIGURE 3.38**  Plain and face milling cutters: (a) plain milling and (b) face milling.

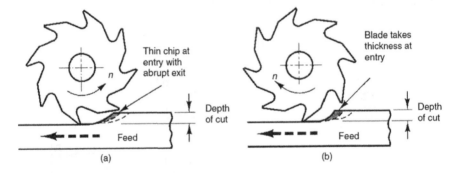

**FIGURE 3.39**  Up-milling and down-milling: (a) up-milling (conventional cut) and (b) down-milling (climb cut).

### 3.4.1.1  Peripheral Milling

In peripheral milling, the cutting occurs by the teeth arranged on the periphery of the milling cutter, and the generated surface is a plane parallel to the cutter axis. Peripheral milling is usually performed on a horizontal milling machine. For this reason, it is sometimes called horizontal milling. The appearance of the surface and also the type of chip formation are affected by the direction of cutter rotation with respect to the movement of the WP. In this regard, two types of peripheral milling are differentiable: up-milling and down-milling.

#### 3.4.1.1.1  Up-Milling (Conventional Milling)

Up-milling is accomplished by rotating the cutter against the direction of the feed of the WP (Figure 3.39a). The tooth picks up from the material gradually; that is, the chip starts with no thickness and increases in size as the teeth progress through the cut. This means that the cycle of operation to remove the chip is first a sliding action at the beginning, and then a crushing action takes place, which is followed by the actual cutting action. In some metals, up-milling leads to strain hardening of the machined surface and also to chattering and excessive teeth blunting.

Advantages of up-milling include the following:

- It does not require a backlash eliminator.
- It is safer in operation (the cutter does not climb on the work).
- Loads on teeth are acting gradually.
- Built-up edge (BUE) fragments are absent from the machined surface.
- The milling cutter is not affected by the sandy or scaly surfaces of the work.

### 3.4.1.1.2  *Down-Milling (Climb Milling)*

Down-milling is accomplished by rotating the cutter in the direction of the work feed, as shown in Figure 3.39b. In climb milling, as implied by the name, the milling cutter attempts to climb the WP. Chips are cut to maximum thickness at the initial engagement of cutter teeth with the work and decrease to zero at the end of its engagement. The cutting forces in down-milling are directed downward. Down-milling should not be attempted if machines do not have enough rigidity and are not provided with backlash eliminators (Figure 3.40). Under such circumstances, the cutter climbs up on the WP, and the arbor and spindle may be damaged.

Advantages of down-milling include the following:

- Fixtures are simpler and less costly, as cutting forces are acting downward.
- Flat WPs or plates that cannot be firmly held can be machined by down-milling.
- Cutters with higher rake angles can be used, which decreases the power requirements.
- Tool blunting is less likely.
- Down-milling is characterized by fewer tendencies of chattering and vibration, which leads to improved surface finish.

### 3.4.1.2  Face Milling

In face milling, the generated surface is at a right angle to the cutter axis. When using cutters of large diameters, it is a good practice to tilt the spindle head slightly at an angle of 1–3° to provide some clearance, which leads to an improved surface finish and eliminates tool blunting (Figure 3.38b). Face milling is usually performed on vertical milling machines; for this reason, the process is called vertical milling, and is more productive than plain milling.

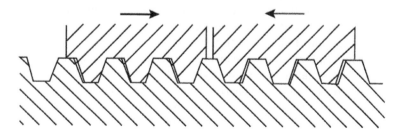

**FIGURE 3.40**   Backlash eliminator in down-milling.

### 3.4.2 MILLING CUTTERS

The milling cutters are selected for each specified machining duty. The milling cutter may be provided with a hole to be mounted on the arbor of the horizontal milling machine or provided with a straight or tapered shank for mounting on the vertical or horizontal milling machine. Figure 3.41 visualizes commonly used milling cutters during their operation. These include the following:

1. Plain milling cutters are either straight or helical. Helical milling cutters are preferred for large cutting widths to provide smooth cutting and improved surface quality (Figure 3.41a). Plain milling cutters are mainly used on horizontal milling machines.
2. Face milling cutters are used for the production of horizontal (Figure 3.41b), vertical (Figure 3.41c), or inclined (Figure 3.41d) flat surfaces. They are used on vertical milling machines, planer-type milling machines, and vertical milling machines with the spindle swiveled to the required angle $\alpha$, respectively.
3. Side milling cutters are clamped on the arbor of the horizontal milling machine and are used for machining the vertical surface of a shoulder (Figure 3.41e) or cutting a keyway (Figure 3.41f).
4. Interlocking (staggered) side mills (Figure 3.41g) mounted on the arbor of horizontal milling machines are intended to cut wide keyways and cavities.
5. Slitting saws (Figure 3.41h) are used on horizontal milling machines.
6. Angle milling cutters are used on horizontal milling machines for the production of longitudinal grooves (Figure 3.41i) or for edge chamfering.
7. End mills are tools of a shank type, which can be mounted on vertical milling machines (or directly in the spindle nose of horizontal milling machines). End mills may be employed in machining keyways (Figure 3.41j) or vertical surfaces (Figure 3.41k).
8. Key-cutters are also of the shank type, which can be used on vertical milling machines. They may be used for single-pass milling or multipass milling operations (Figure 3.41l and m3).
9. Form-milling cutters are mounted on horizontal milling machines. Form cutters may be either concave, as shown in Figure 3.41n, or convex, as in Figure 3.41o.
10. T-slot cutters are used for milling T-slots and are available in different sizes. The T-slot is machined on a vertical milling machine in two steps:
    • Slotting with end mill (Figure 3.41j)
    • Cutting with T-slot cutter (Figure 3.41p)
11. Compound milling cutters are mainly used to produce compound surfaces. These cutters realize high productivity and accuracy (Figure 3.41q).
12. Inserted tool milling cutters have a main body that is fabricated from tough and less expensive steel. The teeth are made of alloy tool steel, HSS, carbides, ceramics, or cubic boron nitride (CBN) and mechanically attached to the body using setscrews and in some cases, are brazed. Cutters of this type are confined usually to large-diameter face milling cutters or horizontal milling cutters (Figure 3.41q).

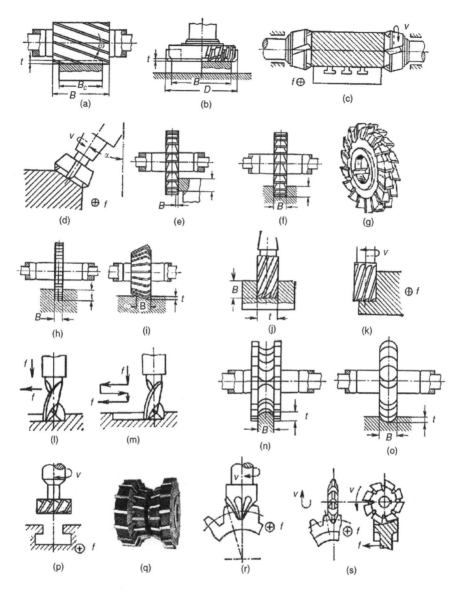

**FIGURE 3.41** Different types of milling cutter during operation. (a) plain milling cutters; (b) vertical face milling cutters; (c) horizontal face milling cutters; (d) inclined milling cutters; (e) side milling cutters; (f) keyway cutters; (g) interlocking side mills; (h) slitting saws; (i) angle milling cutters; (j) slotting end mill; (k) milling vertical surfaces; (l) key-cutters for single-pass milling; (m) key-cutters for multipass milling; (n) concave-milling cutters; (o) convex milling cutters; (p) T-slot cutter; (q) compound milling cutters; (r, s) gear milling cutters.

13. Gear milling cutters are used for the production of spur and helical gears on vertical or horizontal milling machines (Figure 3.41r and s). Gear cutters are form-relieved cutters, which are used to mill contoured surfaces. They are sharpened at the tooth face. Hobbing machines and gear shapers are used to cut gears for mass production and high-accuracy demands.

### 3.4.3 General-Purpose Milling Machines

Milling machines are employed for machining flat surfaces, contoured surfaces, and complex and irregular areas, slotting, threading, gear cutting, and production of helical flutes, twist drills, and spline shafts to close tolerances.

Milling machines are classified by application into the following categories:

- General-purpose milling machines, which are used for piece and small-lot production.
- Special-purpose milling machines, which are designed for performing one or several distinct milling operations on definite WPs. They are used in mass production.

The general-purpose milling machines are extremely versatile and are subdivided into these types:

1. Knee type
2. Vertical bed type
3. Planer type
4. Rotary table

#### 3.4.3.1  Knee-Type Milling Machines

The special feature of these machines is the availability of three Cartesian directions of table motion. This group is further subdivided into plain horizontal, universal horizontal, vertical, and ram-head knee-type milling machines. The name "knee" has been adopted because it features a knee that mounts the worktable and travels vertically along the vertical guideway of the machine column. In plain horizontal milling machines, the spindle is horizontal, and the table travels in three mutually perpendicular directions. Universal horizontal milling machines (Figure 3.42) are similar in general arrangement to plain horizontal machines. The principal difference is that the table can be swiveled about its vertical axis through ±45°, which makes it possible to mill helical grooves and helical gears. In contrast to horizontal milling machines, vertical-type milling machines have a vertical spindle and no overarm (Figure 3.43). The overarm serves to hold the bearing bracket supporting the outer end of the tool arbor in horizontal machines. Ram-head milling machines (Figure 3.44) differ from the universal type in that they have an additional spindle that can be swiveled about both the vertical and the horizontal axes. In ram-head milling machines, the spindle can be set at any angle in relation to the WP being machined. In modern machines, a separate drive for the principal movement (cutter), feed movement (WP), rapid traversal of the worktable in all directions, and a single lever control for changing speeds and feeds are provided. Units and components of milling machines are widely unified. Horizontal knee-type milling machine specifications are as follows:

- Dimensions of table working surface
- Maximum table travel in the three Cartesian directions

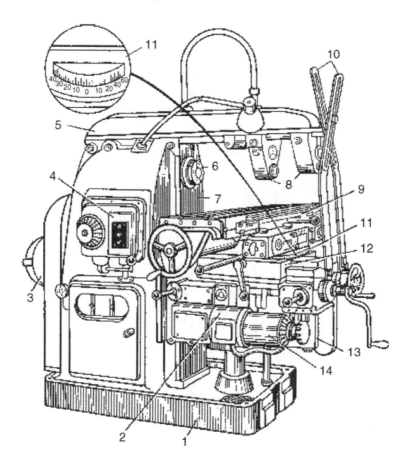

**FIGURE 3.42** Universal horizontal-spindle milling machine.

- Maximum angle of table swivel
- Arbor diameter
- Maximum distance between arbor axis and the overarm underside
- Number of spindle speeds
- Number of feeds in the three directions
- Power and speed of main motor
- Power and speed of feed motor
- Overall dimensions and net weight

Figure 3.42 visualizes the main parts of the horizontal universal milling machine. These are base (1), column (7), knee (13), saddle (12), table swivel plate with graduation (11), worktable (9), overarm (5), holding bearing bracket (8), main motor (3), spindle (6), speed gearbox (4), feed gearbox (2), feed control mechanism (14), and braces (10) to link the overarm with the knee for high-rigidity requirements in heavy-duty milling machines.

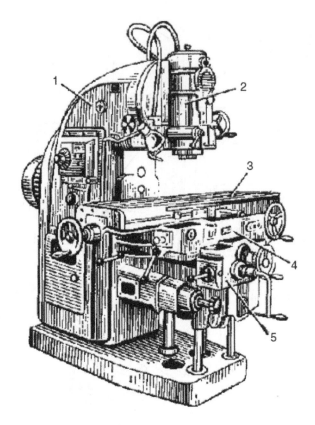

**FIGURE 3.43** Vertical milling machine.

### 3.4.3.2  Vertical Bed-Type Milling Machines

These machines are rigid and powerful; hence, they are used for heavy-duty machining of large WPs (Figure 3.45). The spindle head containing a speed gearbox travels vertically along the guideways of the machine column and has a separate drive motor. In some machines, the spindle head can be swiveled. The work is fixed on a compound table that travels horizontally in two mutually perpendicular directions. The adjustment in the vertical direction is accomplished by the spindle head.

### 3.4.3.3  Planer-Type Milling Machines

These are intended for machining horizontal, vertical, and inclined planes as well as form surfaces by means of face, plain, and form-milling cutters. These machines are of single or double housing, with one or several spindles; each has a separate drive. Figure 3.46 shows a single-housing machine with two spindle heads traveling vertically and horizontally.

### 3.4.3.4  Rotary-Table Milling Machines

These are also called continuous milling machines, as the WPs are set up without stopping the operation. Rotary-table machines are highly productive; consequently, they are frequently used for both batch and mass production. The WPs being

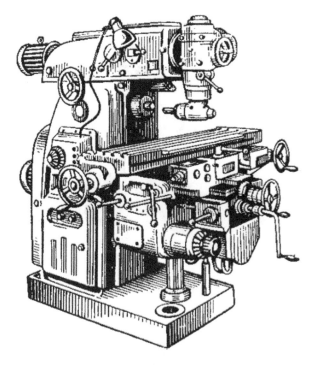

**FIGURE 3.44**  Ram-head milling machine.

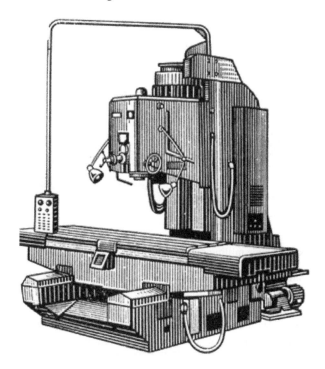

**FIGURE 3.45**  Vertical-bed general-purpose milling machine.

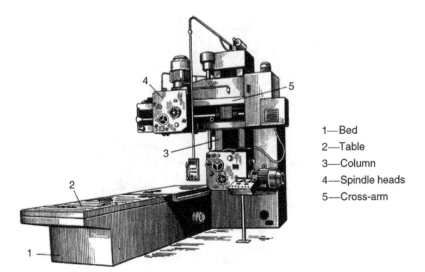

1—Bed
2—Table
3—Column
4—Spindle heads
5—Cross-arm

**FIGURE 3.46**   Planer-type general-purpose milling machine.

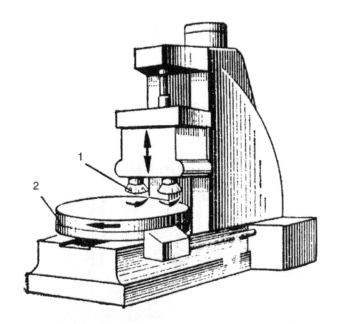

**FIGURE 3.47**   Rotary-table milling machine.

machined are clamped in fixtures installed on the rotating table (2) (Figure 3.47). The machines may be equipped with one or two spindle heads (1).

When several surfaces are to be machined, the WPs are indexed in the fixtures after each complete revolution of the table. The machining cycle provides as many table revolutions as the number of surfaces to be machined.

### 3.4.4   Holding Cutters and Workpieces on Milling Machines

#### 3.4.4.1   Cutter Mounting

The nose of milling machine spindles has been standardized. It is provided with a locating flange ɸ H7/h6 and a steep taper socket of 7:24 (1:3.4286) corresponding to an angle of 16° 35.6′ (Figure 3.48) to ensure better location of the arbor and end mill shanks. Rotation is transmitted to the cutter through the driving key secured to the end face of the spindle. Large face milling cutters are mounted directly on the spindle flange and are secured to the flange by four screws, whereas rotation is transmitted to the cutter through the driving keys on the spindle (Figure 3.48). Plain and side milling cutters are mounted on an arbor whose taper shank is drawn up tight into the taper socket of the spindle (2) with a draw-in bolt 1 (Figure 3.49). Milling arbors are long or short (stub arbors). The outer end of the long arbor (3) is supported by an overarm support (5) in horizontal milling machines, and the cutter (4) is mounted at the required position on the arbor by a key (or without a key in the case of slitting saws) and is clamped between collars or spacers (6) with a large nut.

The system shown in Figure 3.50 is used in duplex-bed milling machines. On the stub arbors, the shell end mill or the face milling cutters are driven by either a feather key, as shown in Figure 3.50a, or an end key (Figure 3.50b). End mills, T-slot cutters, and other milling cutters of tapered shanks are secured with a draw-in bolt directly

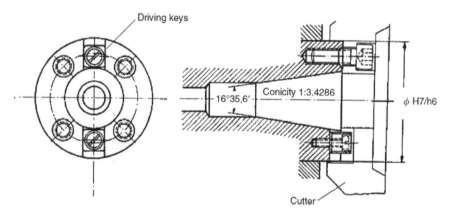

**FIGURE 3.48**   Typical nose of milling machine spindle.

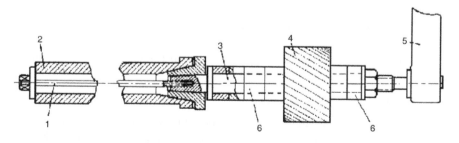

**FIGURE 3.49**   Milling machine arbor.

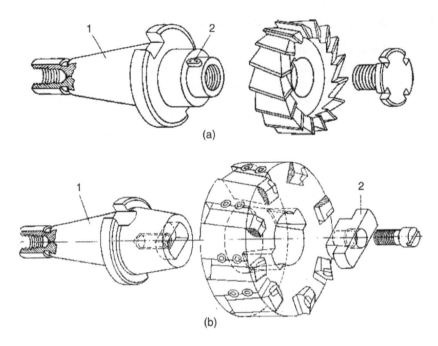

**FIGURE 3.50** Mounting of end mills and face milling cutters on duplex-bed milling machine.

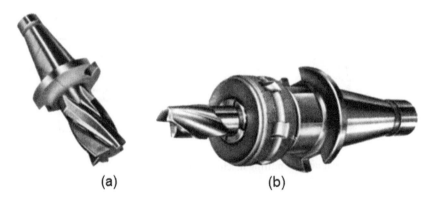

**FIGURE 3.51** Mounting of (a) tapered and (b) straight-shank milling cutters.

in the taper socket of the spindle by means of adaptors (Figure 3.51a). Straight shank cutters are held in chucks (Figure 3.51b).

### 3.4.4.2 Workpiece Fixturing
Large WPs and blanks that are too large for a vise are clamped directly on the worktable using standard fastening elements such as strap clamps, support blocks, and T-bolts (Figure 3.52). Small WPs and blanks are clamped most frequently in

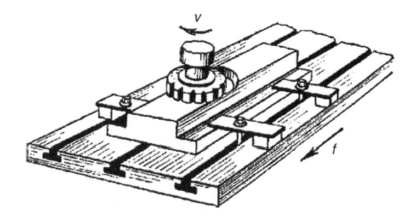

**FIGURE 3.52** Clamping of large WPs directly on the worktable.

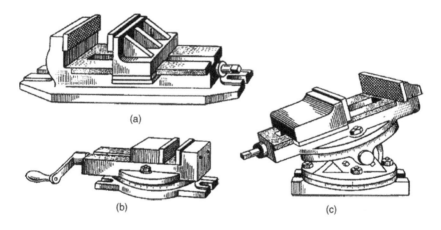

**FIGURE 3.53** Vises for clamping of small WPs on milling machines: (a) plain vise, (b) swivel vise, and (c) universal vise.

general-purpose plain, swivel, or universal milling vises fastened to the worktable (Figure 3.53). Shaped jaws are sometimes used instead of the flat type to clamp parts of irregular shapes. For more accurate and productive work, expensive milling fixtures are frequently used.

Figure 3.54 shows a simple milling fixture for a bearing bracket. A full-form and a flatted locator, firmly fitted into the base plate, are used to locate the WP from two previously machined holes. The clamping is effected by two solid clamps. To achieve correct alignment and hence, increased accuracy, a tool-setting block is used to locate the cutter with respect to the WP. Figure 3.54 illustrates how the height of the cutter is set up using setting blocks and a 0.7 mm feeler. The main body of the fixture is frequently made of CI because of its ability to absorb vibrations initiated by the milling operation. However, welded and other steel constructions are also used for various specialized purposes.

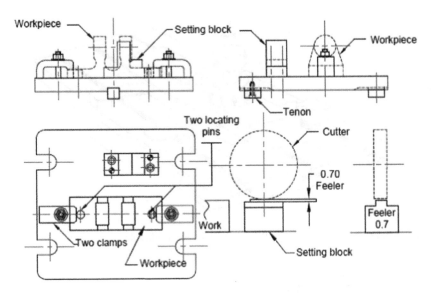

**FIGURE 3.54**  Simple milling fixture for a bearing bracket. (From Mott, L. C., *Engineering Drawing and Construction,* Oxford University Press, Oxford, 1976. With permission.)

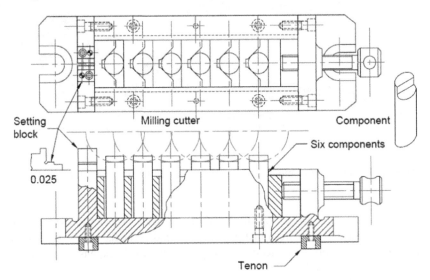

**FIGURE 3.55**  Special fixture for milling six cylindrical WPs. (From Mott, L. C., *Engineering Drawing and Construction,* Oxford University Press, Oxford, 1976. With permission.)

Figure 3.55 illustrates a vise used as a fixture for milling six cylindrical WPs in one clamp. The setting block is designed for a feeler gage of 0.025 mm, the thickness of which should be stamped on the setting block in some suitably prominent position. In this type of fixture, it is essential that when the components are unloaded, all the swarf must be removed; otherwise, the component subsequently loaded into the fixture will not seat correctly.

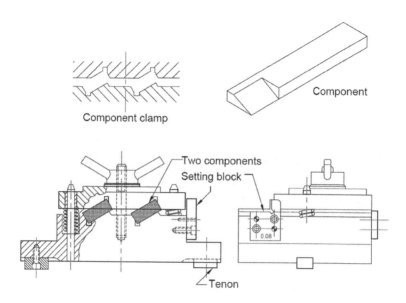

**FIGURE 3.56** Special milling fixture for mounting two rectangular components. (From Mott, L. C., *Engineering Drawing and Construction,* Oxford University Press, Oxford, 1976. With permission.)

Figure 3.56 shows a WP and fixture of more specialized nature designed by the U.S. Naval Gun Factory. Two rectangular components are to be milled together. They are located and clamped between two mating surfaces. The holding plate is positioned by two spring-loaded dowels and a central fixing stud. A setting block is doweled and screwed to the fixture. It is designed for use with a feeler gauge of 0.08 mm thickness. The disadvantage of this setup is that the arbor is unsupported at its free end, and therefore, only light cuts are taken. Duplex milling machines enable WPs to be machined from both sides at once to ensure high accuracy and enhance productivity.

### 3.4.5   Dividing Heads

Dividing heads are attachments that extend the capabilities of milling machines. They are mainly employed on knee-type milling machines to enhance their capabilities toward milling straight and helical flutes, slots, grooves, and gashes whose features are equally spaced about the circumference of a blank (and less frequently, unequally spaced). Such jobs include milling spur and helical gears, spline shafts, twist drills, reamers, milling cutters, and others. Therefore, dividing heads are capable of indexing the WP through predetermined angles. In addition to the indexing operation, the dividing head continuously rotates the WP, which is set at the required helix angle during the milling of helical slots and helical gears. There are several versions of dividing heads:

- Plain dividing heads (Figure 3.57) are mainly used for indexing milling fixtures.
- Universal dividing heads.

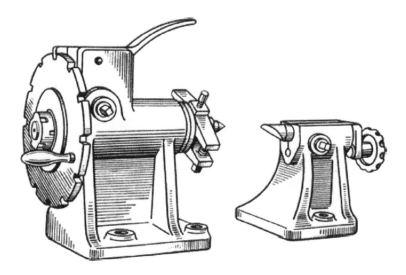

**FIGURE 3.57**   Plain milling dividing head.

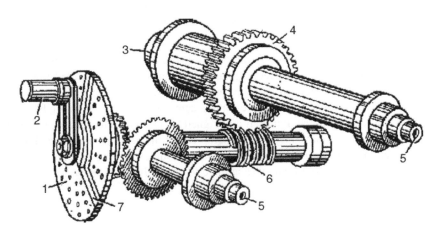

**FIGURE 3.58**   Isomeric gearing diagram of a universal dividing head.

- Optical dividing heads are commonly used for precise indexing and also for checking the accuracy of marking graduation lines on dial scales. Their main drawback is that they cannot be used in the milling of helical gears.

### 3.4.5.1   Universal Dividing Heads

The most widely used type of dividing head is the universal dividing head. Figure 3.58 illustrates an isometric view of the gearing diagram of a universal dividing head in a simple indexing mode. Periodical turning of the spindle (3) is achieved by rotating the index crank (2), which transmits the motion through a worm gearing 6/4 to the WP (gear ratio 1:40; that is, one complete revolution of the crank corresponds to 1/40 revolution of the WP). The index plate (1), having several concentric circular rows of accurately and equally spaced holes, serves for indexing the index crank (2) through the required angle. The WP is clamped in a chuck screwed on the spindle (3). It can also be clamped between two centers.

The dividing head is provided with three index plates (Brown and Sharpe) or two index plates (Parkinson). The plates have the following number of holes:

Brown and Sharpe
*Plate 1:* 15, 16, 17, 18, 19, and 20
*Plate 2:* 21, 23, 27, 29, 31, and 33
*Plate 3:* 35, 37, 39, 41, 43, 47, and 49
Parkinson
*Plate 1:* 24, 25, 28, 30, 34, 37, 38, 39, 41, 42, and 43
*Plate 2:* 46, 47, 49, 51, 53, 54, 57, 58, 59, 62, and 66

### 3.4.5.2 Modes of Indexing

The universal dividing head can be set up for simple or differential indexing, or for milling helical slots.

1. Simple Indexing
    The index plate (1) is fixed in position by a lock pin (4) to be motionless (Figure 3.59). The work spindle (3) is rotated through the required angle by rotating

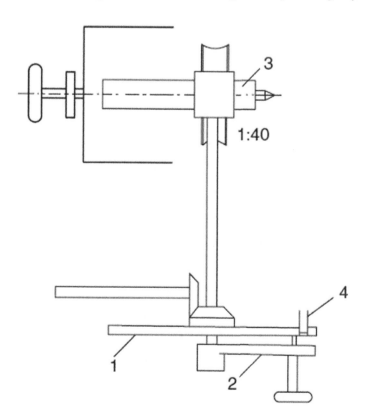

**FIGURE 3.59** Simple indexing.

the index crank (2). For determining the number of index crank revolutions $n$ to give the number of divisions $Z$ on the job periphery (assuming a worm/worm gear ratio of 1:40), the kinematic balance equation is given by

$$n = \frac{40}{Z} \qquad\qquad (3.9)$$

### Illustrative Example 1

It is required to determine the suitable index plates (Brown and Sharpe) and the number of index crank revolutions $n$ necessary for producing the following spur gears of teeth number 40, 30, and 37.

#### SOLUTION

$$Z = 40 \text{ teeth}$$

$$n = \frac{40}{40} = 1\,\text{rev}$$

The crank should be rotated one complete revolution to produce one gear tooth. Any index plate and any circle of holes can be used.

$$Z = 30 \text{ teeth}$$

$$n = \frac{40}{30} = 1\frac{1}{3}\,\text{rev}$$

$$= 1 + \frac{6}{18}$$

Then, choose plate 1 (Brown and Sharpe) and select the circle of 18 holes. The crank should be rotated one complete revolution plus six holes out of 18.

$$Z = 37 \text{ teeth}$$

$$n = \frac{40}{37} = 1 + \frac{3}{37}\,\text{rev}$$

Choose plate 3, and select hole circle 37. The crank should be rotated one complete revolution plus three holes out of 37.

To avoid errors in counting the number of holes, the adjustable selector (Figure 3.60) on the index plate should be used.

2. Differential Indexing

This is employed where simple indexing cannot be effected; that is, where an index plate with the number of holes required for simple indexing is not available.

In differential indexing, a plunge (5) is inserted in the bore of the work spindle (Figure 3.61) while the index plate is unlocked. The spindle drives the plate through change and bevel gears while the crank through the worm is driving the spindle.

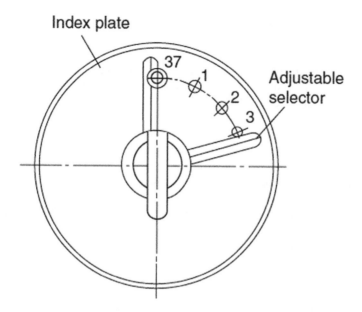

**FIGURE 3.60** Counting with adjustable sector.

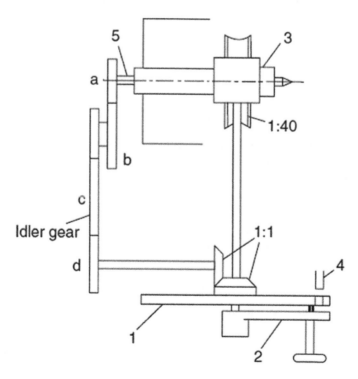

**FIGURE 3.61** Differential indexing.

Hence, the required turn of the work spindle is obtained as the sum of two turns (Figure 3.61):

- A turn of the index crank (2) relative to the index plate (1)
- A turn of the index plate itself, which is driven from the work spindle through change gears $(a/b) \times (c/d)$ to provide the correction

Depending on the setup, the index plate rotates either in the same direction with the index crank or in the opposite direction. An idler gear should be used if the crank and plate move in opposite directions to each other (Figure 3.61).

To perform a differential indexing, the following steps are to be considered:

- The number of revolutions of the index crank is set up in the same manner as in simple indexing but not for the required number of divisions $Z$. Another number $Z'$, nearest to $Z$, makes it possible for simple indexing to be carried out.
- The error of setup using $Z'$ is compensated for by means of a respective setting up of the differential change gears $a$, $b$, $c$, and $d$ (Figure 3.61). The change gears supplied to match the three plate system (Brown and Sharpe) are 24(2), 28, 32, 40, 44, 48, 56, 64, 72, 86, and 100 teeth.
- The numbers of teeth of the change gears $a$, $b$, $c$, and $d$ are determined from the corresponding kinematic balance equation:

$$\frac{40}{Z'} + \frac{1}{Z}\frac{a \cdot c}{b \cdot d} = \frac{40}{Z} \tag{3.10}$$

from which

$$\frac{a \cdot c}{b \cdot d} = \frac{40}{Z'}(Z' - Z) \tag{3.11}$$

It is more convenient to assume that $Z' > Z$ to avoid the use of an idler gear. If $Z' < Z$, then an idler gear must be used (Figure 3.61).

## Illustrative Example 2

Select the differential change gears and the index plate (Brown and Sharpe), and determine the number of revolutions of the index crank for cutting a spur of $Z = 227$ teeth.

### SOLUTION

Assume $Z' = 220$, $Z' < Z$; therefore, idler is required:

$$n = \frac{40}{Z'} = \frac{40}{220} = \frac{2}{11} = \frac{6}{33}$$

$$\frac{a}{b} \times \frac{c}{d} = \frac{40}{Z'}(Z' - Z)$$

$$= \frac{2}{11}(220 - 227) = -\frac{2 \times 7}{11}$$

$$= -\frac{8}{4} \times \frac{7}{11} = -\frac{64}{32} \cdot \frac{28}{44}$$

$a = 64$, $b = 32$, $c = 28$, and $d = 44$ teeth with an idler gear.

3. Setting the Dividing Head for Milling Helical Grooves

In milling helical grooves and helical gears, the helical movement is imparted to the WP through a reciprocating movement along its axis and rotation of the WP about the same axis. The WP receives reciprocation together with the worktable, and rotation from the worktable lead screw through a set of change gears. The table is set to the spindle axis at an angle $\omega_h$ equal to the helix angle of the groove being cut. The table is swiveled clockwise for left-hand grooves and counterclockwise for right-hand grooves (Figure 3.62).

$$\tan \omega_h = \frac{\pi D}{t_{hel}} \tag{3.12}$$

where

$t_{hel}$ = lead of helical groove (mm)
$D$ = diameter of the WP (mm)

The kinematic balance is based on the fact that for every revolution of the blank, it travels axially a distance equal to the lead of the helical groove to be milled. This balance is obtained by setting up the gear train that links the lead screw to the work spindle; therefore,

$$t_{hel} = t_{ls} \times \frac{d_1}{c_1} \cdot \frac{d_1}{a_1} \times 1 \times 40 \tag{3.13}$$

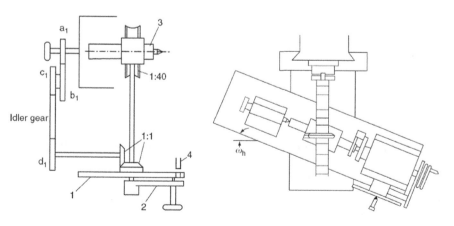

**FIGURE 3.62** Setting the dividing head for milling helical grooves.

from which

$$\frac{a_1}{b_1} \cdot \frac{c_1}{d_1} = 40 \frac{t_{ls}}{t_{hel}}$$                              (3.14)

where

$t_{ls}$ = lead of worktable lead screw (mm)

$(a_1/b_1) \cdot (c_1/d_1)$ = change gears (Figure 3.62)

### Illustrative Example 3

It is required to mill six right-hand helical flutes with a lead of 600 mm; the blank diameter is 90 mm. If the pitch of the table lead screw is 7.5 mm, give complete information about the setup.

#### SOLUTION

Indexing:

$$n = \frac{40}{6} = 6\frac{2}{3} = 6\frac{12}{18} \text{ crank revolution}$$

Choose plate 1 (Brown and Sharpe) and select hole circle 18. The crank should be rotated six complete revolutions plus 12 holes out of 18.

Helix:

$$\tan \omega_h = \frac{\pi D}{t_{hel}} = \frac{\pi \times 90}{600} = 0.471$$

then

$$\omega_h = 25.23°$$

The milling table should be set counterclockwise at an angle of 25.23°.

Change gears:

$$\frac{a_1}{b_1} \cdot \frac{c_1}{d_1} = 40 \times \frac{t_{ls}}{t_{hel}}$$

$$= 40 \times \frac{7.5}{600} = \frac{24}{56} \cdot \frac{28}{24}$$

Then, $a_1 = 24$, $b_1 = 56$, $c_1 = 28$, and $d_1 = 24$ teeth.

## 3.5  SHAPERS, PLANERS, AND SLOTTERS AND THEIR OPERATIONS

### 3.5.1  Shaping, Planing, and Slotting Processes

These processes are used for machining horizontal, vertical, and inclined flat and contoured surfaces, slots, grooves, and other recesses by means of special single-point tools. The difference between these three processes is that in planing, the work is reciprocated and the tool is fed across the work, while in shaping and slotting, the tool is reciprocating and the work is fed across the cutting tool. Moreover, the tool travel is horizontal in shaping and planing and vertical in the case of slotting (Figure 3.63).

The essence of these processes is the same as that of turning, where metals are removed by single-point tools similar in shape to lathe tools. A similarity also exists in chip formation. However, these operations differ from turning in that the cutting action is intermittent, and chips are removed only during the forward movement of the tool or the work. Moreover, the conditions when shaping, planing, and slotting tools are less favorable than in turning, even though the tools have the opportunity to cool during the return stroke, when no cutting takes place. That is because these tools operate under severe impact conditions. For these conditions, the related machine and tools are designed to be more rigid and strongly dimensioned, and the cutting speed in most cases does not exceed 60 m/min. Consequently, tools used in these processes should not be shock sensitive like ceramics and CBN. It is sufficient to use low-cost and easily sharpened tools such as HSS and carbides.

The limited cutting speed and the time lost during the reverse stroke are the main reasons behind the low productivity of shaping, planing, and slotting compared

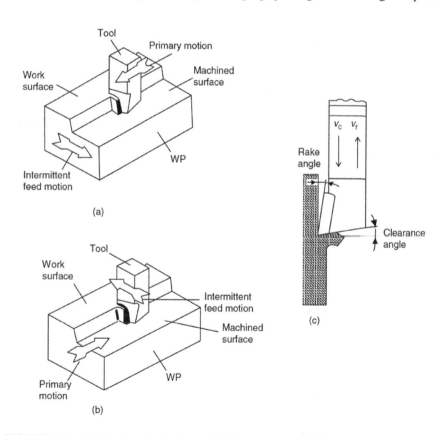

**FIGURE 3.63** (a) Shaping, (b) planing, and (c) slotting operations.

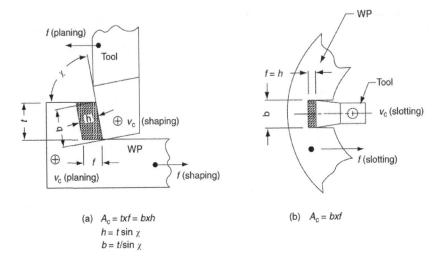

(a)   $A_c = txf = bxh$
      $h = t \sin \chi$
      $b = t/\sin \chi$

(b)   $A_c = bxf$

**FIGURE 3.64**  Kinematics and machining parameters in (a) shaping and planning and (b) slotting.

with turning. However, in planing, not only the productivity but also the accuracy is enhanced due to the possibility of using multiple tooling in one setting. Figure 3.64 illustrates the kinematics and machining parameters in shaping, planing, and slotting. The basic machining parameters are the average speed during the cutting stroke $v_{cm}$, the feed $f$, the depth of cut $t$, and the uncut cross-section area $A_c$. The feed is the intermittent relative movement of the tool (in planing) or the WP (in shaping and slotting) in a direction perpendicular to the cutting motion and expressed in millimeters per stroke. The feed movement is always actuated at the end of the return stroke when the tool is not engaged with the work. The depth of cut is the layer removed from the WP in millimeters in a single pass and is measured perpendicular to the machined surface. The uncut chip cross section in square millimeters is given by the following equation for shaping and planing:

$$A_c = b \cdot h = t \cdot f \text{ mm}^2 \tag{3.15}$$

where
  $b$ = chip contact length (mm)
    = $t/\sin x$
  $h$ = chip thickness (mm)
    = $f \sin x$
  $x$ = setting angle (frequently $x = 75°$)

and the following equation for slotting

$$A_c = b \cdot f \text{ mm}^2 \tag{3.16}$$

where $b$ is the slot width (mm).

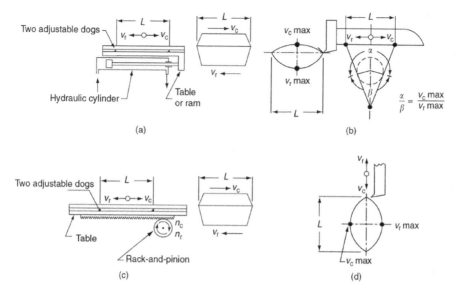

**FIGURE 3.65** Shaper, planer, and slotter mechanisms. (a) Hydraulic shapers, planers, and slotters ($v_c < v_r$), (b) shapers and slotters of lever arm mechanism ($v_c < v_r$), (c) planers of rack and pinion mechanism ($v_c < v_r$), (d) small slotters of simple crank mechanism ($v_c < v_r$).

$v_{cm}$ in meters per minute can be calculated depending on the type of machine mechanism. It should not exceed the permissible cutting speed, which depends upon:

- Machining conditions (depth of cut, feed, tool geometry, and related conditions)
- Tool material used
- Properties of WP material

### 3.5.1.1 Determination of $v_{cm}$ in Accordance with the Machine Mechanism

1. Machines Equipped by the Quick-Return Motion (QRM) Mechanisms ($v_{cm} < v_{rm}$)

$$v_{cm} = \frac{nL(1+Q)}{1000}\,\text{m / min} \tag{3.17}$$

where

$L$ = selected stroke length (mm)
$n$ = selected number of strokes per minute
$Q = v_{cm}/v_{rm} < 1$ ($v_{cm}$ and $v_{rm}$ are the mean cutting and the mean reverse speed, respectively)

Equation 3.17 is applicable to:

- Hydraulic shapers, planers, and slotters, where $v_c$ and $v_r$ are constant (Figure 3.65a)
- Shapers and slotters (Figure 3.65b) of lever arm mechanism
- Planers of rack and pinion mechanism (Figure 3.65c)

2. Machines of Crank Mechanism ($v_{cm} = v_{rm}$)

These machines are applicable to small-size slotters (Figure 3.65d).

$$v_{cm} = v_{rm} \frac{2nL}{1000} \, \text{m} / \text{min} \tag{3.18}$$

where

$L$ = selected stroke length (mm)

$n$ = selected number of strokes per minute

Hydraulic shapers, planers, and slotters are becoming increasingly popular for the following characteristics:

- Greater flexibility of speed (infinite variable)
- Smoother in operation
- Ability to slip in case of overload, thus eliminating tool and machine damage
- Possibility of changing speeds and feeds during operation
- Providing a constant speed all over the stroke

### 3.5.2 SHAPER AND PLANER TOOLS

Shaper and planer tools are strongly dimensioned single-point tools designed to withstand the operating impact loads. Figure 3.66 shows typical tools that are used for different machining purposes. These include the following:

a. Straight-shank roughing tool
b. Bent-type roughing tool
c. Side-cutting tool
d. Finishing tool
e. Broad-nose finishing tool
f. Slotting tool
g. Tee-slot tool
h. Gooseneck tool

Figure 3.67 compares straight and gooseneck tools. The gooseneck tools are used to reduce digging in and scoring the WP, and better surface quality is thereby achieved. The tendency to gouge will be lessened if the tool nose is leveled up with the base of the tool shank. For eliminating chatter, and accordingly achieving an acceptable surface quality, the tool overhang should be kept as small as possible. Shaper and planer tools have rake angles of 5–10° for HSS tools and between 0 and −15° for carbide tools. The cutting edge inclination angle is normally 10°, while a nose radius of 1–2 mm is used in the case of roughing tools.

### 3.5.3 SHAPERS, PLANERS, AND SLOTTERS

#### 3.5.3.1 Shapers

A shaper machine is commonly used in single-piece and small-lot production as well as in repair shops and tool rooms. Due to its limited stroke length, it is conveniently

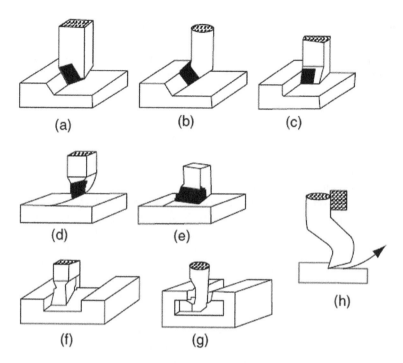

**FIGURE 3.66** Shaper and planer tools: (a) straight-shank roughing tool; (b) bent-type roughing tool; (c) side-cutting tool; (d) finishing tool; (e) broad-nose finishing tool; (f) slotting tool; (g) tee-slot tool; (h) gooseneck tool.

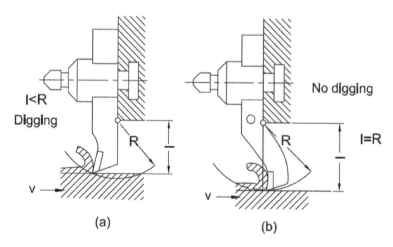

**FIGURE 3.67** Performance of straight and gooseneck tools. (a) Straight, (b) gooseneck.

adapted to small jobs and best suited to surfaces comprising straight-line elements and contoured surfaces when the shaper is equipped with a tracing attachment. It is also applicable for cutting keyways and splines on shafts. Although the shaping process is inherently slow, it is quite popular because of its short setup time, inexpensive tooling, and ease of operation. In comparison to a planer, it occupies less floor space, consumes less power, costs less, is easier to operate, and is about three times quicker in action, as stroke length and inertia forces are lower. Its stroke length is limited to 750 mm, as the accuracy decreases for longer strokes due to ram overhanging. Figure 3.68 shows a typical shaper. The column (1) houses the speed gearbox, the crank, and the slotted arm mechanism. The power is, therefore, transmitted from the motor (2) to the ram (3). Ram travel is the primary reciprocating motion, while the intermittent cross travel of the table is the feed motion. The tool head (5), carrying the clapper box and the tool holder (6), is mounted at the front end of the ram and is fed manually or automatically. The slot with the clamp (7) serves to position the ram in setting up the shaper.

The tool head has a tool slide and feed screw rotated by a ball crank handle (8) for raising and lowering the tool to adjust the depth of the cut. A swivel motion of the tool head enables it to take angular cuts to machine inclined surfaces. The WP is either clamped directly on the table or held in a machine vise. By means of a ratchet and pawl mechanism (9) driven from the crank and slotted arm mechanism, the table is fed crosswise in a horizontal plane. The table is raised or lowered by the elevating screw (10). A support bracket (11) is provided to clamp the table rigidly during operation. The number of ram strokes per minute is set by shifting levers (12).

Figure 3.68b shows a simplified gearing diagram of a mechanical shaper. Rotation of the main motor is transmitted to the six-speed gearbox (A). The pinion, $Z = 25$, drives the bull gear, $Z = 102$. The rotation movement of the bull gear is converted to reciprocating motion of the slotted arm (B) linked to the ram (3). The stroke length of the ram can be varied by adjusting the radius of the crankpin on the bull gear. This adjustment alters the speed of the ram (see Equation 3.17). Figure 3.69 visualizes the table feed mechanism of the mechanical shaper. Rotation of the crank gear (1), mounted on the driving shaft (9) (driven by the gear, $Z = 35$ [Figure 3.68b]), rocks the pawl carrier (4) and pawl (5) of the ratchet and pawl arrangement through the connection (3). In its forward stroke, the pawl engages a tooth of the ratchet wheel (6), which is fastened to the table lead screw (7). This causes the ratchet wheel and lead screw to rotate a fraction of a revolution. On the return stroke, the pawl slips over the ratchet teeth. Accordingly, the table is fed upon rotation of the lead screw. Radial adjustment of the cluster (2) on the crank gear (1) varies the amount by which the ratchet wheel is moved at each stroke and consequently, changes the table displacement per stroke (feed). Maximum feed is obtained when the cluster is adjusted to its maximum radius. The table feed direction can be reversed by reversing the ratchet (5).

### 3.5.3.2 Planers

Planers are intended for machining large-size WPs because of their capacity for long table travel (1–15 m) and robust construction. They are used to machine plane surfaces that may be horizontal, vertical, or at an angle. Angular surfaces are often easier to machine on planers. Some of the work formerly done on planers is done now on planer-type milling machines using large face milling cutters. However, it is found that milling cutters tend to be glazed, and the machined component is work hardened and hence becomes difficult to hand-scrape. Therefore, plane surfaces that require hand-scraping are preferably machined on planers. Both the productivity

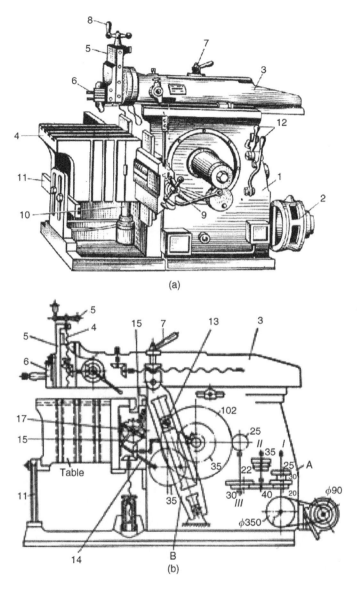

**FIGURE 3.68** Typical mechanical shaper: (a) General view and (b) gearing diagram.

and the accuracy of planers are considerably enhanced, because it is possible to take multicuts on the WP in a single stroke. Generally, it is usual to mount two tool holders on the cross-rail and one each side of the column. The setting time, therefore, is around five to six times longer than that of a shaper. It is also possible to machine a large number of small parts by setting them properly on the planer table. Planers produce large work at the lowest cost in comparison to any other machine tool. The operation of a planer requires a high degree of mental effort and mechanical skill. Heavy cuts can be performed on planers. A depth of cut up to 18 mm and a feed

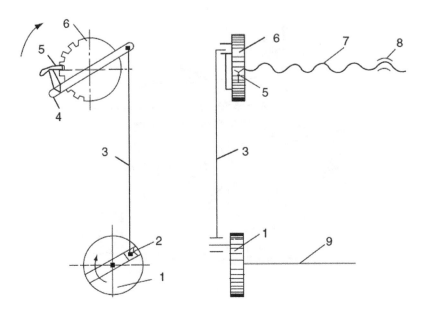

**FIGURE 3.69**   Table feed mechanism of a mechanical shaper.

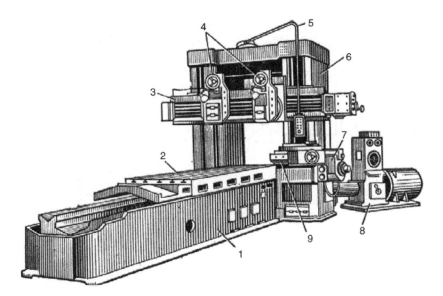

**FIGURE 3.70**   Double housing planer.

rate of 1.5–3 mm/stroke can be taken for roughing, while a depth of 0.25–0.5 mm may be used for finish cuts. Straightness of 8 μm/m and surface roughness $R_a$ of the order of 1 μm can be attained (Jain, 1993). Planers may be of either the open-side or the housing type. A double housing planer, illustrated in Figure 3.70, operates in the following manner.

The table (2) carrying the WP reciprocates on the bed (1). The table is powered by a variable-speed dc motor (8) through a reduction gearbox and a rack and pinion drive (Figure 3.71). The housing (6) mounts the side tool head (9), while the cross-rail (3) is raised and lowered from a separate motor on the housings to accommodate WPs of different heights set up on the table. The upper tool heads (4) may be traversed by a lead screw (feed motion). The side tool head is traversed vertically (feed motion) by the feed gearbox (7) to machine vertical surfaces. All tool heads operate independently of each other. The control panel and the suspended cable (5) are shown in Figure 3.70. The tool heads (4) may be swiveled to machine an inclined surface. Like all reciprocating machine tools, planers are equipped with a clapper box to raise the tools on the return stroke. As the tool and holder are quite heavy, air cylinders are employed to lift the tool from the WP on the return stroke.

### 3.5.3.3 Slotters

Slotters are commonly used for internal machining of blind holes or vertical machining of complicated shapes that are difficult to machine on horizontal shapers. They are useful for machining keyways and cutting internal and external teeth on large gears.

As illustrated in Figure 3.72, the job is generally supported on a round table (3) that has a rotary feed in addition to the usual table movement in cross-directions. The ram (1) travels vertically along the ways of the column (2). The ram stroke of a slotter ranges from 300 to 1800 mm. The slotters are generally very robust machines, and there is a possibility of tilting the ram up to ±15° from vertical to permit machining of dies with relief. The rams are either crank driven or hydraulically driven. Ram

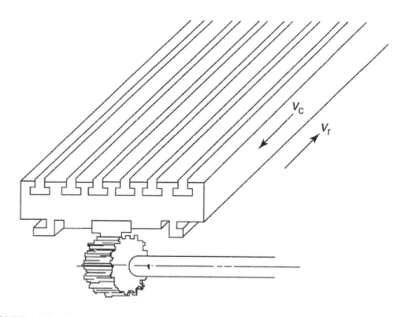

**FIGURE 3.71** Rack-and-pinion mechanism of a planer.

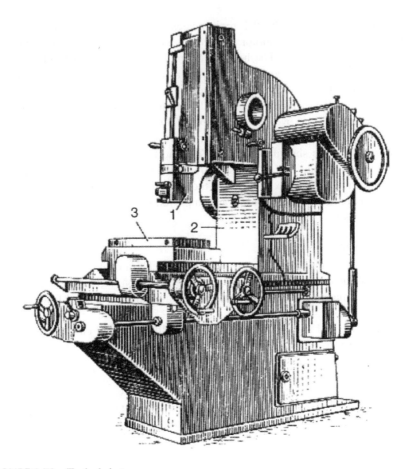

3   1
2

**FIGURE 3.72**   Typical slotter.

speeds are usually from 2 to 40 m/min. Longitudinal and transverse feeds range from 0.05 to 2.5 mm/stroke. The cutting action takes place on the downward stroke.

## 3.6   BORING MACHINES AND OPERATIONS

### 3.6.1   BORING

Boring is the machining process in which internal diameters are generated in true relation to the centerline of the spindle by means of single-point tools. It is the most commonly used process for enlarging and finishing holes or other circular contours. Although most boring operations are performed on simple straight-through holes, the process may be also applied to a variety of other configurations. Tooling can be designed for boring blind holes, holes with bottle configurations, circular-contoured cavities, and bores with numerous steps, undercuts, and counterbores. The process is not limited by the length-to-diameter ratio of holes. Boring is sometimes used after drilling to provide drilled holes with greater dimensional accuracy and improved surface finish. It is used for finishing large holes in castings and forgings that are too large to be produced by drilling.

### 3.6.2 BORING TOOLS

The boring tools can be mounted in either a stub-type bar, held in the spindle, or a long boring bar that has its outer end supported in a bearing. This provides rigid support for the boring bar and permits accurate work to be done.

#### 3.6.2.1 Types of Boring Tools

Figure 3.73 illustrates a number of typical boring tools:

a. A single-point cutter mechanically secured to a boring bar. When the tool becomes worn, it is removed for sharpening and reset again. Resetting sharpened tools is tedious and requires a fair degree of skill.
b. An adjustable single-point cutter advanced for wear compensation.
c. Boring tools clamped in a universal boring head that is attached to the end of the boring bar. The head is designed to accommodate a variety of tool configurations.
d. A fixed cutter, held by a stub boring bar, which is simple and widely used.
e. A blade-type boring tool, where the cutter is inserted through the body, thus providing two cutting edges, which enable a substantially higher increase

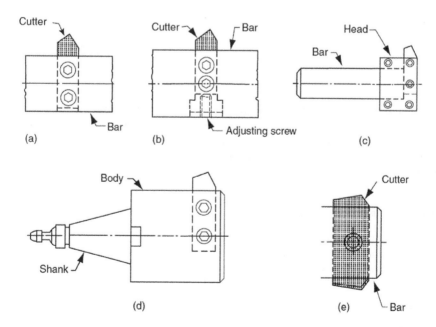

**FIGURE 3.73** Typical boring tools: (a) single-point cutter; (b) adjustable single-point cutter; (c) boring tools in a universal boring head; (d) fixed cutter, held by a stub boring bar; (e) blade-type boring tool.

**FIGURE 3.74**   Adjustable tools in stub. (From DeVlieg Machine Co., Michigan, USA.)

of feed rate than is possible when only one cutting edge is used. The main advantage of this tool is that it equalizes the force imposed on the bar during operation. Therefore, it is possible to maintain closer tolerances with bars having maximum unsupported length than when using a boring tool that has only one cutting edge. The main disadvantage is that the blade cannot be adjusted to compensate for wear and therefore must be removed for sharpening and then reset. The boring bar illustrated in Figure 3.74 is the same as (d) in Figure 3.73, but the cutter is adjustable for wear compensation.

### 3.6.2.2   Materials of Boring Tools

1. For low cutting speeds, HSS is more suitable than carbide.
2. Carbides are used almost exclusively for precision boring, when the maximum rigidity is maintained in the setup.
3. Ceramics are increasingly applied for precision boring operation at high cutting speeds.

Ceramic inserts are characterized by reduced tool wear. Ceramics have the ability to bore hard materials (steel of 60–65 HRC), thus eliminating the need for subsequent grinding. Ceramics are not recommended for boring interrupted cuts and refractory metals. Also, they are not recommended for aluminum alloys, because they develop BUE.

### 3.6.3   Boring Machines

Boring is performed on almost every type of machine that has facilities for rotating a spindle or a WP. Most boring operations are done in conjunction with turning and cutting on numerical control (NC) and computer numerical control (CNC) machines, and so on, discussed in other chapters of this book. However, in this section, general-purpose horizontal boring machines and jig-boring machines are discussed.

### 3.6.3.1   General-Purpose Boring Machines

In these machines, the WP remains stationary; the tool rotates and may simultaneously perform a feed motion. The boring machine is designed to machine relatively large, irregular, and bulky WPs that cannot be easily rotated. Among the operations performed on this machine are boring, facing, drilling, counterboring, counterfacing, external and internal thread cutting, and milling. A horizontal boring machine is especially suitable for work where several parallel bores with accurate center distances are to be produced. Because of its flexibility, this machine is especially suited to work in which other machining operations are performed in conjunction with boring.

A typical general-purpose boring machine is shown in Figure 3.75. The cutting tool is mounted either in the spindle (13) or on the facing slide (8). The rotation of the spindle and faceplate (7) is the principal movement that is effected by the main motor (11) through the speed gearbox housed in the headstock (9). The spindle can also be fed axially, so that drilling and boring can be done over a considerable distance without moving the work. The WP is installed either directly on the table (6) or in a fixture. The table is moved longitudinally or transversally on the cross slide (5). The table and the cross slide are located on a saddle (4), which moves longitudinally on the bed (3).

The headstock (9) moves vertically along the column (10) simultaneously with the spindle rest (2), which is moving vertically along the end support column (1). The spindle travels axially when boring or cutting internal thread, and so on. The facing slide is moved radially on the faceplate to perform facing operations. The rotational speed of the faceplate is much lower than that of the spindle. The table feed and its rapid reverse are powered by the motor (12). In some setups, the work is fed toward the tool, while in other cases, the tool is fed toward the work.

### 3.6.3.2 Jig-Boring Machines

Jig borers are extra-precise vertical boring machines intended for precise boring, centering, drilling, reaming, counterboring, facing, spot facing, and so on in addition to layout work. They are mainly designed for use in tool making, jigs and fixtures, and machining of other precision parts. No jigs whatsoever are required in these machines. A jig-boring machine contains similar features to a vertical milling

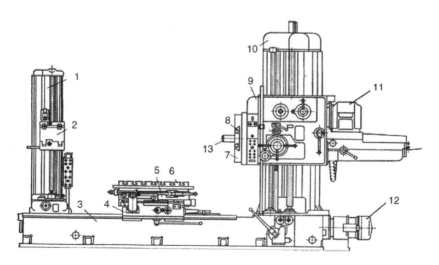

**FIGURE 3.75** Typical general-purpose boring machine.

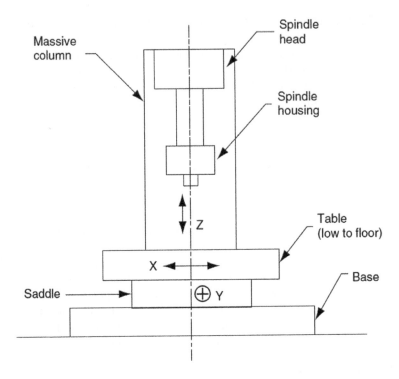

**FIGURE 3.76**  Schematic of a jig-boring machine.

machine, except that the spindle and its bearings are constructed with very high precision, and the worktable permits extra-precise movement and control.

As illustrated schematically in Figure 3.76, the jig borer is generally built lower to the floor and is of much more rigid and accurate construction than any other machine tool. The table and saddle ensure the longitudinal and cross movements, $X$ and $Y$. The machine has a massive column that supports and accurately guides the spindle housing in the vertical direction, thus achieving the third position adjustment, $Z$. Jig-boring machines are rigid enough to perform heavy cuts and sensitive enough for précising. They are equipped with special devices ensuring accurate positioning of the machine operative units, including a precision lead screw and nut, and are supplemented by vernier dials and precision scales in combination with optical readout devices, inductive transducers, and also optical and electrical measuring devices. To prevent the influence of ambient temperature changes on the machining accuracy, jig borers are installed in special environmental enclosures with temperature maintained at a level of 20 °C. Currently, jig-boring machines are often replaced by CNC machining centers to do similar work.

## 3.7  BROACHING MACHINES AND OPERATIONS

### 3.7.1  BROACHING

Broaching is a cutting process using a multitoothed tool (broach) having successive cutting edges, each protruding to a greater distance than the preceding one

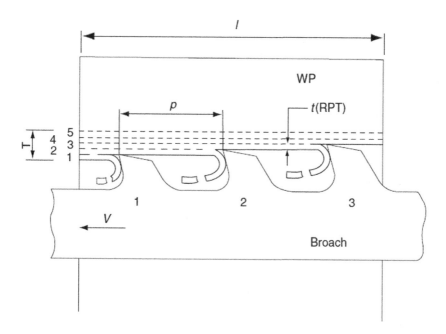

**FIGURE 3.77**   Cutting action of broaching process.

in the direction perpendicular to the broach length. In contrast to all other cutting processes, there is no feeding of the broach or the WP. The feed is built into the broach itself through the consecutive protruding of its teeth. Therefore, no complex motion of the tool relative to the WP is required, and the tool is moved past the WP with a rectilinear motion $v_c$ (Figure 3.77). Equally effective results are obtained if the tool is stationary and the work is moved. The total depth of the material removed in one stroke $T$ is the sum of rises of teeth of the broach. $T$ may be as deep as 6 mm broached in one stroke. If more depth is to be broached, two broaches may be used to perform the task. Broaching is generally used to machine through holes of any cross-sectional shape, straight and helical slots, external surfaces of various shapes, and external and internal toothed gears (Figure 3.78). To permit the broaching of spiral grooves and gun-barrel rifling, a rotational movement should be added to the broach. Broaching usually produces better accuracy and finish than drilling, boring, or reaming operations. A tolerance grade of IT6 and a surface roughness $R_a$ of about 0.2 μm can be easily achieved by broaching. Broaching dates back to the early 1850s, when it was originally developed for cutting keyways in pulleys and gears. However, its obvious advantages quickly led to its development for mass-production machining of various surfaces and shapes to tight tolerance. Today, almost every conceivable form and material can be broached. Broaches must be designed individually for a particular job. They are very expensive to manufacture ($15,000–$30,000 per tool). It follows that the broaching can only be justified when a very large batch size (100,000–200,000) is to be machined. However, sometimes the WP is designed so that an inexpensive standard broach can be used.

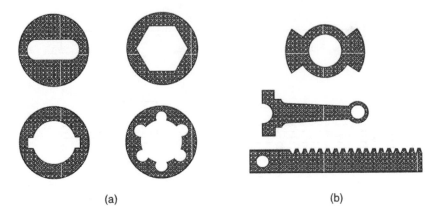

(a)                                                    (b)

FIGURE 3.78  Typical parts produced by internal and external broaching: (a) internal broaching and (b) external broaching. (From Kalpakjian, S. and Schmid, S. R., *Manufacturing Processes for Engineering Materials,* Prentice-Hall, New York, 2003.)

### 3.7.1.1  Advantages and Limitations of Broaching

Advantages

- Broaching is a process in which both roughing and finishing operations are completed in one pass, giving a high rate of production.
- It is a fast process; it takes only seconds to accomplish a task that would require minutes with any other method. Rapid loading and unloading of fixtures keeps the total production time to the minimum.
- Automation is easily arranged.
- Internal and external surfaces can be machined within the close tolerance that is normally required for interchangeable mass production.
- As all the performance is built into the tooling, little skill is needed to operate a broaching machine.
- Broaches have an exceptionally long life (10,000–20,000 parts per sharpening) as each tooth passes over the work only once per pass.

Limitations

- Broaches are costly to make and sharpen. Hence, broaching is adopted only in cases of mass production.
- Standard broaches are available; however, most broaches are expensive, as they are made especially to perform only one job.
- Special precautions may be necessary when broaching cast and forged parts to control the variations in stock. Operations for removing excess stock may be necessary, which add to the overall cost of manufacturing.
- Surfaces to be machined must be parallel to the axis of the broach.
- Broaching is impractical in the following cases:
  a.  A surface that has obstructions across the path of the broach travel
  b.  Blind holes and pockets

  c.  Fragile WPs, because they cannot withstand broaching forces without distortion or breakage

### 3.7.2  THE BROACH TOOL

#### 3.7.2.1  Tool Geometry and Configuration

Figure 3.79 illustrates the broach tooth terminology. Each individual tooth has the basic wedge form.

- Depending on the material being cut, the rake (hook) angle ranges from 0° to 20°. The small clearance (back-off) angle is usually 3°–4° for roughing teeth and 1°–2° for finishing teeth.
- The rise per tooth (RPT) (superelevation) is the difference in height of two consecutive teeth (Figure 3.77). It is selected depending upon the material to be machined and the type (form) of the broach (Table 3.3).
- The pitch is the distance between two consecutive teeth of a broach. It depends upon the following factors:
- Length of cut $l$
- Material of WP and its mechanical properties
- RPT (superelevation)

The pitch $p$ can be expressed empirically by

$$p = 3\sqrt{\text{RPT} \cdot l \cdot X} \qquad (3.19)$$

where
  $X$ = chip space number
   = 3–5 for brittle WP materials
   = 6–10 for ductile and soft WP materials

A relatively large pitch and tooth depth are required for roughing teeth to accommodate greater chip volume in the chip gullet, especially when machining materials

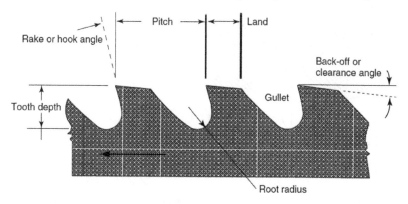

**FIGURE 3.79**  Broach tooth terminology.

**TABLE 3.3**
**RPT of Broaches**

|  | RPT (μm) | | | |
| --- | --- | --- | --- | --- |
| Type of Broach | Steel | CI | Aluminum | Bronze-brass |
| Round | 15–30 | 30–100 | 20–50 | 50–120 |
| Spline | 25–100 | 40–100 | 20–100 | 50–120 |
| Square and hexagon | 15–80 | 80–150 | 20–100 | 50–200 |
| Keyway | 50–200 | 60–200 | 50–80 | 80–200 |

From Arshinov, V. and Alekseev, G., *Metal Cutting Theory and Cutting Tool Design*, Mir Publishers, Moscow, 1970. With permission.

produce continuous chips. For semifinishing and finishing teeth, the pitch is reduced to about 60% of that of roughing teeth to reduce the overall length of the broach. The calculated pitch, according to Equation 3.19, should not be greater than $l/2$ ($l$ is the length to be cut) to provide better guidance of the tool and to prevent the broach from drifting. To prevent possible chattering and to obtain better surface finish, the pitch $p$ should be made nonuniform as shown in Figure 3.80. To avoid the formation of long chips, especially when broaching profiles and circular shapes, chip breakers are uniformly cut into the cutting edges of the broach in a staggered manner. Chip breakers are not necessary when broaching brittle materials that produce discontinuous chips. They are not used for finishing teeth and small-size broaches. The use of chip breakers reduces the pitch and consequently, the overall length of the broach. As a result, the productivity is enhanced, and the tool cost may be reduced.

### 3.7.2.1.1 Broach Configuration

Figure 3.81 illustrates the terminology of a pull-type internal broach for enlarging circular holes. The cutting teeth on the broach have three regions: roughing,

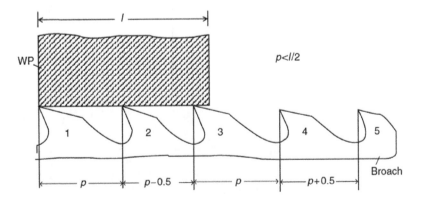

**FIGURE 3.80** Nonuniform pitching to prevent chattering, and engagement of more than two teeth to ensure guidance.

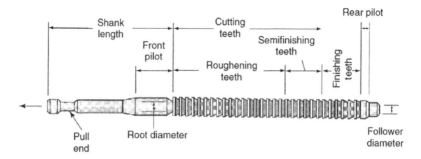

**FIGURE 3.81**   Solid-pull broach configuration.

**FIGURE 3.82**   Shell broach. (From Degarmo, E. P. et al., *Materials and Processes in Manufacturing*, 8th Edition, Prentice-Hall, New York, 1997.)

semifinishing, and finishing teeth. On some round broaches, burnishing teeth are provided for finishing or sizing. These teeth have no cutting edge but are rounded. Their diameters are oversized by 25–30 μm larger than the finished hole. Irregular shapes are produced by starting from circular broaching in the WP originally provided with a drilled, bored, cored, or reamed hole. The pull end provides a means of quickly attaching the broach to the pulling mechanism. The front pilot aligns the broach in the hole before it begins to cut, and the rear pilot keeps the tool square with the finished hole as it leaves the WP. It also prevents sagging of the broach. The follower end is ground to fit in the machine follower rest.

Internal broaches are also made of shells mounted on an arbor (Figure 3.82). Shell broaches are superior to solid broaches in that worn or broken shells can be replaced without discarding the entire broach. Shell construction, however, is initially more expensive than a solid broach of comparable size. The disadvantage of shell broaches is that some accuracy and concentricity are sacrificed.

Regarding the application of broaching force, two types of broaching are distinguished (Figure 3.83):

1. Pull broaching, as the name implies, involves the broach being pulled through the hole (Figure 3.83a). In this case, the main cutting force is applied to the front of the broach, subjecting the body to tension. Most internal broaching is done with pull broaches. Because there is no problem of buckling, pull broaches can be longer than push types for the same broaching depth. Pull broaches can be made to long lengths, but cost usually limits the length to approximately 2 m. Broaches longer than 2 m are shell broaches, because the cost is less for replacing damaged or worn sections than for replacing the entire broach.

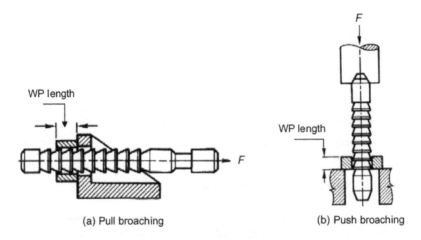

(a) Pull broaching                                          (b) Push broaching

**FIGURE 3.83**   (a) Pull broaching and (b) Push broaching.

2. Push broaching applies the main cutting force to the rear of the broach, thus subjecting the body to compression (Figure 3.83b). A push broach should be shorter than a pull broach, and its length does not usually exceed 15 times its diameter to avoid buckling.

### 3.7.2.2   Broach Material

The low cutting speeds used in most broaching operations (2–12 m/min) do not lend themselves to the advantage of carbide tooling. Accordingly, most broaches are made of alloy steel and HSS of high grades (Cr-V-grade), which have less distortion during heat treatment. This is an important factor in the manufacture of long broaches. Titanium-coated HSS broaches are becoming more common due to their prolonged tool life. Recently, carbide-tipped K-type (cobalt group) tools have been employed to machine CI, thus allowing higher cutting speeds, increased durability, and improved surface finish. However, carbide-tipped broaches are seldom used for machining steels and forged parts, as the cutting edges tend to chip in the first stroke due to lack of rigidity of the work fixture/tool combination.

### 3.7.2.3   Broach Sharpening

Broach sharpening is essential, as dull tools require more force, leading to less accuracy and broach damage. Dull internal broaches have the tendency to drift during cutting. The clearance angle of the sizing teeth of a broach is made as small as possible (1°–2°) to minimize the loss of size when it is sharpened. Also, the finishing or sizing teeth are commonly provided with a land of a small width of 50–200 μm to limit the size loss due to sharpening. Most broaches are sharpened by grinding the hook faces of the broach. The lands must not be reground, because this would change the size of the broach (Figure 3.84). After sharpening, the tooth

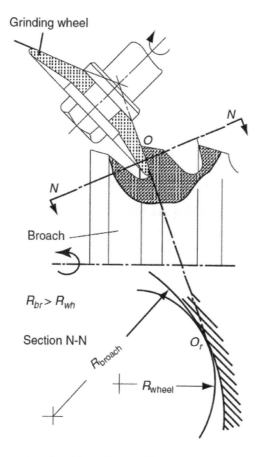

**FIGURE 3.84**   Sharpening of tool face of a round broach.

characteristics, such as rake angle, clearance angles, tooth depth, root radius, RPT, and pitch, should not be altered.

### 3.7.3   BROACHING MACHINES

In comparison with other types of machine tools, broaching machines are notable for their simple construction and operation. This is due to the fact that the shape of the surface produced in broaching depends upon the shape and arrangement of the cutting edges on the broach. The only cutting motion of the broaching machine is the straight-line motion of the ram. Broaching machines have no feed mechanisms, as the feed is provided by a gradual increase in the height of the broach teeth. Hydraulic drives, developed in the early 1920s, offered pronounced advantages over the various early mechanical driving methods. Most broaching machines existing today are of hydraulic drive and are accordingly characterized by smooth running and safe operation. The choice between vertical and horizontal machines is determined primarily by the length of stroke required and the available floor space. Vertical machines seldom have strokes greater than 1.5 m because of ceiling limitations. Horizontal machines can have almost any stroke length; however, they require greater floor space.

The main specifications of a broaching machine are as follows:

- Maximum pulling or pushing force (capacity) (tons)
- Maximum stroke length (m)
- Broaching speed (m/min)
- Overall dimensions and total weight

### 3.7.3.1  Horizontal Broaching Machines

Currently, horizontal machines are finding increasing favor among users because of their long strokes and the limitation that ceiling height places on vertical machines. About 47% of all broaching machines are horizontal units (ASM International, 1989). Horizontal internal broaching machines are used mainly for particular types of work such as automotive engine blocks. The pulling capacity ranges from 2.5 to 75 tons, strokes are up to 3 m, and cutting speeds are limited to less than 12 m/min. Broaching that requires rotation of the broach, as in rifling and spiral splines, is usually done on horizontal internal broaching machines. Horizontal machines are seldom used for broaching small holes.

Horizontal surface broaching machines may be hydraulically or electromechanically driven. In these machines, the broach is always supported in guides. Surface hydraulic broaching machines are built with capacities up to 40 tons, strokes up to 4.5 m, and cutting speeds up to 30 m/min. These machines have been basically used in the automotive industry to broach a great variety of CI parts for nearly 30 years. On the other hand, the electromechanically driven horizontal surface broaching machines are available with higher capacities, stroke lengths, and cutting speeds (up to 100 tons, 9 m, and 30 m/min, respectively). Carbide-tipped broaches are used to machine CI blocks of internal combustion engines (ICEs).

### 3.7.3.2  Vertical Broaching Machines

These machines are almost all hydraulically driven. They are used in every major area of metal working. Depending on their mode of operation, they may be pull-up, pull-down, or push-down units. Figure 3.85 schematically illustrates a pull-down vertical broaching machine in which the work is placed on the worktable. These machines are capable of machining internal shapes to close tolerances by means of special locating fixtures. They are available with pulling capacities from 2 to 50 tons, strokes from 0.4 to 2.3 m, and cutting speeds up to 24 m/min. When cutting strokes exceed existing factory ceiling clearances, expensive pits must be dug for the machine so that the operator can work at the factory floor level.

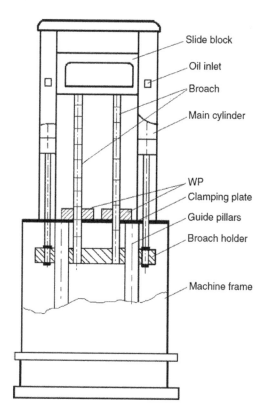

Slide block
Oil inlet
Broach
Main cylinder

WP
Clamping plate
Guide pillars
Broach holder

Machine frame

**FIGURE 3.85** Schematic of a pull-down vertical broaching machine. (Adapted from ASM International, *Machining,* Vol. 16, *Metals Handbook,* ASM International, Materials Park, OH, 1989.)

### 3.7.3.3  Continuous Horizontal Surface Broaching Machines

In this type of machine, the broaches are usually stationary and mounted in a tunnel on the top of the machine, while the work is pulled past the cutters by means of a conveyor (Figure 3.86). Fixtures are usually attached to the conveyor chain, so that the WPs can be provided automatically by the loading chute at one end of the bed and removed at the other end. The key to the productivity of this type of machine is the elimination of the return stroke by mounting the WPs on a continuous chain. In rotary continuous horizontal broaching machines, the broaches are also stationary, while the work is passed beneath or between them. The work is held in fixtures on a rotary table. These machines are also used in mass production, as there is no loss of time due to the noncutting reciprocating strokes.

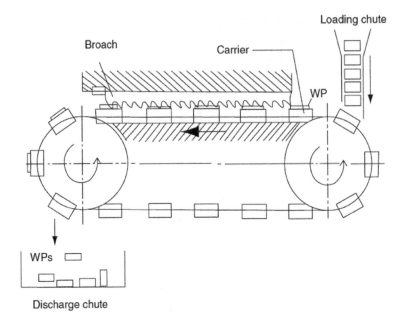

**FIGURE 3.86** Continuous horizontal surface broaching machine. (Adapted from ASM International, *Machining,* Vol. 16, *Metals Handbook,* ASM International, Materials Park, OH, 1989.)

## 3.8 REVIEW QUESTIONS

3.8.1    The produced accuracy of planers is superior to that of shapers. Explain.

3.8.2    What is the maximum table travel of a large planer? What is the maximum ram travel in a shaper?

3.8.3    What are the main features that limit the stroke of a shaper?

3.8.4    How does form turning differ from ordinary turning?

3.8.5    Why is the table spindle hollow?

3.8.6    What will happen to a WP held between centers if the centers are not exactly in line?

3.8.7    How does a steady rest differ from a follower rest?

3.8.8    Why should the projected length of a lathe tool be minimized?

3.8.9    List the methods of taper turning on a lathe.

3.8.10   Why it is desirable to use a heavy depth of cut and light feeds rather than the opposite?

3.8.11   Why is the cutting speed in planing, shaping, and slotting limited to 50–60 m/min?

3.8.12   On what diameter is the rpm of the work based for a facing cut, assuming given work and tool material?

3.8.13   Why are vertical boring machines better suited than a facing lathe to machining large WPs?

3.8.14    Mark true or false.
- a. To enhance productivity, efficient cutting tools such as carbide, CBN, and ceramic tools are employed on shapers and planers.
- b. Surfaces to be hand-scraped would be better produced on planer-type milling machines.
- c. Planers produce large work at the lowest cost in comparison to any other machine tools.
- d. Broaching machines are simpler in basic design than other machine tools.

3.8.15    Define these terms: boring, broaching, counterboring, countersinking, reaming, and spot facing.

3.8.16    When a large-diameter hole is to be drilled, why is a smaller-diameter hole often drilled first?

3.8.17    Explain why a gooseneck tool is highly recommended to use as a shaping or planing tool.

3.8.18    What is unique about broaching compared with other basic machining operations?

3.8.19    Why is broaching more practically suited for mass production?

3.8.20    What is the main point to be considered in pitching the broach teeth that reduce chattering?

3.8.21    Why are broaching speeds usually relatively low compared with those of other machining operations?

3.8.22    What are the advantages of a shell-type broach?

3.8.23    Can continuous broaching machines be used for broaching holes? Explain why or why not.

3.8.24    State some ways to improve the efficiency of a planer. Do any of these apply to a shaper?

3.8.25    How does the process of shaping differ from planing?

3.8.26    How does a gang drilling machine differ from a multiple-spindle drilling machine?

3.8.27    It is required to drill eight equally spaced holes $\phi$ 10 mm in a bolt-hole circle of 160 mm. The holes must be $\pm 1°$ from each other around the bolt-hole circle. Calculate the tolerance between hole centers.
- Do you think a typical multiple-spindle drill set up could be used, or would a drilling jig be better in this situation?

3.8.28    What might happen when holding work by hand during drilling?

3.8.29    In a turning operation, a cutting speed of 55 m/min has been selected. At what rpm should a 15 mm diameter bar be rotated?

3.8.30    Describe the relative characteristics of climb milling and up-milling, mentioning the advantages of each.

3.8.31    Which type of milling (up or down) do you think uses less power under the same cutting conditions?

3.8.32    Why does the use of climb milling make it easier to design a milling fixture than up-milling?

3.8.33    What are the advantages of a helical-tooth cutter over a straight-tooth cutter for slab milling?

3.8.34   Explain the steps required to produce a T-slot by milling.
3.8.35   Why would a plain horizontal-knee milling machine be unsuitable for milling helical flutes?
3.8.36   What is the basic principle of a universal dividing head?
3.8.37   The input end of a universal dividing head can be connected to the lead screw of the milling machine table—for what purpose?
3.8.38   What is the purpose of indexing plates on a universal dividing head?
3.8.39   Explain how a standard universal dividing head having a hole circle 21, 24, 27, 30, and 32 would be operated to cut 18 gear teeth.

## REFERENCES

Arshinov, V & Alekseev, G 1970, *Metal cutting theory and cutting tool design*, Mir Publishers, Moscow.
ASM International 1989, *Machining, vol. 16, metals handbook*, ASM International, Materials Park, OH.
Degarmo, EP, Black, JT & Kohser, RA 1997, *Materials and processes in manufacturing*, 8th edn, Prentice-Hall, New York.
DeVlieg Machine Co., Ferndale, MI
Jain, RK 1993, *Production technology*, 13th edn, Khanna Publishes, Delhi, India.
Kalpakjian, S & Schmid, SR 2003, *Manufacturing processes for engineering materials*, 4th edn, Prentice-Hall, New York.
Mott, LC 1976, *Engineering drawing and construction*, Oxford University Press, Oxford.

# 4 General-Purpose Abrasive Machine Tools

## 4.1 GRINDING MACHINES AND OPERATIONS

### 4.1.1 GRINDING PROCESS

Grinding is a metal removal process that employs an abrasive grinding wheel (GW) whose cutting elements are grains of abrasive materials of high hardness and high refractoriness. Grinding is generally among the final operations performed on manufactured products. It is not necessarily confined to small-scale material removal; it is also used for large-scale material removal operations and specifically competes economically in this domain with some machining processes such as milling and turning. The development of abrasive materials and better fundamental understanding of the abrasive machining have contributed to the placing of grinding among the most important basic machining processes.

Because the abrasives employed are very hard, abrasive machining is used for:

- Finishing hard materials and hardened steels
- Shaping hard nonmetallic materials such as carbides, ceramics, and glass
- Cutting off hardened shafts, masonry, granite, and concrete
- Removing weld beads
- Cleaning surfaces

The sharp-edged and hard grains are held together by bonding material. Projecting grains (Figure 4.1) abrade layers of metal from the work in the form of very minute chips as the wheel rotates at high speeds of up to 60 m/s. Due to the small cross-sectional area of the chip and the high cutting speed, grinding is characterized by high accuracy and good surface finish. Consequently, it is usually employed as a finishing operation. However, it is also used in snagging.

The chip formation in grinding is similar to that in milling. In spite of the small size of the layer being cut in grinding, the chip has a comma form similar to that obtained by milling. However, in grinding, not all the grains participate equally in the metal removal as they do in milling.

Along with the general features of other typical methods of machining, the grinding process has certain specific features of its own, such as the following:

- In contrast to the teeth of a milling cutter, individual grains of a GW have an irregular and nondefinite geometry. They are randomly spaced along the periphery of the GW.

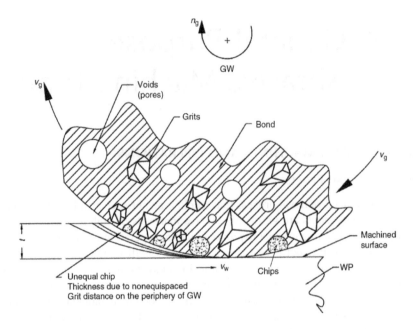

**FIGURE 4.1**   Cutting principles and main variables of a surface grinding process.

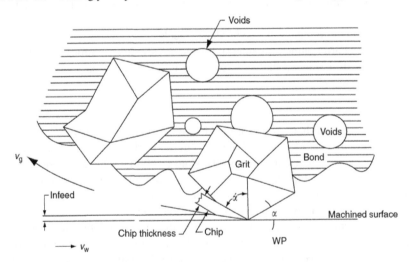

**FIGURE 4.2**   Schematic to illustrate the constituents of a GW.

- The radial positions of the (protruding) grains on the wheel periphery vary, which causes the grains to cut layers of material in the form of chips of different volumes (Figure 4.1).
- The grains of the GW are characterized by high negative rake angles of −40° to −80°; consequently, the shear angles are very small (Figure 4.2).
- Owing to the minute chip thickness and the highly abrasive negative rakes of the grinding operation, the specific cutting energy in grinding is

considerably larger than that of operations using tools of definite geometry. Grinding is thus not only time-consuming but also power-consuming and is hence a costly operation.

- The GW has a self-sharpening characteristic. As the grains wear during grinding, they either fracture or are torn off the wheel bond, exposing new sharp grains to the work.
- The cutting speeds of GWs are very high, typically 30 m/s, which together with the minute chip removal of the grains, provides high dimensional and form accuracy along with high surface quality.

These features make the grinding process more complicated than the other kinds of machining processes and offer considerable difficulties in both theoretical and experimental investigations. However, grinding possesses certain advantages over other metal-cutting methods:

- It cuts hardened steels easily. Parts requiring hard surfaces are first machined to shape in annealed condition with only a small amount left as the grinding allowance, considering the tendency of material to warp during hardening operation.
- Very accurate dimensions and smoother surfaces can be achieved in a very short time.
- Very little pressure is required, thus permitting very light work to be ground that would otherwise tend to spring away from the tool. This permits the use of magnetic chucks for holding the work in many grinding operations.

*Machining variables of a surface grinding process.* Figure 4.1 illustrates the main variables of a surface grinding process. The main rotary motion is performed by the GW ($v_g$), whereas the feed motion is performed by the WP ($v_w$). The depth of cut $t$ (feed/stroke) is fed by the GW perpendicular to the machined surface.

The cutting speed $v_g$ is given by

$$v_g = \frac{\pi D n_g}{1000} \, \text{m/min} \tag{4.1}$$

where
$n_g$ = rotational speed of the GW (rpm)
$D$ = outside diameter of the GW (mm)

The feed motion of the WP, $v_w$, is considerably smaller than the main cutting speed of the GW, $v_g$. Typical values of the ratio $v_w/v_g$ range from 1/20 (for rough grinding) to 1/120 (for finishing). The depth of cut $t$ (feed/stroke) ranges from 10 μm (for finish cuts) to 100 μm (for roughing).

## 4.1.2  GRINDING WHEELS

GWs of all shapes are composed of carefully sized abrasive grains held together by a bonding material. Pores between the grains and the bond allow the grains to act as single-point tools and at the same time provide chip clearance to prevent clogging of the GW (Figure 4.2). GWs are produced using the appropriate grain size of abrasive with the required bond, and the mixture is sintered into shape. GWs are distinguished by their shapes, sizes, and manufacturing characteristics.

### 4.1.2.1 Manufacturing Characteristics of Grinding Wheels

A number of variables are considered that influence the performance of a GW; these are:

#### 4.1.2.1.1 Abrasive Materials

The abrasives for grinding wheels are generally harder than the material of a single-point tool. In addition to hardness, friability is an important characteristic of abrasives. Friability is the ability of abrasive grains to break down into smaller pieces; this property of abrasives enhances the self-sharpening characteristic, which is important in maintaining the sharpness of the GW. High friability indicates low fracture resistance.

Aluminum oxide ($Al_2O_3$) has lower friability than silicon carbide (SiC); thus, it has less tendency to fragment and self-sharpen. The shape and size of a grain also affect its friability. Small grains of negative rakes are less friable than plate-like grains.

The four types of abrasive materials used in manufacturing of GWs are produced synthetically. They are classified into:

a. *Conventional Abrasives as follows:*
   * Aluminum oxide, $Al_2O_3$ (corundum), which has high hardness (Knoop number = 2100) and toughness and is mainly used for grinding metals and alloys of high tensile strength, such as steel, malleable iron, and soft bronze.
   * Silicon carbide, SiC (carborundum), which is harder than $Al_2O_3$ (Knoop number = 2500). It is more friable (more brittle) and is mainly used for grinding materials that have low strength, such as cast iron (CI), aluminum, cemented carbides, and so on. Silicon carbides are available in black (95% SiC) or green (98% SiC). Carbide-tipped tools that are sharpened by SiC dull more rapidly than $Al_2O_3$ when grinding steels.
b. *Super Abrasives:*
   * CBN, which has been manufactured by the General Electric Company since 1970 under the trade name of Borazon. Its properties are similar to those of diamond. CBN is very hard (Knoop number = 4500). It is used for manufacturing wheels intended for grinding extrahard materials at high speeds. CBN is 10–20 times more expensive than $Al_2O_3$.
   * Diamond, which is the hardest of all materials (Knoop number = 7500). It has been synthetically produced since 1955. Synthetic diamonds are friable. Diamond has a very high chemical resistance as well as a low coefficient of thermal expansion. Diamond abrasive wheels are

extensively used for sharpening carbide and ceramic cutting tools. Diamonds are used for truing and dressing other types of abrasive wheels. Diamonds are best suited for nonferrous metal and are not recommended for machining steels.

Table 4.1 shows the characteristics of abrasive materials used in GW manufacturing and their applications.

### 4.1.2.1.2 Abrasive Grain Size

The size of an abrasive grain is identified by the grit number, which is a function of sieve size. The smaller the sieve size, the larger the grit number. The sieve sizes (mesh numbers) of abrasives are grouped into four categories:

| | |
|---|---|
| Coarse | 10, 12, 14, 16, 20, and 24 |
| Medium | 30, 36, 46, 56, and 60 |
| Fine | 70, 80, 90, 100, 120, 150, and 180 |
| Very fine | 220, 240, 280, 320, 400, 500, and 600 |

The choice of the grain size is determined by the nature of the grinding operation, the material to be ground, and the relative importance of the stock removed rate to the finish required. Coarse and medium sizes are normally used for roughing and semifinishing operations. Fine and very fine grains are used for finishing operations and also used for making form GWs.

### 4.1.2.1.3 Wheel Grade

The wheel grade designates the force holding the grains. It is a measure of the strength of the bond. The wheel grade depends upon the type and amount of the bond, the structure of the wheel, and the amount of abrasive grains. Because strength and hardness are directly related, the grade is also referred to as the hardness of the bonded abrasive.

**TABLE 4.1**

**Characteristics of Abrasive Materials and Their Applications**

| Abrasive | Knoop Number | Uses |
|---|---|---|
| $Al_2O_3$ | 2100–3000 | Softer and tougher than SiC, used for steels and high-strength materials |
| SiC | 2500–3000 | Nonferrous, nonmetallic materials, CI, carbides, hard metals, and good finish |
| CBN | 4000–5000 | Hard and tough tool steel, stainless steel, aerospace alloys, hard coating |
| Diamond | 7000–8000 | Nonferrous metals, sharpening carbide, and WC tools |

From Raw, P. N., *Metal Cutting and Machine Tools,* Tata McGraw-Hill, New Delhi, 2000.

The grade is designated by letters, as follows:

| | |
|---|---|
| Very soft | A, B, C, D, E, F, and G |
| Soft | H, I, J, and K |
| Medium | L, M, N, and O |
| Hard | P, Q, R, and S |
| Very hard | T, U, V, W, X, Y, and Z |

Soft grades are generally used for machining hard materials, and vice versa. When grinding hard materials, the grit is likely to become dull quickly, thus increasing the grinding force, and tends to knock off the dull grains easily. In contrast, when a hard grade is used to machine soft material, the grits are retained for a longer period of time, which prolongs the GW's life. Table 4.2 illustrates the recommended wheel grades for different materials and operations. During grinding, and depending on the machining variables, the wheel behaves as if it were harder or softer than its nominal or selected grade, as illustrated in Table 4.3.

### 4.1.2.1.4    Wheel Structure

The structure of a GW is a measure of its porosity. Some porosity (Figure 4.1) is essential to provide clearances for the grinding chips; otherwise, they would interfere with the grinding process. The wheel loses its cutting ability due to loading by chips.

**TABLE 4.2**

**Grinding Wheel Hardnesses for Different Materials and Operations**

| | Wheel Hardness | | | |
|---|---|---|---|---|
| WP Material | Cylindrical Grinding | Surface Grinding | Internal Grinding | Deburring |
| Steel up to 80 kg/mm² | L, M, N | K, L | K, L | |
| Steel up to 140 kg/mm² | K | K, J | J | |
| Steel more than 140 kg/mm² | J | L, J | I | |
| Light alloys | J | I, K | I | O, P, Q, R |
| CI | K | I | J | |
| Bronze, brass, and copper | L, M | J, K | J | |

From Raw, P. N., *Metal Cutting and Machine Tools,* Tata McGraw-Hill, New Delhi, 2000.

**TABLE 4.3**

**Effect of Machining Variables on Wheel Grade**

| Variable | Wheel Grade Appears |
|---|---|
| Increasing work speed ($v_w$) | Soft |
| Increasing wheel speed ($v_g$) | Harder |
| Increasing work diameter ($d$) | Harder |
| Increasing wheel diameter ($D$) | Harder |

Wheels of open or porous structure are used for high metal removal rates that produce rough surfaces, whereas those of dense or compact structure are used for precision grinding at low material removal rates (MRRs). Wheel structure is designated by numbers from 1 (for extra-dense) to 15 (for extra-compact).

### 4.1.2.1.5 Wheel Bond

The wheel bond holds the grains together in the wheel with just the right strength to permit each grain on the cutting face to perform its work effectively. As the grains become dull, they may be either broken, forming new cutting edges, or torn out, leaving the bond. Thus, the bond acts like a tool post that supports the abrasive grains. When the amount of bond is increased, the size of the posts connecting each grain is increased. The seven standard GW bonds are:

*Vitrified bond* (V). This bond is of refractory clay, which vitrifies or fuses into glass. About 70% of GWs are made of vitrified bond, as its strength and porosity yield high stock removal rates. Moreover, vitrified bonds are not affected by water, oils, or acids. However, they are brittle and sensitive to impact, but they can withstand velocities up to 2000 m/min.

*Resinoid bond* (B). This bond is stronger and more elastic than a vitrified bond. However, it is not resistant to heat and chemicals. Because the bond is an organic compound, wheels with resinoid bonds are also called "organic wheels." Resinoid bonded wheels can be used for rough grinding, parting off, and high-speed grinding at 3500 m/min.

*Silicate bond* (S). This is a soda silicate bond ($NaSiO_3$) that releases abrasive grains more rapidly than a vitrified bond. It is used to a limited extent where the heat generated in grinding must be kept to a minimum, as in a very large GW bond for tool sharpening.

*Rubber bond* (R). This is the most flexible bond, as the principal constituent is natural or synthetic rubber. It is not as porous and is widely used in thin cut-off large wheels, portable snagging wheels, and centerless regulating wheels (RWs).

*Shellac bond* (E). This bond is frequently used for strong, thin wheels having some elasticity. They tend to produce a high polish and thus have been used in grinding such parts as camshafts and mill rolls. Thin cut-off wheels may be shellac bonded.

*Oxychloride bond* (O). This magnesium oxychloride bond is used to a limited extent in certain wheels and segments used on disk grinders.

*Metallic bond* (M). These are made of Cu- or Al-alloys. Metallic bonds are used in diamond and CBN wheels, especially for electrochemical (EC) grinding applications. The depth of the abrasive layer can be up to 6 mm.

### 4.1.2.1.6 Grinding Wheel Marking

A standard marking system has been adopted by the American National Standards Institute (ANSI). It is implemented by all GW manufacturers today. This system involves the use of numbers and letters in the sequence indicated in Figure 4.3: abrasive type–grain size–grade–structure–bond.

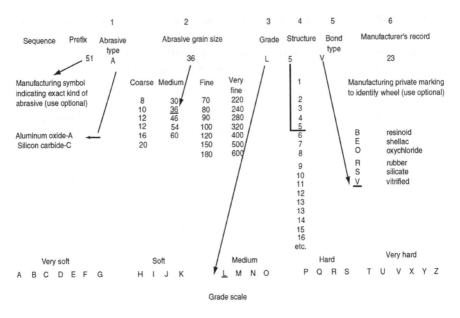

**FIGURE 4.3** GW marking according to ANSI.

The wheel selected in Figure 4.3 is, therefore, designated as:

$$51 \, (\text{optional}) \, A \, 36 \, L \, 5 \, V \, 23 \, (\text{optional})$$

Moreover, the maximum allowable peripheral speed should be printed on the GW. Because GWs are brittle and operated at high speeds, precautions must be carefully followed in their handling, storage, and use. Failure to follow warnings and instructions printed on individual wheel labels may result in serious injury or even death. In general, the following guidelines are considered when selecting a GW marking:

- Choose $Al_2O_3$ for steels and SiC for CI, carbides, and nonferrous metals.
- Choose a hard grade for soft materials and a soft grade for hard materials.
- Choose a large grit size for soft ductile materials and a small grit for hard brittle materials.
- Choose a small grit for a good surface finish and a large grit for maximum metal removal rate.
- Choose an open structure for rough cutting and a compact one for finishing.
- Choose a resinoid, rubber, or shellac bond for a good surface finish and a vitrified bond for maximum removal rate.
- Do not choose vitrified bonded wheels for cutting speeds higher than 32 m/s.
- Choose softer grades for surface and internal cylindrical grinding and harder grades for external cylindrical grinding.
- Choose harder grades on nonrigid grinding machines.
- Choose softer grades and friable abrasives for heat-sensitive materials.

## 4.1.2.2 Grinding Wheel Geometry

GW shapes must permit proper contact between the wheel and the surfaces to be ground. Figure 4.4 illustrates eight standard shapes of GWs, whose applications are as follows:

Shapes 1, 3, and 5 are intended for grinding external or internal cylindrical surfaces and for plain surface grinding.

Shape 2 is intended for grinding with the periphery or the side of the wheel.

Shape 4 is of a safely tapered shape to withstand breakage during snagging.

Shape 6 is a straight cup intended for surface grinding.

Shape 7 is a flaring cup intended for tool sharpening.

Shape 8 is a dish type intended for sharpening cutting tools and saws.

Each has a specific grinding surface. Grinding on other faces is improper and unsafe. Figure 4.5 shows a variety of standard face contours for straight GWs.

## 4.1.2.3 Mounting and Balancing of Grinding Wheels and Safety Measures

A. *Mounting:* Proper and reliable clamping of the GW on its spindle is a prime requisite both for operator safety and to ensure high accuracy and surface finish.

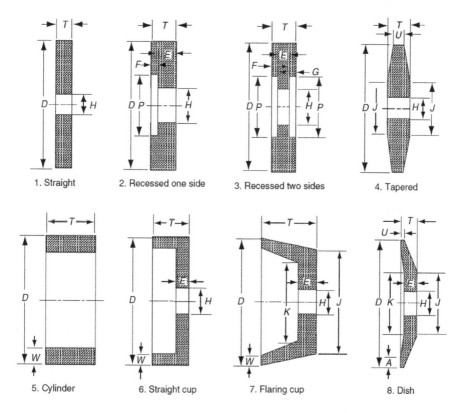

**FIGURE 4.4** Standard shapes of GWs.

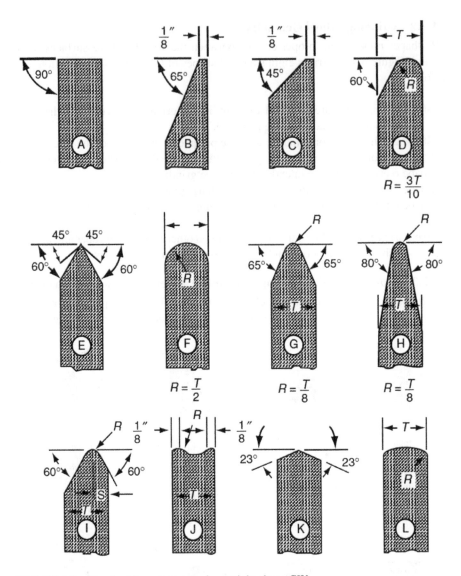

**FIGURE 4.5**   Standard face contours of a straight-shape GW.

Figure 4.6 shows different methods of wheel mounting, which depends upon the type and construction of the grinder and the shape and size of the GW. Wheels of small diameter, used in chucking-type internal grinding, are either seated on the spindle nose (Figure 4.6a) or cemented or glued on the spindle stem (Figure 4.6b).

GWs with small bores (all shapes except shape 5 [Figure 4.4]) are directly clamped by flanges on the spindle (Figures 4.6c through e). Rubber or leather washers of 0.5–3 mm thickness must be inserted between the flanges and the wheel to ensure that the clamping pressure is evenly disturbed. Figure 4.6f shows the recommended

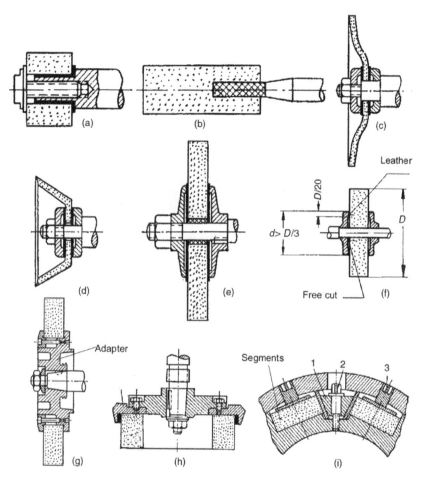

**FIGURE 4.6** Methods of mounting GWs: (a) seated on the spindle; (b) cemented on the spindle stem; (c, d and e) clamped by flanges; (f) proportions of flanges; (g) mounted on an adaptor; (h) secured on a special chuck; (i) segmental wheels clamped in chucks.

proportions of flanges relative to the wheel diameter D. GWs of large mounting holes are mounted on an adaptor (Figure 4. 6g), which in turn, is mounted on the spindle.

Cylindrical wheels (shape 5 [Figure 4.4]) are secured on a special chuck, either by cementing with bakelite varnish or by pouring molten sulfur, Babbitt, or lead into the gap between the wheel and the chuck flange (Figure 4.6h). The surfaces of the wheel and chuck being jointed must be rough, cleaned of all dirt, and degreased.

Segmental wheels are held in their chucks either by cementing or by mechanical clamping using tapered keys (1) and screws (2 and 3, Figure 4.6i).

B. *Balancing:* Because of the high rotational speeds involved, GWs must never be used unless they are in good balance. A slight imbalance produces vibrations that cause waviness errors and harm the machine parts. This may cause wheel breakage, leading to serious damage and injury. Static unbalance of a GW is necessary due to

**FIGURE 4.7**   Revolving-disk wheel balancing stand.

the lack of coincidence between its center of gravity and its axis of rotation. Lack of balance is measured at the manufacturing plant in special balancing machines and is eliminated. The user balances GWs either on a balancing stand or directly in the grinder. In the first case, and before mounting the wheel on the spindle, each wheel with its sleeve should be balanced on an arbor that is placed on the straight edges or revolving disks for a balancing stand (Figure 4.7). The wheel is balanced by shifting three balance weights (1) in an annular groove of the wheel sleeve (or mounting flange). The wheel is rotated until it no longer stops its rotation at a specific position. Certain grinders are equipped with a mechanism for balancing the wheel during operation without stopping the wheel spindle rotation.

C. *Safety measures:* Any unsafe practice in grinding can be hazardous for operation and deserves careful attention. Various important aspects in this respect are:

- Mounting of GWs. The wheel should be correctly mounted and enclosed by a guard. The wheel bore should not fit tightly on the sleeve.
- Wheel speed. The printed speed on the GW should not be exceeded.
- Wheel inspection. Before mounting the wheel, it should be checked for damage, cracks, and other defects. A ringing test should be performed. This is good enough for vitrified bonded wheels.
- Wheel storage. When not used, the wheels should be stored in a dry room and placed on their edges in racks.
- Wheel guards should always be used during grinding.
- Dust collection and health hazard precautions. When grinding dry, provisions for extracting grinding dust should be made. Operators should wear safety devices to protect themselves from abrasives and dust.
- Adequate power is necessary; otherwise, the wheels slow down and develop flat spots, causing the wheel to run out of balance.
- Wet grinding. The wheel should not be partly immersed, as this would throw the wheel seriously out of balance.

#### 4.1.2.4 Truing and Dressing of Grinding Wheels

In the grinding process, the sharp grains of the GW become rounded and hence lose their cutting ability. This condition is termed GW-glazing. Along with grain wear (glazing), another factor that reduces the cutting ability is the loading of voids between the grains with the chips and waste of the grinding process, resulting in a condition known as wheel loading. Loading especially occurs when grinding ductile and soft materials.

A worn and loaded wheel ceases to cut. Its cutting ability can be restored by dressing or truing.

Dressing is a sharpening operation, which removes the worn and dull grits and embedded swarf to improve the cutting action. Truing is an allied operation, using the same tools, done to restore the correct geometrical shape of the wheel when it has been lost due to nonuniform wear. Truing makes the face of the wheel concentric and its sides plane and parallel or forms the wheel true for grinding special contours. It also restores the cutting ability of a worn wheel, as in dressing. Dressing a wheel does not necessarily make it true; however, the distinction between truing and dressing is a difficult one. There is some difference between diamond truing and crush dressing. The abrasive grit, being crystalline, tends to fracture along the most highly stressed crystallographic plane. Diamond truing tends to chip the grits along planes that make a small angle with respect to the direction of motion of the grit. Crush dressing may cause shear fractures along planes that make a large angle with respect to the direction of motion of the grit (Figure 4.8). As shown in Figure 4.8b, the crushed grit is likely to have more favorable cutting angles than diamond-trued grit (Figure 4.8a). A crush-dressed GW will have free-cutting properties but will not produce a finish on the work equal to that of a wheel that is diamond-trued.

In diamond truing and dressing, a single diamond (0.25–2 carat) is held in a steel holder. The GW is rotated at a normal speed, and a small depth, typically of 25 μm, is given while moving the diamond across the face of the wheel in an automatic feed. The diamond tool is pointed in the same direction during wheel rotation to prevent gouging of the wheel face. It is placed at the height of the wheel axis or 1–2 mm below it. Figure 4.9 shows the setting angle in two planes for truing and dressing operations. For the best results in diamond truing and dressing, the maximum rate of

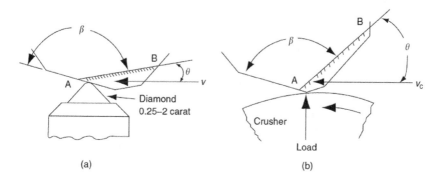

**FIGURE 4.8** (a) Diamond truing and (b) crush dressing of GWs.

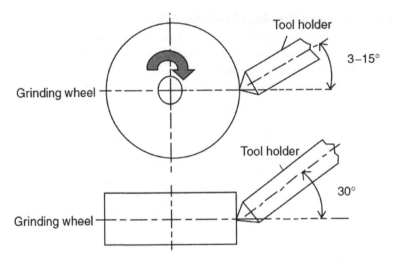

**FIGURE 4.9** Diamond truing and dressing of GWs.

traverse should be 0.05–0.4 m/min and the infeed 5–30 μm/pass, with two or three roughing and one or two finishing passes. The lower the rates of longitudinal traverse and infeed, the smoother the active surface of the wheel will be.

Wheel truing and dressing that do not require diamond make use of:

- Solid cemented carbide rollers
- Rollers of cemented carbide grains in a brass matrix
- Steel rollers and star-type dressers
- Abrasive wheels of black SiC with a vitrified bond of diameter 60–150 mm and width of 20–32 mm

Wheel truing and dressing without diamond are less efficient and do not require the expensive diamond tool. Of all the dressing tools that do not require diamond, abrasive wheel dressers are the most widely employed. They have a grain size three to five steps coarser and five or six grades harder than the wheel that is to be dressed or trued. Three to five passes are made in dressing or truing; the traverse feed is 0.5–0.9 m/min, and the infeed is 10–30 μm/pass. The last (finishing) passes are made without infeed and at reduced traverse feed (0.4–0.5 m/min). Ample coolant is applied in all dressing and truing methods that do not use diamond.

Diamond grinding wheels are not conventionally dressed; they are trued only when their shape is no longer sufficiently accurate. Metal-bonded diamond wheels are trued with a green SiC dressing stick having a vitrified bond and a grain size of 16 or 12, and of hard grades. Resinoid-bonded diamond wheels are trued with pumice. Truing is done at the working speed of the wheel with a coolant being applied. However, in more recent developments, metal-bonded diamond GWs can be dressed nonconventionally by electrodischarge (ED) and electrochemical (EC) techniques, which erode away a very small layer or a metal bond, thus exposing new diamond cutting edges.

### 4.1.3 Grinding Machines

A distinguishing feature of a grinding machine is the rotating GW. Grinding machines handle workpieces (WPs) that have been previously machined on other machine tools, which leaves a small grinding allowance. Such an allowance depends upon the required accuracy, the size of the work, and the preceding machining operations to which it has been subjected. Grinding machines are available for various WP geometries and sizes. Modern machines are computer-controlled, with features such as automatic WP loading and unloading, clamping, cycling, gauging, and dressing, thereby reducing labor cost and producing parts accurately and repetitively. According to the shape of the ground surface, general-purpose grinding machines are classified into surface external cylindrical, internal cylindrical, and centerless grinding machines. In addition to the general-purpose grinding machines, there are other important types of single-purpose grinding machines, such as thread-gear, spline, contour, milling rolls, and tool grinders. The general-purpose machines and their related operations are discussed briefly in the following sections.

#### 4.1.3.1 Surface-Grinding Machines and Related Operations

These machines are used to finish flat surfaces. The most widely used types are:

1. *Horizontal-spindle reciprocating-table grinders.* Figure 4.10 illustrates a typical horizontal-spindle reciprocating-table grinder, on which a straight-shaped wheel (7) is commonly used. The bed (1) contains the drive

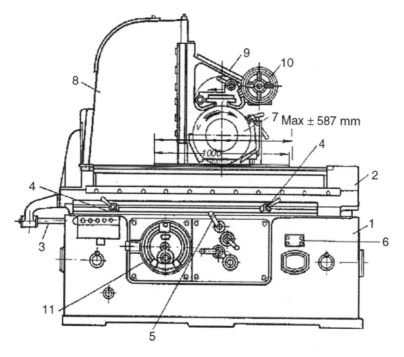

**FIGURE 4.10** Typical surface grinder.

mechanisms and the main table hydraulic cylinder. The table (2), actuated by the piston rod (3) of the hydraulic cylinder, reciprocates along ways on the bed to provide the longitudinal feed of the WP. T-slots are provided in the table surface for clamping WPs directly onto the table or for clamping grinding fixtures or a magnetic chuck. Nonmagnetic materials are held by a vise or special fixtures. The table stroke is set up by adjustable dogs (4). By means of a lever (5), the dogs reverse the table travel at the ends of the stroke. Push-button controls (6) start and stop the machine. A column (8) secured to the bed guides the vertical slide (9), which can be raised or lowered with the GW manually by the hand wheel (11). The vertical slide has horizontal ways to guide the wheel horizontally crosswise for traverse grinding. This slide is actuated by hand using a wheel (10) or by a hydraulic drive housed in the slide. The GW rotates at a constant speed; it is powered by a special built-in motor. Operations that can be performed on horizontal-spindle reciprocating-table grinders are:

a.  Traverse grinding, in which the table reciprocates longitudinally $(v_w)$ and is periodically fed laterally after each stroke at a rate $f_2$ that is less than the GW width. The wheel is fed down to provide the infeed $f_1$ after the entire surface has been ground (Figure 4.11a).

b.  Plunge grinding, in which the wheel is fed perpendicular to the work surface at a rate $f_1$, while the work reciprocates, as in grinding a groove (Figure 4.11b).

c.  Creep feed grinding (CFG), which is used for large-scale metal removal operations. The work is fed very slowly past the wheel, and the tool depth (d = 1–6 mm) is accomplished in a single path. The wheels are mostly of a softer grade with a capability for continuous dressing using a diamond roll to improve the surface finish (Figure 4.11c). The machines used for CFG commonly have special features such as high power of up to 225 kW, high stiffness, high damping factor, variable and well-controlled spindle and worktable speeds, and ample grinding fluids. Although a single pass generally is sufficient, a second pass may be necessary to improve the surface finish.

2. *Vertical-spindle reciprocating-table grinders.* In these machines, a cup, ring, or segmented wheel grinds the work over its full width using the end face of the wheel in one or several strokes of the table. The tool is fed down periodically at the infeed rate *f* (Figure 4.12).

3. *Horizontal-spindle rotary-table grinders.* The reciprocating cross-feed motion $f_1$ is transmitted in these machines to either the GW or the unit; the feed $f_2$ is table actuated per table revolution (Figure 4.13a). The worktable rotates at a speed $v_w$.

4. *Vertical-spindle rotary-table grinders.* These machines are similar to the previous type, except that the spindle is vertical. The configuration of these machines allows a number of pieces to be ground in one setup (Figure 4.13b).

### 4.1.3.2  External Cylindrical Grinding Machines and Related Operations

This type of machine is mainly used for grinding external cylindrical surfaces, which may be parallel or tapered, or filets, grooves, shoulders, or other formed surfaces of

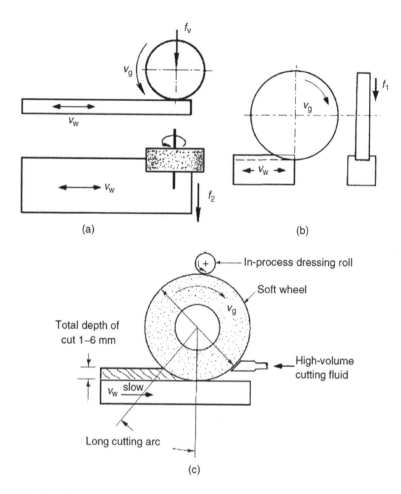

**FIGURE 4.11** Operations performed on horizontal-spindle reciprocating-table grinders. (a) Transverse grinding, (b) plunge grinding, (c) CFG.

revolution. Typical applications include crankshaft bearings, spindles, shafts, pins, and rolls for rolling mills. The rotating cylindrical WP reciprocates laterally along its axis. However, in machines used for long shafts, the GW reciprocates. The latter design configuration is called a roll grinder and is capable of grinding heavy rolls as large as 1.8 m in diameter.

Center-type cylindrical grinders are subdivided into:

1. Universal-type grinders, which make it possible to swivel the GW by swiveling the headstock. This enables steep tapers to be ground. Due to their versatility, universal cylindrical grinders are best suited to tool room applications.
2. Plain-type grinders, in which the worktable can be swiveled through an angle of only ±6°. This type is basically designed for heavy repetitive single work. It is not very versatile and is used for grinding tapers with small included angles.

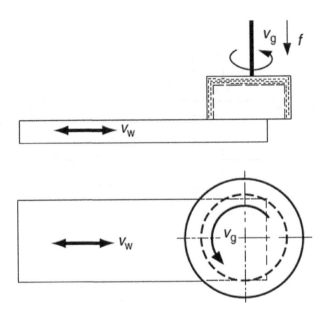

**FIGURE 4.12**   Operations performed on vertical-spindle reciprocating-table grinders.

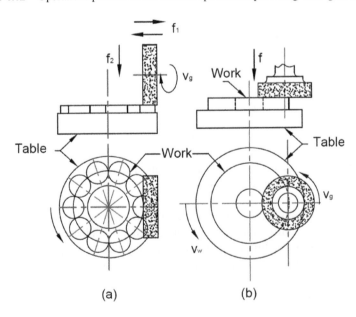

**FIGURE 4.13**   Operations performed on rotary-table grinders. (a) Horizontal spindle, (b) vertical spindle.

Similarly to surface grinders, described before, the table assembly of cylindrical grinders is reciprocated using a hydraulic drive. The table speed, therefore, is infinitely varied, and the stroke is controlled by means of adjustable trip dogs.

   Infeed is provided by the movement of the wheel head crosswise to the table axis. Most grinders have automatic infeed with retraction when the desired size has been reached. Such machines are also equipped with an automatic diamond wheel truing device that dresses the wheel and resets the measuring element before grinding is started on each piece. Similarly to engine lathes, cylindrical grinders are identified by the maximum diameter and length of the WP that can be ground.

These machines are generally equipped with computer control, simultaneously reducing labor cost and producing parts accurately and repetitively. Computer-controlled grinders are capable of grinding noncylindrical parts and cams. Moreover, in these machines, the WP spindle speed can be synchronized such that the distance between the WP and the wheel axes is varied continuously to produce accurate longitudinal profiles.

Two methods of cylindrical grinding are illustrated in Figure 4.14:

1. Traverse cylindrical grinding, in which the wheel has two movements: rotation about its axis and infeed into the work to remove the grinding allowance (usually an intermittent crosswise motion at the ends of the traverse stroke). The work rotates about its axis and also traverses longitudinally past the wheel so as to extend the grinding action over the full length of the work (Figure 4.14a). The longitudinal traverse should be about $\frac{1}{4} - \frac{1}{2}$ of the wheel width per revolution of the work. For a fine surface finish, it should be held to the smaller value of this range. The depth of cut (infeed) varies according to the finish required. It ranges from 50 to 100 μm for rough cut and 6–12 μm for finish cut. The grinding allowance ranges from 125 to 250 μm for short parts and from 400 to 800 μm for long parts subjected to hardening treatment. About 70% of the grinding allowance is allocated for roughing and 30% for finish grinding.

2. Plunge-cut cylindrical grinding, in which there is no traversal motion of either the wheel or the work. The GW extends over the entire length of the surface being ground on the work ($B > l$), which rotates about its axis. The wheel rotates and, at the same time, is continuously fed into the work at a rate of 2.5–20 μm per revolution of the work (Figure 4.14b). This method is used in form grinding of relatively short work at high output. In any plunge-cut operation (cylindrical or surface), the wheel is fed normal to the

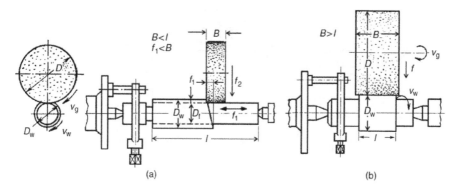

**FIGURE 4.14** External cylindrical grinding operations: (a) traverse; (b) plunge-cut.

work surface (infeed). The feed $f$, which is the depth of the layer of material removed during one work revolution or stroke in the case of surface grinding or stroke, will initially be lower than the nominal feed setting on the machine. The difference results from the machine-tool elements, the GW, and the WP. It occurs due to the forces generated during the grinding operation. Thus, on completion of the estimated number of work revolutions or strokes required, some additional work material has to be removed. The removal of this material, called sparking out, is achieved by continuing the grinding operation with no further application of feed until metal removal becomes insignificant (no further sparks appear). Therefore, when calculating grinding time, an additional time $t_s$ should be considered for sparking out.

### 4.1.3.3 Internal Grinding Machines and Related Operations

In internal grinding, a small GW is used to grind the inside diameter of a WP, such as bushings, bearing races, and heavy housings. It is usually of the traverse type; however, the plunge-cut technique may also be used. Two difficulties in internal grinding are encountered:

1. The GW and consequently, the machine spindle should be small to suit the small internal holes. The reduced rigidity of small spindles makes it impossible to take heavy cuts. Furthermore, the rotational speed of the small GW must be very high (up to 150,000 rpm) to operate at the recommended cutting speeds. Therefore, high-speed drives for the GW with special spindle mounting are required.
2. In internal grinding, conventional methods of coolant supply are not efficient. A method of internal coolant delivery is illustrated in Figure 4.15. The coolant is pumped to the GW through the axial hole A in the wheel spindle, and the radial holes B and C are drilled in the spindle nose and the sleeve on which the wheel is mounted. Due to the action of the centrifugal force, the coolant passes through the pores of the wheel to its periphery. The

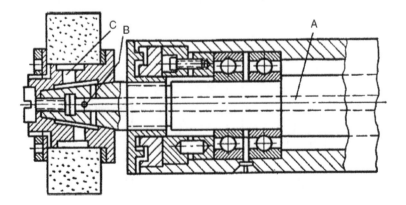

**FIGURE 4.15**  GW cooling for internal grinding operations.

coolant is applied intensively in the grinding zone, where it also washes the waste products of grinding out of the wheel. This method raises the output by 10–20%, improves the surface finish by one class, avoids burning, and reduces the GW wear.

There are two types of internal cylindrical grinding machines:

a. *Chucking-type machine.* Used in grinding comparatively small WPs. In addition to the primary cutting motion of the GW $v_g$, the following feed motions are encountered (Figure 4.16a):
- Work feed $v_w$ due to the work rotation
- Traverse feed $f_1$ as a reciprocating motion of the work or GW
- Infeed $f_2$ as a periodic crosswise motion of the GW

b. *Planetary-type machine.* Designed to grind holes in large irregular parts that are difficult to mount and rotate (Figure 4.16b). In this case, the work is stationary, while the wheel rotates not only around its own axis $v_g$ but also around the axis ($v_w$) of the hole being ground. In addition to these two motions, traders feed $f_1$ and infeed $f_2$, as in the chucking type, are affected.

### 4.1.3.4 Centerless Grinding Machines and Related Operations

As the name implies, the work is not supported between centers but is held against the face of the GW, a supporting rest, and the RW. Therefore, centerless grinding does not require center holes, a driver, and other fixtures for holding the WP. During cutting, the WP (1) is supported on the work-rest blade (2) by the action of the GW (3). The RW (4) of infinite variable speed holds the WP against the horizontal force controlling its size and imparting the necessary rotational and longitudinal feeds of the WP (Figure 4.17).

The wheels rotate clockwise, and the work driven by the RW, having approximately the same peripheral speed of typically 20–30 m/min, rotates counterclockwise. To increase friction between the work and the RW, the latter has a fine grain size of mesh number 100–180 and is rubber bonded and of a sufficiently hard grade (R or S). Resinoid-bonded RWs are also employed. In comparison to the RWs, the GWs run at a much higher speed (2000 m/min) and accomplish the cutting action. To ensure that a true cylindrical surface is ground on the work, it is set above the

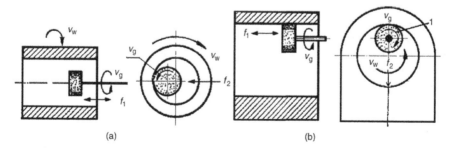

**FIGURE 4.16** Internal cylindrical grinding operations: (a) chucking and (b) planetary.

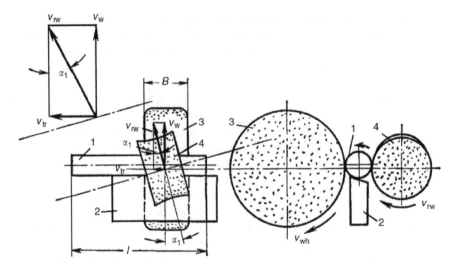

**FIGURE 4.17**  Centerless grinding operations.

centers of the RWs and GWs by 0.15–0.25 of the work diameter, but not over 10 mm, to avoid chattering.

The material of the work-rest blade mainly depends upon the type of WP material to be ground. For machining mild steel, the material of the blade should be CI. For nonferrous and small-diameter jobs, HSS is recommended as a blade material, and for machining stainless steels, either sintered carbides or hard bronze is to be selected as a blade material.

In comparison to cylindrical grinding, centerless grinding has the following advantages:

- The rate of production is much greater than with cylindrical grinding.
- The work is supported rigidly along the whole length, ensuring better stability and accuracy.
- Less grinding allowance is required, as the work centers itself during operation.
- The process is suitable for long jobs.
- Work of very small diameter can be ground using external centerless grinding.
- Because centering is unnecessary, no time is lost in job setting, and the cost of providing centers is eliminated.
- The machine is easy and economical to maintain.
- The production cost is considerably lower.
- The process can often be made automatic.
- Very little skill is required of the operator.

Due to the advantages listed, centerless grinding plays an important role in the field of production technology. The process is applicable to WPs 0.1–150 mm in diameter and from short jobs to precise bars of about 6 m in length required for Swiss-type automatics. Centerless grinders are now capable of wheel surface speeds on the order of 10,000 m/min using CBN abrasive wheels. The accuracy that can be obtained from centerless grinding is of the order of 2–3 μm, and with suitable selected wheels, high degrees of finishes are obtained.

The major disadvantages are as follows:

- Special machines are required that can do no other type of work.
- The work must be round; that is, flat surfaces or keyways cannot be worked on.
- In grinding tubes, there is no guarantee that internal and external diameters are concentric.
- A most common defect of centerless grinding is lobbing (unevenly ground surface). It occurs during grinding of steel bars whose surfaces have some high and some low spots due to hot or cold rolling.

Centerless grinding may be external or internal.

### 4.1.3.4.1 External Centerless Grinding

Basically, three methods of centerless grinding are commonly used in practice.

a. *Through-Feed Centerless Grinding*

In this method, the axial traverse motion is imparted to the work by the RW, because the latter is inclined at a small angle $\alpha_I$ with respect to the axis of the GW (Figure 4.18a) or because the work-rest blade is inclined to an angle $\alpha_1$ (Figure 4.18b).

The peripheral velocity $v_p$ of the RW is resolved into peripheral speed of the work $v_w$ and the rate of the work traverse $v_{tr}$ (mm/min) (Figure 4.17), which can be calculated by the equation

$$v_{tr} = \pi d_R n_R \sin \alpha_1 \tag{4.2}$$

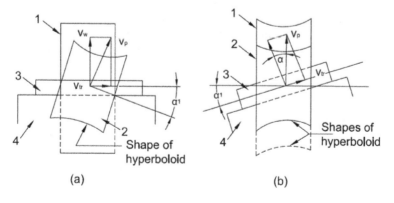

(a)    (b)

**FIGURE 4.18** Ensuring linear contact between WP and wheels in through-feed centerless grinding by providing wheels of hyperboloid of revolution profiles: (a) RW inclination and (b) work-rest inclination.

where

$\alpha_1 = 0.5°$–$1.5°$ for finish grinding

  = $1.5°$–$6°$ for rough grinding

$d_R$ = diameter of the RW

$n_R$ = rotational speed of the RW

Equation 4.2 does not consider the effect of slip. The contact between the WP and the GWs and RWs must be linear in through-feed grinding. For this reason, the face of the RW (in the case of RW inclination) or the faces of both RWs and GWs (in the case of work-rest inclination) are trued by diamonds to the shape of a hyperboloid of revolution (Figure 4.18).

Advantages of through-feed grinding are as follows:
- The method becomes automatic by employing magazine feeds for bars and hoppers for small jobs.
- Long bars can be ground easily without any deflection being produced.

Disadvantages of through-feed grinding are as follows:
- This method is used for straight cylindrical parts; if there is a head on the WP or it is tapered, then the process cannot be employed.
- Form grinding cannot be produced by this method.

b. *Infeed (Plunge-Cut) Centerless Grinding*

This method of grinding is used when the WP is of headed, stepped, or tapered form (Figure 4.19). In this case, there is no axial movement of the WP, which has a rotating movement only. The WP (3) is supported on the work-rest blade (4). After approaching the end stop (5), the cross feed is actuated by a method similar to plunge-cut grinding in which the grinding or RW is fed in a direction square to the WP axis by a precise feed movement.

In some cases, the infeed is ensured by the use of an RW (2) of special shape. Its periphery consists of sectors I, II, and III (Figure 4.20). Sectors I and III are circular arcs of different radii. Sector II is an Archimedean spiral, which enables infeed to be actuated without movement of the wheel heads. The whole grinding cycle takes place during one revolution of the

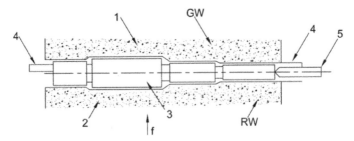

f-Infeed after approaching the end stop

**FIGURE 4.19**   Schematic of infeed centerless grinding.

RW. The WP (3) is automatically loaded between the wheels (1 and 2) from the top, and the axis of the RW (or the work-rest blade, 4) is inclined slightly at an angle $\alpha_1 = \frac{1}{2}°$ to provide for a fixed axial position of the WP by holding it against the locating stop (not shown in Figure 4.20). Infeed is done at various rates (sector II). At the beginning of the process, a large part of the allowance is removed with a high rate of infeed, and then this rate is reduced. At the end of operation, the WP is ground for several revolutions without infeed for sparking out (sector III). The RW has a longitudinal slot (A) into which the finished WP drops after rolling off the work-rest blade and is removed outside the grinding zone.

c. *End-Feed Centerless Grinding*

This method is essentially an intermediate method between through-feed and infeed centerless grinding. It is employed for headed components that are too long to be ground by the infeed method. It is used when the length of WPs is greater than the width of the GW and unable to pass between the wheels for through-feed (Figure 4.21).

The work (3) is fed as in case of the infeed method, and after a certain portion of length has been ground, the axial movement takes place until the whole of length has been ground after approaching the end stop (5). In this method, the angle of inclination of the RW is typically 2–3°, which is larger than that used in the infeed method.

#### 4.1.3.4.2  Internal Centerless Grinding

This process was recently developed for grinding internal surfaces of short or long tube work. The arrangement is schematically illustrated in Figure 4.22. The WP

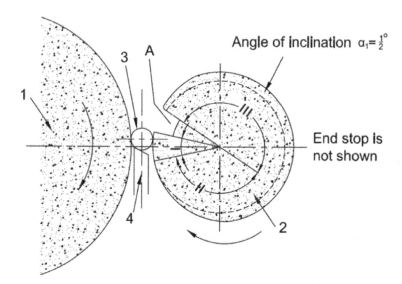

Angle of inclination $\alpha_1 = \frac{1}{2}°$

End stop is
not shown

**FIGURE 4.20**  Automatic-infeed centerless grinding provided by a profiled RW with a loading recess.

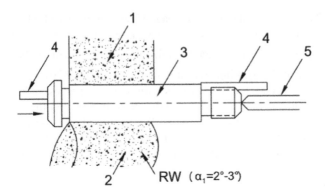

**FIGURE 4.21** Schematic of end-feed centerless grinding.

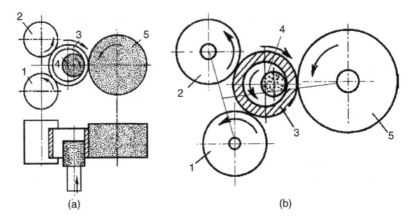

**FIGURE 4.22** Internal centerless grinding: (a) on-center principle (used for thin-walled components) and (b) above-center principle.

is supported by two steel rollers (1 and 2) and an RW (5). Roller 1 is a supporting roller, and roller 2 is a pressure roller. The GW (4) and the WP (3) rotate in the same direction, while the RW (5) rotates in the opposite direction. The GW is generally smaller than the RW. The process may work either on the on-center principle (Figure 4.22a) or on the above-center principle (Figure 4.22b). The on-center method is used for thin-walled components; however, it tends to duplicate the errors of the outside diameter and those of roundness and waviness, which can to some extent be corrected by the above-center arrangement. In internal centerless grinding, as the roundness of internal surface depends to a great extent upon the external surface, the latter must be ground first.

## 4.2  MICROFINISHING MACHINES AND OPERATIONS

These are operations by which a product receives the final machining stage that applies to the service for which it is intended. These remove a very small amount of metal, and hence, the surface finish obtained is specified in the range of microfinishes. These operations include honing, microhoning (superfinishing), lapping, polishing, and buffing. The first three operations are discussed briefly in the following sections.

### 4.2.1 HONING

Honing is a controlled, low-speed sizing and surface-finishing process in which stock is abraded by the shearing action of a bonded abrasive honing stick.

In honing, simultaneous rotating and reciprocating action of the stick (Figure 4.23a) results in a characteristic cross-hatch lay pattern (Figure 4.23b). For some applications, such as cylinder bores, angles between cross-hatched lines are important and may be specified within a few degrees. Because honing is a low-speed operation, metal is removed without the increased temperature that accompanies grinding, and thus, any surface damage caused by heat (heat-affected zone [HAZ]) is avoided.

In addition to removing stock, honing involves the correction of errors from previous machining operations. These errors include:

- Geometrical errors such as out-of-roundness, waviness, bell mouth, barrel, taper, rainbow, and reamer chatter
- Dimensional inaccuracies
- Surface character (roughness, lay pattern, and integrity)

Honing corrects all these errors with the lowest possible amount of material removal; however, it cannot correct hole location or perpendicularity errors. The most frequent application of honing is the finishing of internal cylindrical holes. However, numerous outside surfaces also can be honed. Gear teeth, valve components, and races for antifriction bearings are typical applications of external honing. The hone is allowed to float by means of two universal joints so that it follows the axis of the hole

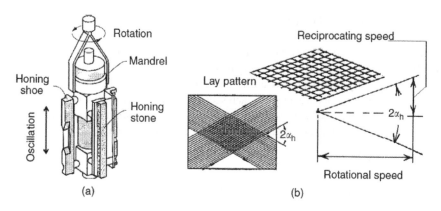

**FIGURE 4.23** Honing operation: (a) honing head with honing sticks and (b) cross-hatched angle. (Adapted from ASM International, *Machining,* Vol. 16, *Metals Handbook,* ASM International, Materials Park, OH, 1989.)

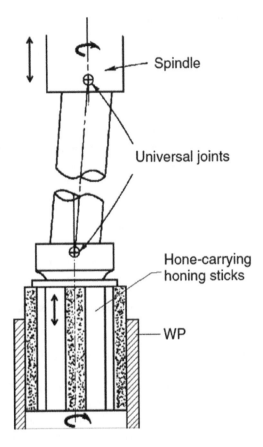

**FIGURE 4.24**   Floating hone using two universal joints to permit the bore and the tool to align.

(Figure 4.24). Due to the fact that the tool floats, the honing sticks are able to exert an equal pressure on all sides of the bore regardless of the machine vibrations, and therefore, round and straight bores are produced. As the tool reciprocates through the bore, the pressure and the resulting penetration of grit are greatest at high spots, and consequently, the waviness crests are abraded, making the bore straight and round. After high spots are leveled, each section of the bore receives equal abrading action. The hole axis is usually in the vertical position to eliminate gravity effects on the honing process; however, for long parts, the axis may be horizontal.

Advantages of Honing

- It is characterized by rapid and economical stock removal with a minimum of heat and distortion.
- It generates round and straight holes by correcting form errors caused by previous operations.
- It achieves high surface quality and accuracy.

### 4.2.1.1  Process Capabilities

#### 4.2.1.1.1  Materials

Although CI and steel are the most commonly honed materials, the process can also be used for finishing materials ranging from softer metals like Al- and Cu-alloys to extremely hard materials like case-nitrided steels or sintered carbides. The process can also be used for finishing ceramics and plastics.

#### 4.2.1.1.2  Bore Size and Shape

Bores as small as 1.6 mm in diameter can be honed. The maximum bore diameter is governed by the machine power and its ability to accommodate the WP. Machines powered by motors of up to 37 kW are available, which can hone bores up to about 1200 mm in diameter. Honing bores up to 760 mm in diameter is a common practice (ASM International, 1989). Although most internal honing is done on simple, straight-through holes, blind holes with a slight taper can also be honed. It is not feasible to hone the sides of a blind hole flush with the bottom. Bores having keyways can be honed and so can male or female splines (ASM International, 1989).

#### 4.2.1.1.3  Stock Removal

In honing, a general rule is to remove twice as much stock as the existing error in the WP. For example, if a cylinder is 50 μm out of round or tapered, a removal of 100 μm will be required for complete cleanup. The work in preceding operations is usually planned so that the amount of stock removed in honing is minimized. On the other hand, stock removal of up to 6.4 mm may be practical for rough honing in some applications. For instance, as much as 2.5 mm is honed from the inside diameter of hydraulic cylinders, because stock removal through honing is more practical and economical than attaining close preliminary dimensions by grinding or boring. Another example occurs in finishing bores of long tubes, where even larger amounts, as much as 6 mm, may be removed by honing, because it is the only practical method. Such tubes are finished by honing immediately after drawing. Honing is performed at a rate of 32 cm³/min from soft steel tubes; for tubes steel-hardened to 60 HRC, the rate is reduced to 16 cm³/min (ASM International, 1989).

Rough honing is employed before finish honing when large amounts of stock are to be removed and specific finishes are required. Sticks containing abrasives of 80 grit or even coarser are used for rough honing to maximize the removal rate. Finish honing is accomplished by abrasives of 180–320 grit or finer.

#### 4.2.1.1.4  Dimensional Accuracy and Surface Finish

Internal honing to tolerances of 2.5–25 μm is common. Surface roughness $R_a$ of 0.25–0.38 μm can be easily obtained by rough honing, and roughness of less than 0.05 μm can be achieved and reproduced in finish honing. Figure 4.25 compares typical ranges of surface roughness obtained by honing with other common microfinishing processes.

#### 4.2.1.1.5  Honing Sticks

The same ANSI-designation system of GWs is applied to honing sticks. Honing sticks commonly used may be vitrified, resinoid, or metallic bonded. The bond must

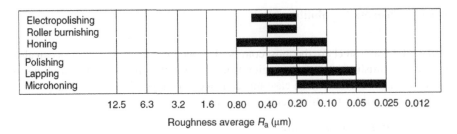

**FIGURE 4.25** Average surface roughness of common microfinishing operations. (Adapted from ASM International, *Machining*, Vol. 16, *Metals Handbook*, ASM International, Materials Park, OH, 1989.)

be strong enough to hold the grit; however, it must not be so hard as to rub the bore and hence retard the cutting action.

The grit size selection depends generally on the desired rate of material removal and the degree of surface finish required. Guide rules for selecting the type of abrasive materials are as follows:

- $Al_2O_3$ is widely used for steels.
- SiC is generally used for CI and nonferrous materials.
- CBN is used for all steels (soft and hard), Ni- and Co-base super alloys, stainless steels, Br-Cu-alloy, and Zr.
- Diamonds are used for chromium plating, carbides, ceramics, glass, CI, brass, bronze, and surfaces nitrided to depths greater than 30 μm.

### 4.2.1.2 Machining Parameters

Parameters affecting the performance of honing process are:

1. *Rotation speed.* The choice of the optimum surface speeds is influenced by:
   - *Material being honed*—higher speed can be used for metals that shear easily.
   - *Material hardness*—harder material requires lower speed.
   - *Surface roughness*—rougher surfaces that mechanically dress the abrasive stick permit higher speed.
   - *Number and width of sticks in the hone*—speed should be decreased as the area of abrasive per unit area to hone increases.
   - *Finish requirement*—higher speed usually results in finer surface finish.

   Depending on the material to be honed, the rotational surface speed typically varies from 15 to 90 m/min. Experience with a particular application may indicate advantages for higher or lower speeds. Rotation speeds as high as 183 m/min have been used successfully. However, a reduction of surface rotation speed can reduce the number of rejects (ASM International, 1989). Excessive speeds contribute to decreased dimensional accuracy, overheating of the WP, and glazing of the abrasive stick. Overheating causes breakdown of honing fluid and distortion of the WP.

2. *Reciprocation speed.* Reciprocation speed commonly ranges from 1.5 to 30 m/min for a variety of metals and alloys.

3. *Control of cross-hatch angle.* The cross-hatch angle $2\alpha_h$ (Figure 4.23b) obtained on a honed surface is given by

$$\tan \alpha_h = \frac{v_{rc}}{v_{rt}} \qquad (4.3)$$

where
$v_{rc}$ = reciprocation speed (m/min)
$v_{rt}$ = surface rotation speed (m/min)
$\alpha_h$ = half cross-hatch angle

When the rotation and reciprocation speeds are equal, the cross-hatch angle is 90°.

For some applications (engine cylinder bores), the cross-hatch angle is an important feature that should be noted in specifications. The cross-hatch scratch pattern left on the wall of cylindrical surfaces tends to retain lubricating fluids and thus reduce the wear in mating components. In the majority of applications, although an angle of 30° is commonly recommended, any angle within the range 20–45° is usually suitable.

4. *Honing pressure.* This is selected depending on the hardness and toughness of the material, characteristics of the honed surface (plain or interrupted by keyways), type of stick, and so on. Insufficient pressure results in a subnormal rate of metal removal and rough surface finish. Excessive removal rate and rough finish can cause an increased stick cost as well as decreased productivity due to time lost by frequent tool exchange.

5. *Honing fluids.* Lubrication is more critical in honing than in most other material removal operations. Honing fluids are necessary to act as lubricants and coolants and to remove swarf. No single honing fluid possesses all the requirements needed for the honing process. Therefore, mixtures of two or more liquids are commonly used.

Water-based solutions are superior as coolants, but they are poor lubricants, have insufficient viscosity to prevent chatter, and cause rust. Because of this, water-based solutions are seldom used as honing fluids.

Mineral seal oil is effective and widely used for honing. It has a higher viscosity and flash point than kerosene. It is less likely to cause skin irritation. Mineral oils used for other machining operations have also proved satisfactory when one part oil is diluted with four parts kerosene.

### 4.2.1.3 Honing Machines

For the production of few parts, honing may be performed on drill presses or engine lathes on which arrangements can be made for simultaneous rotating and reciprocating motions. The stroking can be done manually or by power, depending on the equipment capabilities. On the other hand, the production honing is done with machines built for the purpose. These vertical machines are available in a wide range of sizes and designs. Some horizontal machines operate by manual stroking. In power stroking, the

WP is usually held stationary in a rigid fixture, while the hone is rotated and hydraulically powered for stroking, which is considered beneficial for heavier WPs.

### 4.2.2 SUPERFINISHING (MICROHONING)

Superfinishing (microhoning) is an abrading process that is used for external surface refining of cylindrical, flat, and spherical-shaped parts. It is not a dimension-changing process but is mainly used for producing finished surfaces of superfine quality. Only a slight amount of stock is removed (2–30 μm), which represents the surface roughness (Figure 4.26). The process of honing involves two main motions, whereas superfinishing requires three or more motions. As a result of these motions, the abrasive path is random and never repeats itself.

The primary distinction between honing and superfinishing is that in honing, the tool rotates, while in superfinishing, the WP always rotates. The operating principle of the superfinishing process is illustrated in Figure 4.27. The bonded abrasive stone, whose operating face complies with the form of the WP surface, is subjected to very light pressure. A short, high-frequency (HF) stroke, superimposed on a reciprocating traverse, is used for superfinishing of long lengths.

#### 4.2.2.1 Machining Parameters

The following parameters affect the superfinishing process considerably:

1. Abrasive stones

    Two types are mainly used: $Al_2O_3$ for carbon and alloy steels and SiC (more friable) for very soft and tough steels as well as for CI and most non-ferrous metals.

    *Grit size.* The grit size is selected from a wide range (60–1000) to suit the machining situation, which varies from rough superfinishing to fine or extra-fine finishing.

    *Grade.* This varies from J (soft), used for extremely hard alloys, to P (very hard), used for extremely soft materials, CI, and nonferrous metals.

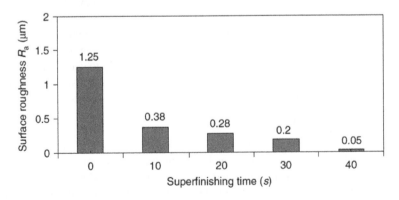

**FIGURE 4.26** Gradually improving a rough surface by superfinishing. (Adapted from ASM International, *Machining*, Vol. 16, *Metals Handbook*, ASM International, Materials Park, OH, 1989.)

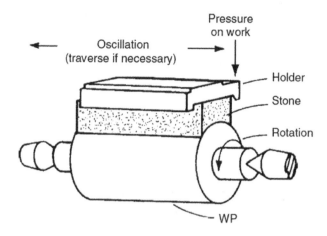

**FIGURE 4.27** Principles of superfinishing process.

*Width.* This ranges from 60% to 80% of the part diameter, but not more than 25 mm.

*Number of stones.* For parts over 150 mm diameter, several stones are arranged.

*Stone length.* This length is somewhat lower than the work length but not more than three times the width of the stone. For superfinishing of longer work, an additional traverse movement is needed.

2. Work speed

*Roughing.* 12–15 m/min.

*Finishing.* 30–60 m/min (for very fine finish, higher speeds of 120 m/min may be applied). At lower work speeds, the superfinishing process generally develops a distinct cross-hatched pattern, which may be desirable in many applications despite its low surface reflectivity. At higher speeds, this pattern disappears, and a brighter surface is developed.

3. Stroke length and speed of stone reciprocation

The fast reciprocation of stones in a short stroke is a main characteristic that sets superfinishing apart from honing. Some machines employ a single stroke length of 4.76 mm, while others provide a variable stroke length over a range of about 2–5 mm. The actual linear speed of oscillation is a function of the stroke length (amplitude) and the rate of reciprocation (frequency). Typical extreme reciprocation speeds are 3–20 m/min.

4. Stone pressure

For normal work, $p_h = 1.5$–$3.0$ kg/cm$^2$

For roughing, $p_h = 3.0$–$6.0$ kg/cm$^2$

For extra-fine finishing, $p_h = 1.0$–$2.0$ kg/cm$^2$

### 4.2.3 LAPPING

The usual definition of lapping is the random rubbing of the WP against a CI lapping plate (lap) using loose abrasives carried in an appropriate vehicle (oil) to improve fit and finish. It is a low-speed, low-pressure abrading process. In general, the surface quality that can be obtained by lapping is not easily or economically obtained by other processes. Moreover, the life of the moving parts that are subjected to wear can be increased by eliminating hills and valleys that create a maximum percentage at bearing area.

Lapping is a final machining operation that realizes the following major objectives:

- Extreme dimensional accuracies
- Mirror-like surface quality
- Correction of minor shape imperfections
- Close fit between mating surfaces

It does not require holding devices, and consequently, no WP distortion occurs. Also, less heat is generated than in most other finishing operations. Therefore, metallurgical changes are totally avoided. The temperature increase of the surface is only 1–2 °C above ambient.

### 4.2.3.1  Machining Parameters

The following parameters have an effect on the lapping process:

1. *Abrasive type:*
   - Diamond is used for lapping tungsten carbide (WC) and precious stones.
   - $B_4C$ is used for lapping dies and gauges. It is 10–25 times more expensive than SiC and $Al_2O_3$.
   - SiC is intended for rapid stock removal. It is mainly used for lapping hardened steels, CI, and nonferrous metals.
   - $Al_2O_3$ is intended for improved finish. It is used for lapping soft steels.
2. *Grit size and abrasive grading.* Grit size (mesh number) generally ranges from 50 to 3800; however, more frequently, grit size from 100 to 1000 is used, depending on the degree of surface finish required. Soft materials require finer grains to obtain a good finish. Commercially available abrasives of certain grit size may contain finer or coarser grit than the specified size. Abrasives increase in cost as their grading becomes closer. The use of a low-cost, loosely graded commercial abrasive is not recommended for reasons of economy.
3. *Vehicle.* This prevents scoring of the lapped surfaces and varies from clean water to heavy grease. It is selected to suit the work, method, and type of surface finish required. For machine lapping, an oil-based type is recommended; however, a commercial mixture of kerosene and machine oil can be used. Grease-based vehicles are recommended for lapping soft metals.
4. *Speed.* Speeds of 1.5–4 m/s are commonly used for machine lapping.
5. *Pressure.* A pressure of 0.1–0.2 $kg/cm^2$ is used for soft materials, while 0.7 $kg/cm^2$ is recommended for lapping hard materials. If the preceding values are exceeded, rapid breakdown and scoring of the WP results.

### 4.2.3.2  Lapping Machines

Lapping machines usually fall into one of the two categories: individual-piece and equalizing lapping machines.

### 4.2.3.2.1 Individual-Piece Lapping

This is the most effective lapping method for hard metals and other hard materials. It is used to produce optically flat surfaces and accurate surface plates. When both sides of a flat WP are lapped simultaneously, extreme parallelism is achieved.

Individual-piece lapping is performed using a lap that is softer than the WP, so that the abrasives become embedded in the lap. The lap is usually made of close-grain soft CI that is free from porosity and defects. When CI is not suitable as a lap material, steel, brass, copper, or aluminum may be used. Wood is sometimes used for certain applications.

The vast majority of individual-piece lapping installations are of the following categories:

- Specialized single- or double-plate machines, such as ball or pin laps
- Single-sided flat or double-sided planetary laps
- A cup-lapping machine for lapping spherical surfaces
  1. *Double-plate lapping machines for cylindrical WPs*
     Figure 4.28 visualizes a typical vertical lapping machine used for lapping cylindrical surfaces in production quantities. The laps are two opposing CI circular plates that are held on vertical spindles of the machine. The WPs are retained between laps in a slotted-holder plate and rotate and slide in and out to break the pattern of motion by moving over the inside and outside edges of the laps, which prevents grooving. The lower lap is usually rotated and drives the WPs. The upper one is held stationary but is free to float so that it can adjust to the variations in WP size. The lower lap regulates the speed of rotation. To avoid damage to the surface being lapped, the holder plate or carrier is made of soft material (copper, laminate fabricate base, and so on).

     An alternative design of this machine is illustrated in Figure 4.29. Accordingly, the retainer is arranged eccentrically between the two laps and has a separate drive. In this design, both the upper and lower laps are rotating. The abrasive with vehicle is provided to the laps before starting. Oil or kerosene is then added during the cycle to prevent drying of the vehicle, which could result in surface scratching. The best

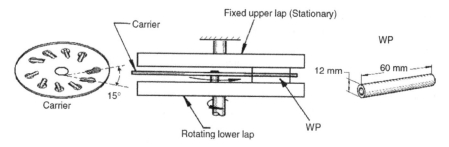

**FIGURE 4.28** Typical vertical lapping machine for cylindrical surfaces. (Adapted from ASM International, *Machining,* Vol. 16, *Metals Handbook,* ASM International, Materials Park, OH, 1989.)

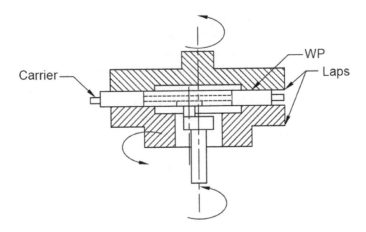

**FIGURE 4.29**  Two-plate lapping machine with two rotating laps and eccentrically rotating plate holder.

lapping practice is to load as many parts as possible to reduce the pressure applied on each part and slowdown the operation, which provides easier control of tolerances.

*Achieved accuracy and surface finish.* Fine surface finishes of 0.025 μm $R_a$ and metal removal of 2.5–10 μm are feasible when CI laps are used. A diametral tolerance as low as ±0.5 μm, out-of-roundness of 0.13 μm, and taper less than 0.25 μm have been achieved. Such accuracies depend greatly on the accuracies achieved in prior machining operations.

*Applications.* Machine lapping between plates, as described earlier, is an economical and productive (100 parts/h) method of lapping cylindrical surfaces. The machine can be used for lapping parts such as plug gauges, piston pins, hypodermic plungers, ceramic pins, small valve pistons, cylindrical valves, small engine pistons, roller and needle bearings, diesel injector valves, plungers, and miscellaneous cylindrical pins. Either hard or soft materials can be lapped, provided that they are rigid enough to accept the pressure of laps. Because the hardness slows the operation, soft materials lap more rapidly than hard ones. Additionally, hard materials provide easier control of tolerances.

*Limitations.* A part with diameter greater than its length is difficult or impossible to machine lap between plates. Parts with shoulders require special fixtures. Parts with keyways, flats, or an interrupted surface are difficult to lap, because the variations in lapping pressure that occur are likely to fall out of round. If the relief extends over the entire length of the part, this method of lapping cannot be used at all.

Thin-wall tubing can be lapped, provided that the deflection due to lapping pressure is insignificant. Parts that are hollow on one end and solid on the other present problems in obtaining roundness and straightness. Plugging the hollow end of the part will sometimes solve the problem.

The outside edges of the laps lap at a faster rate than the inside edges. Therefore, it is expected that the cylindrical WP will become tapered. One method of overcoming

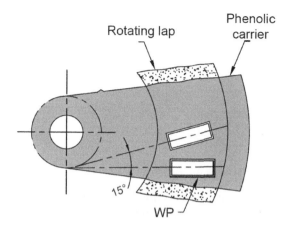

**FIGURE 4.30** Lapping setup that minimizes taper for production quantity of cylindrical parts. (Adapted from ASM International, *Machining*, Vol. 16, *Metals Handbook*, ASM International, Materials Park, OH, 1989.)

this problem consists of using short cycles while the WPs are reversed in their slots. In addition, they are mixed between slots. Taper can also be minimized by positioning the work-holder so that parts in slots are at 15° angle to a radius, as illustrated in Figure 4.30.

### Example

The valve needle (high-alloy tool steel of 60–65 HRC) shown in Figure 4.31 is to be lapped to achieve the accuracy requirements, where $R_a = 0.05$ μm, tolerance = ±0.13 μm, out-of-roundness = 0.13 μm, and taper = 0.25 μm. Discuss the possible alternatives to achieve the preceding requirement.

### SOLUTION

There are two alternatives for lapping:

1. For small quantities, a ring lap of CI is used (Figure 4.31). Each needle is chucked by its stem and rotated in a lathe at 650 rpm. The CI lap is stroked back and forth over the needle until grinding marks vanish. The needle is coated with lapping compound (CrO mixed with spindle oil).
2. For lot and mass production, the part is finished on a two-plate lapping machine (Figure 4.29). Before being machine-lapped, parts are carefully ground for roundness and classified into groups according to their diametral variations of ±2.5 μm, ±5 μm, and so on. A laminated phenolic work-holder is designed to be eccentric to the laps to provide an Figure 4.29). The short cycles are stopped to measure the parts with an electro limit gauge. If the desired size has not yet been attained, more lapping compound is added, and lapping proceeds until the required finish is achieved.

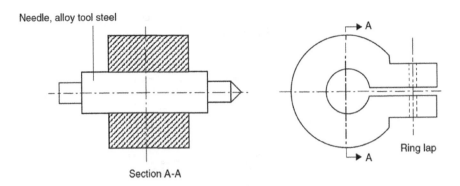

Section A-A

**FIGURE 4.31**  Lapping of valve needle using a ring lap. (Adapted from ASM International, *Machining,* Vol. 16, *Metals Handbook,* ASM International, Materials Park, OH, 1989.)

2. *Lapping of flat surfaces*

Flat surfaces can be lapped by either manual or mechanical methods.

a. *Manual lapping*

Manual lapping is used only for limited quantities or when special requirements must be met. Hand rubbing of a flat WP on a plate lap charged with an abrasive compound is the simplest method of flat lapping. The lap, usually made of iron, has regularly spaced grooves of about 1.6 mm depth to retain the lapping medium. The WP is rubbed on the lap in a figure eight or a similar pattern that covers almost the entire lap surface. The lap remains flat for a considerable amount of work. This method of lapping is time-consuming and tedious and requires a high degree of labor skill. Another, somewhat faster method makes use of a vertical drilling press, where the lap is fixed on the machine table and the work is held by the spindle. The WP rotates against the lap, while light pressure is applied by hand. However, this method violates one of the basic rules of lapping, namely, the random and the nonrepeated paths between the lap and the work.

b. *Mechanical lapping*

Mechanical lapping is performed by flat lapping machines. The two general types are single- and dual-face lapping machines. However, dual-face lapping machines are preferred due to their enhanced accuracy. Most dual-face lapping machines are of the planetary type, with the work-holder (carriers) nested between a center drive and a ring drive. These drives can be either gear- or pin-type configurations and must have positive engagement (Figures 4.32 and 4.33). The WP is propelled by the carrier in a serpentine path between lap plates on which abrasives have been charged or continuously fed in the form of slurry. In the planetary fixed-plate machine (Figure 4.32), the bottom lap is fixed, and the top lap is restrained from rotating. It is allowed to float to bear on the largest pieces and laps all the pieces to the same size. The

**FIGURE 4.32**   Planetary fixed-plate double-face lapping machine for flat surfaces. (From Hoffman Co., Carlisle, PA, USA.)

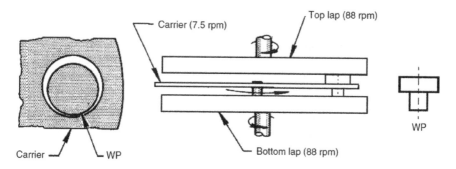

**FIGURE 4.33**   Dual-face lapping machine using two bonded abrasive laps. (Adapted from ASM International, *Machining,* Vol. 16, *Metals Handbook,* ASM International, Materials Park, OH, 1989.)

part is dragged between the plates by the carrier, and all the power is directed to the flat, thin carrier plates, exerting high forces on their thin teeth, which may cause edge chipping on fragile parts.

Figure 4.33 illustrates another dual-face lapping machine having two bonded abrasive laps (400 grit SiC) that are rotated in opposite directions at 88 rpm. The head is air-actuated to provide the lapping pressure to the top lap. The WP carrier is eccentrically mounted over the bottom lap and rotates at 7.5 rpm. The viscous cutting oil is fed to the laps during operation. The laps are dressed two or three times during an 8 hour shift. Figure 4.34 illustrates some typical shapes that can be machined on flat lapping machines. Symmetrical components (a) and (b) do not require work-holders. Asymmetrical components (c)

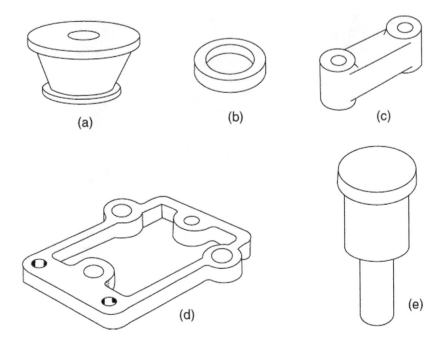

**FIGURE 4.34** Typical shapes lapped on flat lapping machine: (a, b) symmetrical components do not require work holders; (c, d) asymmetrical components require work holders; (e) parts require holders to keep them from tipping. (Adapted from ASM International, *Machining,* Vol. 16, *Metals Handbook,* ASM International, Materials Park, OH, 1989.)

and (d) require work-holders. Parts similar to (e) require holders to keep them from tipping.

*Tolerance, roughness, flatness, and parallelism.* Achieved tolerance of parts having parallel shapes can be from ±2.5 μm (for small parts) to ±25 μm (for large parts). It is difficult to maintain accuracy for parts of uneven configuration. Such parts may require fixtures that determine the level of accuracy attainable. The flatness may attain a value of 0.3 μm, and the achievable surface roughness $R_a$ is 0.05 μm (Figure 4.34).

Flat parallel surfaces can be lapped on either double-lap machines, which lap both sides of the WP in a single operation, or single-lap machines, which require two operations. In the latter case, extraordinary attention is required to such details as cleanliness and lap flatness. Flatness of laps must be kept within the required flatness tolerance of the WP. In the case of lot production, a parallelism of 0.2 μm/mm dictates the use of a dual machine; however, it is possible to produce parts with opposing faces parallel within 0.02 μm/mm. Allowance for stock removal in this operation should be 1.5–2 times the amount of the part out-of-parallelism plus the amount of the variation in part size (ASM International, 1989).

3. *Lapping Machines for Spherical Surfaces*

These are classified into two classes: single- and multiple-piece lapping machines. Single-piece machines have the following two configurations:

a. *A single-spindle machine with a vertical spindle that rotates the lap.* Ferrous WPs are held stationary by a magnetic chuck; those of nonferrous materials are clamped in a fixture. A crank held by the chuck of a lathe, provided by a ball-end crankpin that fits in a drilled hole in the back of the lap (Figure 4.35a), rotates over the spherical surface of the WP. The WP is in line with the spindle of the lathe. The lap should be heavy enough to provide the required lapping pressure.

b. *A two-spindle machine.* One spindle holds and rotates the WP, while the other holds the lap in a floating position and oscillates it through an angle large enough to lap the required area of the surface (Figure 4.35b).

### 4.2.3.2.2 Equalizing Lapping

In this process, two WP surfaces are separated by a layer of abrasives mixed with a vehicle and rubbed against each other. Each piece drives the abrasive, so that the particles act on the opposing surface. Irregularities that prevent the two surfaces from fitting together precisely are thus lapped, and the surfaces are mated (equalized).

In many cases, a part is first lapped by individual-piece lapping and then mated with another part by equalizing lapping. Equalizing lapping enables mating parts such as cylinder heads and blocks of internal combustion engines to be liquid- or gas-tight without the need for gaskets. It also eliminates the need for piston rings when fitting plungers to cylinders. Another common application is the equalizing lapping of tapered valve components (Figure 4.36).

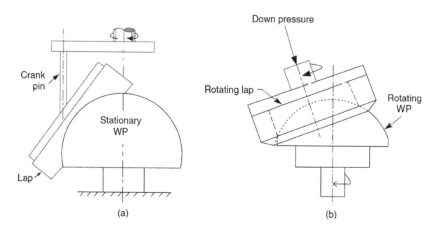

**FIGURE 4.35** Lapping of spherical surfaces: (a) single-spindle machine and (b) two-spindle machine. (Adapted from ASM International, *Machining,* Vol. 16, *Metals Handbook,* ASM International, Materials Park, OH, 1989.)

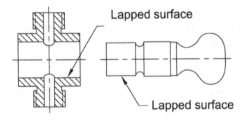

**FIGURE 4.36**  Tapered valve component finished by equalizing lapping. (Adapted from ASM International, *Machining*, Vol. 16, *Metals Handbook,* ASM International, Materials Park, OH, 1989.)

## 4.3  REVIEW QUESTIONS

4.3.1  Why is friability an important grit property in abrasive machining?

4.3.2  What are the commonly used materials for binding GWs?

4.3.3  What are the differences between dressing and truing?

4.3.4  How is a WP controlled in centerless grinding?

4.3.5  Why should a grinding fluid be used in very copious quantities when performing wet grinding?

4.3.6  What is CFG?

4.3.7  How does CFG differ from conventional plunge surface grinding?

4.3.8  What are the common causes of grinding accidents?

4.3.9  What other machine tool does a cylindrical grinding resemble?

4.3.10  A set of granite or wooden stairs shows wear on the treads in the central region where people step when they climb or descend the stairs. The higher the stair step, the less wear on the tread. Explain.
  • Why do the stairs wear?
  • Why are the lower stairs more worn than the upper ones?

4.3.11  Explain the major differences between the specific energies involved in grinding and in machining.

4.3.12  Explain why the same GW might act soft or hard.

4.3.13  A soft-grade GW is generally recommended for hard materials. Explain.

4.3.14  Explain why CFG has become an important process.

## REFER\ENCES

ASM International, *Machining*, Vol. 16, *Metals Handbook*, ASM International, Materials Park, OH, 1989.

Hoffman Co., Carlisle, PA.

Raw, PN 2000, *Metal cutting and machine tools*, Tata McGraw-Hill, New Delhi.

# 5 Thread-Cutting Machines and Operations

## 5.1 INTRODUCTION

The production of screw threads is of prime importance to engineers, because nearly every piece of equipment has some form of screw thread. Machine parts are held together, adjusted, or moved by screw threads of many sizes and kinds. Screw threads are commonly used as fasteners, to transmit power or motion, and for adjustment. Different thread forms (V, square, acme) and thread series (coarse, fine, and so on) are available. The screw threads used in manufacturing should conform to an established standard in order to be interchangeable and replaceable. The following terms (Figure 5.1) are used to describe the geometry of a screw thread:

*Major diameter.* The outside diameter and the largest diameter of the thread.
*Minor diameter.* The inside diameter of the screw and the base of the thread. It is also called the *root diameter.*
*Pitch.* The distance from one point on a thread to the same point on the adjacent thread. The reciprocal of the pitch is the number of threads per inch (tpi).
*Pitch diameter.* The diameter of an imaginary cylinder whose surface would pass through the threads at such points as to make the width of the thread cut equal to the width of the spaces cut by the imaginary cylinder.
*Crest.* The top surface joining the flanks of the thread.
*Root.* The bottom surface joining the flanks of the thread.
*Flank.* The slanted surface joining the crest and the root; the surface that is in contact with the nut.
*Lead.* The amount by which the nut advances along the screw with one turn if the screw is held stationary. For single-pitch threads, the lead is equal to the pitch; for double-pitch threads, the lead is double the pitch; and for triple-pitch threads, the lead is three times the pitch.

The standard and most widely used threads are as follows (Figure 5.2):

1. *The ISO metric thread.* This thread (Figure 5.2a) is based on the recommendation of the ISO technical committee and was published in British Standard Specification No. 3643 in 1963. It was intended that this thread become a British standard. As a part of the International System of Units (SI), pitches are in millimeters, and the system allows for coarse and fine pitches. The thread form is identical with the unified thread.

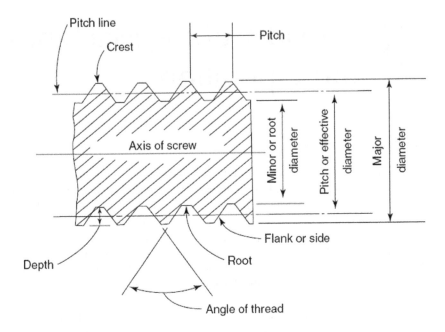

**FIGURE 5.1**  Thread terms.

2. *The unified thread.* This thread standard was published in 1949 as a result of a conference held in Ottawa between the United States, Canada, and Britain in 1945. There are two subtypes of this thread: the unified coarse (UNC) and the unified fine (UNF). The pitches of this thread are in inches (Figure 5.2a).

3. *Whitworth thread.* This thread form was proposed by Sir J. Whitworth in the 1840s. It has been used as the British Standard Whitworth (BSW) ever since (Figure 5.2b).

4. *British Association (BA) thread.* This thread has been used for screws of diameters less than ¼ in. and for electrical fittings and accessories (Figure 5.2c).

5. *Square thread.* As shown in Figure 5.2d, this thread is used for power transmission. It is the most difficult to cut and is not compatible with using split nuts.

6. *Acme thread.* This thread (Figure 5.2e) is often employed instead of the square thread because it is easier to cut; also, it is easier to engage a split nut with an acme thread than with a square one. It has a 29° thread angle.

7. *Trapezoidal metric thread.* Similar to the acme thread, except that it has a 30° thread angle. It may replace the acme thread for lead screws (Figure 5.2e).

Acme, square, and trapezoidal threads are used for power transmissions such as screw jacks, lead screws of lathes, vices, presses, and so on. Acme threads are

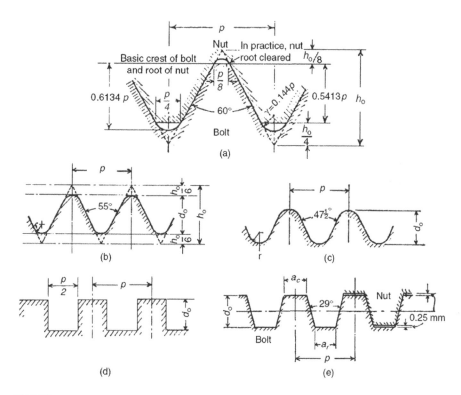

**FIGURE 5.2** Different thread forms: (a) the ISO metric and unified form, (b) Whitworth form ($d_o = 0.64p$, $r = 0.137p$), (c) British Association thread ($d_o = 0.6p$, $r = 2p/11$), (d) square thread ($d_o = 0.5p$), and (e) acme thread ($d_o = 0.5p + 0.25$ mm, $c_1 = 0.371p$, $a_r = a_c - 0.13$).

cheaper to manufacture than square threads but are less efficient. Acme threads are sometimes used with a split nut to facilitate the engagement and disengagement of the nut and to transmit power in any direction, while trapezoidal threads are used to transmit the power in one direction.

Threads can be produced in a number of ways. The manner of producing them depends on many factors, such as the cost and use of the threaded workpiece (WP), the equipment available, the number of parts to be made, the location of the threaded portion, the smoothness and accuracy desired, and the material to be used. Methods of thread manufacturing are shown in Figure 5.3. This chapter deals with thread machining methods; threads produced by rolling and casting methods are beyond the scope of this book.

## 5.2  THREAD CUTTING

During machining by cutting, the tool penetrates into the WP by a depth of cut. Cutting tools have a definite number of cutting edges of known geometry. Moreover, the machining allowance is removed in the form of visible chips. The shape of the

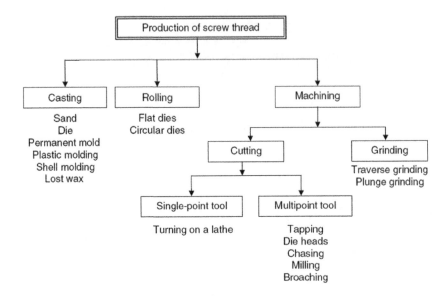

**FIGURE 5.3**  Methods of screw thread production.

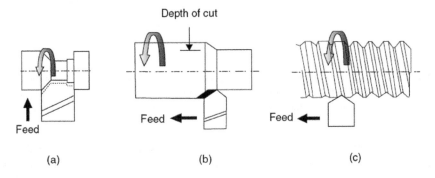

**FIGURE 5.4**  Cutting modes in turning: (a) form cutting, (b) generation cutting, and (c) form and generation cutting.

WP produced depends upon the tool and the relative motions of the WP. Three main arrangements that occur during machining by cutting are as follows:

1. *Form cutting.* The shape of the WP is obtained when the cutting tool completes the final contour of the WP. The WP profile is formed through the main WP rotary motion in addition to the tool feed at a specified depth (Figure 5.4a). For automatic machine tools, a circular-form tool would be used, which has much greater work life, provides more regrinds, and is often easier to manufacture. The quality of the machined surface profile depends on the accuracy of the form-cutting tool used and the tool setting on the work center. Drawbacks of such an arrangement include the

large cutting force and the possibility of vibrations when the cutting profile length is long. Additionally, complex form tools are difficult to produce and hence expensive.

2. *Generation cutting.* In this case, the form of the profile is produced by the cutting edge of the tool, which moves through the required path. The WP is formed by providing the main motion to the WP and moving the tool point in a feed motion. In the turning operation, shown in Figure 5.4b, the WP rotates around its axis while the tool is set at a feed rate to generate the required profile. Much longer profiles can be generated using this method than using the form-cutting method. The work finish is better, and it is easier to generate internal profiles than to form them.

3. *Form and generation cutting.* During thread cutting, the tool, having the thread form (form cutting), is allowed to feed (generation cutting) axially at the appropriate rate while the WP rotates around its axis (main motion) (Figure 5.4c).

Screw threads can be produced by a variety of cutting tools and processes. The simplest of these is the use of a single-point threading tool in an engine lathe, semi-automatics, and automatics. This method is widely used in piece and small-lot production and for cutting coarse-pitch threads. Threads can be produced by hand and in a machine by means of taps and threading dies, which cut internal and external threads, respectively. Solid dies have a low production capacity, and therefore, self-opening dies are currently used. Die heads with radial chasers, circular chasers, and tangential chasers are available. Upon the completion of the thread, the chasers of the self-opening die head automatically withdraw from the work, making reversal of the machine to screw off the die head unnecessary.

Threads can also be produced by milling. Trapezoidal and acme threads with coarse pitch are milled with a disk-type milling cutter, and short threads can be produced with a multithread milling cutter. The axis of this type of cutter is set parallel to the work axis, and its length must be slightly greater than that of the threaded portion of the WP. In thread-milling machines employing this type of cutter, the whole length of the thread is milled in a slightly over one revolution of the WP.

## 5.2.1 Cutting Threads on the Lathe

For cutting an accurate screw thread on the lathe machine, it is necessary to control the relation between the feed movement of the cutting tool and the turns of the WP. This is done by means of the lead screw, which is driven by a train of gears from the spindle (Figure 5.5). Modern lathes are fitted with change gears boxes, by means of which any thread pitches can be cut without working out and setting up the change gears. However, there are some machines on which change gears must be fitted for screw cutting.

When cutting a screw thread, the tool is moved along the bed and is driven by a nut engaging with the lead screw. The lead screw is driven by a train of gears

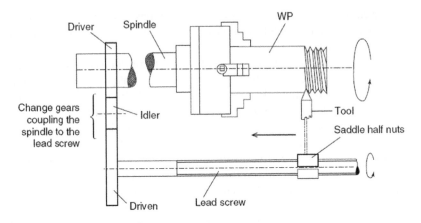

**FIGURE 5.5** Diagrammatic representation of screw cutting on a lathe.

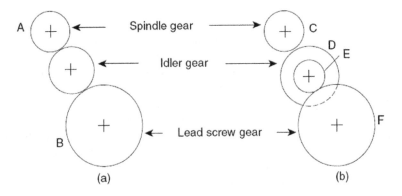

**FIGURE 5.6** Gear trains for thread cutting on a lathe: (a) simple gearing and (b) compound gearing.

from the machine spindle. The gear train may be arranged in one of the following ways:

A. *Simple gear train.* In such a gear train, shown in Figure 5.6a, the following ratio holds:

$$\frac{\text{Turns of lead screw}}{\text{Turns of spindle}} = \frac{\text{Teeth on driver (A)}}{\text{Teeth on driven (B)}}$$

The intermediate gear has no effect on the ratio. It simply acts as a connection that makes the lead screw rotate in the same direction as the machine spindle.

B. *Compound gear train.* In this case, as shown in Figure 5.6b, the gear ratio becomes

$$\frac{\text{Turns of lead screw}}{\text{Turns of spindle}} = \frac{\text{Teeth on C}}{\text{Teeth on D}} \times \frac{\text{Teeth on E}}{\text{Teeth on F}} = \frac{\text{Teeth on drivers}}{\text{Teeth on driven}}$$

Gears supplied with lathes generally range from 20 to 120 teeth in steps of five teeth with two 40s or two 60s. The lead screw on lathes is always single-threaded and of a pitch varying from 5 to 10 mm depending on the size of the machine. For English lathes, the most common screw threads have 2, 4, or 6 tpi.

## Solved Example

Calculate suitable gear trains for the following cases (Chapman, 1981):

   a. 2.5 mm pitch on a 6 mm lead screw
   b. 11 tpi on a 4 tpi lead screw
   c. 7 threads in 10 mm on a 6 mm lead screw
   d. 7/22 in. pitch, 3 start on a lathe with 2 tpi
   e. 2.5 mm pitch on a 4 tpi lead screw
   f. 12 tpi on a lathe having a 6 mm pitch lead screw

### SOLUTION

a. 2.5 mm pitch on a 6 mm lead screw

$$\frac{\text{Drivers}}{\text{Driven}} = \frac{2.5}{6} = \frac{5}{12} = \frac{25}{60}$$

25 teeth driving 60 teeth in a simple train.

b. 11 tpi on a 4 tpi lead screw

$$\frac{\text{Drivers}}{\text{Driven}} = \frac{1/11}{1/4} = \frac{4}{11} = \frac{20}{55} = \frac{40}{110}$$

This gives either 20/55 or 40/110 in a simple train.

c. 7 threads in 10 mm on a 6 mm lead screw
   The pitch of the thread = 10/7

$$\frac{\text{Drivers}}{\text{Driven}} = \frac{10/7}{6} = \frac{10}{42} = \frac{5 \times 2}{7 \times 6} = \frac{50}{70} \times \frac{20}{60}$$

A compound train with 50 teeth and 20 teeth as the drivers and 70 teeth and 60 teeth as the driven.

d. 7/22 in. pitch, 3 start on a lathe with 2 tpi
   Lead of the thread = 3 × 7/22 = 21/22 in.

Pitch of lead screw = ½ in.

$$\frac{\text{Drivers}}{\text{Driven}} = \frac{21/22}{1/2} = \frac{42}{22} = \frac{21}{11} = \frac{3 \times 7}{2 \times 5\frac{1}{5}} = \frac{30}{20} \times \frac{70}{55}$$

A compound train with 30 teeth and 70 teeth as the drivers and 20 teeth and 55 teeth as the driven.

e. 2.5 mm pitch on a 4 tpi lead screw

For cutting metric threads on English lathes:

$$\frac{\text{Drivers}}{\text{Driven}} = \frac{1}{25.4} = \frac{10}{254} = \frac{5}{127}$$

Cutting $p$ mm pitch would require a ratio $p$ as large as $5p/127$, and if the lead screw had $N_t$ threads per inch instead of 1 thread per inch, it would need to turn faster still in the ratio $N_t{:}1$. That is (Chapman, 1981):

$$\frac{\text{Drivers}}{\text{Driven}} = \frac{5pN}{127}$$

where

$N$ = number of threads per inch of the lead screw

$p$ = required pitch (mm)

Hence,

$$\frac{\text{Drivers}}{\text{Driven}} = \frac{5 \times 4 \times 2\frac{1}{2}}{127} = \frac{50}{127}$$

50 teeth driving 127 teeth in a simple train.

f. 12 tpi on a lathe having a 6 mm pitch lead screw

Dealing with English threads (in.) on lathes with metric leadscrew (mm)

$$\frac{\text{Drivers}}{\text{Driven}} = \frac{127}{5pN}$$

$$\frac{\text{Drivers}}{\text{Driven}} = \frac{127}{5 \times 6 \times 12} = \frac{127}{6 \times 60} = \frac{20}{60} \times \frac{127}{120}$$

A compound train with 20 teeth and 127 teeth as the drivers and 60 teeth and 120 teeth as the driven.

## 5.2.2 Thread Chasing

*Thread chasing* is the process of cutting a thread on a lathe with a chasing tool that comprises several single-point tools banked together in a single tool called a *chaser*. Thread chasers are shown in Figure 5.7. Chasing is used for the production of threads that are too large in diameter for a die head. It can be used for internal threads greater than 25 mm in diameter. The chaser moves from the headstock. The chaser is moved radially into the WP for each cut by means of the cross-slide screw. Thread chasing reduces the threading time by 50% compared with single-point threading. However, thread chasing is a relatively slow method of cutting a thread, as a small depth of cut is used per pass. Depending on the size of the thread, 20–50 passes may be required to complete a thread. Multiple threads, square threads, threads on tapers, threads on

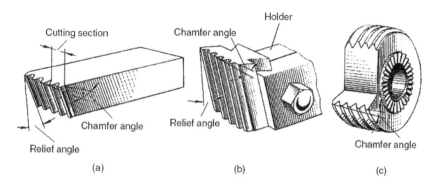

Cutting section  Chamfer angle  Holder

Chamfer angle

Chamfer angle  Relief angle

Relief angle  Chamfer angle

(a)  (b)  (c)

**FIGURE 5.7** Thread chasers: (a) flat (shank type), (b) block, and (c) circular. (From Rodin, P., *Design and Production of Metal Cutting Tools*, Mir Publishers, Moscow, 1968. With permission.)

diameters not practical to thread with a die, threads that are not standard or those that are so seldom cut that buying a tap or die would be impracticable, or threads with a quick lead are all cut by chasing.

Chasing lends itself better to nonferrous materials rather than ferrous ones. Multistart threads can be chased without any indexing of the WP. Taper threads can be generated by chasing, if the chasing attachment is used in conjunction with a taper attachment. For high-speed steel (HSS) cutters, a cutting speed of the order of 40 m/min and upward should be used. Feed varies from 5 to 7.5 cm/min for coarse threads in tough materials to 20–25 cm/min under more favorable conditions. Figure 5.8 shows the different methods of thread chasing.

### 5.2.3 THREAD TAPPING

*Thread tapping* is a machining process that is used for cutting internal threads using a tap having threads of the desired form on its periphery (Figure 5.9). There are hand taps and machine taps, straight shank and bent shank taps, regular pipe taps and interrupted thread pipe taps, solid taps, and collapsible taps. A tap has cutting teeth and flutes parallel to its axis that act as channels to carry away the chips formed by the cutting action. Hand taps are furnished in three sets—taper, plug, and bottoming (Figure 5.10). These three are identical in size, length, and vital measurements, differing only in chamfer at the bottom end. Standard taps are furnished with four flutes and are used for iron and steel. These do not provide sufficient chip room for certain soft metals, such as copper, in which case two- or three-fluted taps should be used. The tap cuts threads through its combined rotary and axial motions. The cost of tapping increases as the work material hardness becomes greater. Fine threads of 360 tpi in 0.33 mm diameter holes and coarse threads such as 3 tpi in 619 mm diameter pipe fitting are possible (*Metals Handbook*, 1989).

Tapping machines are basically drill presses equipped with lead screws, tap holders, and reversing mechanisms. Lead screws convert the rotary motion into a linear one so that the axial motion of the tap into the hole to be threaded conforms with the pitch of the thread. Lead screw control is often used with larger tap sizes to ensure

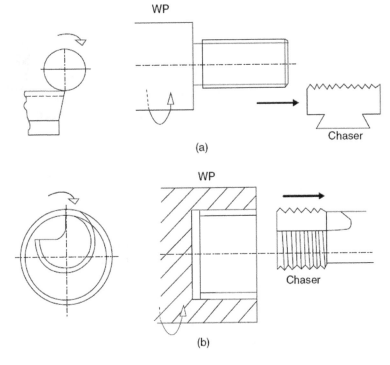

**FIGURE 5.8** Thread chasing methods: (a) right-hand external and (b) right-hand internal.

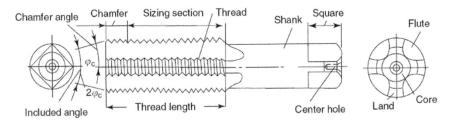

**FIGURE 5.9** Tap nomenclature. (From Rodin, P., *Design and Production of Metal Cutting Tools*, Mir Publishers, Moscow, 1968. With permission.)

high-quality threads. However, such an arrangement has the following two major disadvantages:

- It is necessary to return to the starting point to begin each cycle and to stop the rotation between cycles.
- Changing the taps for different thread sizes requires time-consuming changes in the feed-controlling members.

Tension or compression tapping spindles and attachments provide axial float and compensate for any differences between machine feed and correct tap feed. This provides the possibility to tap different thread pitches at the same time with a single

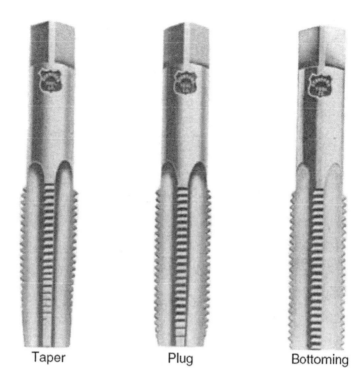

Taper   Plug   Bottoming

**FIGURE 5.10**   Straight flute hand taps. (From Standard Tool Co., Athol, MA.)

machine feed rate. Self-reversing tapping attachments eliminate the need for reversing motors for tap retraction. Nonreversing tapping attachments are generally used with machines equipped with reversing motors. Figure 5.11 shows the components of a tapping attachment. Tapping machines include the following:

1. *Drill presses.* Simple to set up, easy to operate, and can be provided with lead-control devices that regulate the tap feed rates. When a solid tap is used, the drill press must be supplied with a self-reversing tapping attachment or a reversing motor having a tension compression tap holder. With a collapsible tap, the tapping attachment is not required, because the tap automatically collapses at the required depth and returns without stopping or reversing the spindle.
2. *Single-spindle tapping machines.* Used for small to medium production lots. The simpler modes have no lead-control devices but depend on the screw action of the tap in the hole to control the feed (see Figure 5.12).
3. *Multiple-spindle tapping machines.* Used for high-volume production lots. They may have up to 25 spindles that are rotated by a common power source. Holes of different sizes can be tapped simultaneously. Spindles having axial float compensate for differences between the lead of the tap and the feed of the spindle. Thus, different thread pitches can be cut simultaneously on the same machine (see Figure 5.13).

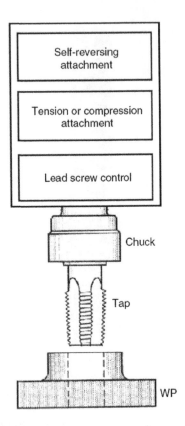

**FIGURE 5.11**   Tapping attachment.

4. *Gang machines.* Permit in-line drilling, reaming, and tapping operations and are generally used for low-volume production lots.

5. *Manual turret lathes.* Used for small production lots. Because the WP rotates, they are more accurate than machines that rotate the tap. The machine capability permits drilling, boring, and tapping on the same machine. A lead-control device is used when tapping on the turret lathe.

6. *Automatic turret lathes.* Tapping may be included among the many other operations of an automatic turret lathe or in a single multiple-spindle bar or chucking-type machine. These machines require long setting times and are therefore used for large production lots. These machines use lead-control devices for regulating the feed.

The selection of a tapping machine depends on the following factors:

- Size and shape of the WP
- Production quantity
- Tolerance
- Surface finish
- Number of related operations
- Cost

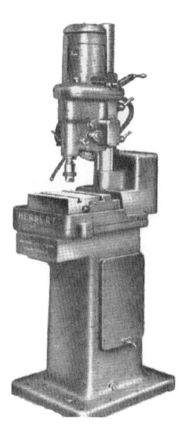

**FIGURE 5.12** Herbert flash tapping machine with automatic cycle. (From Alfred Herbert Ltd., Coventry, UK.)

Generally, small diameters and fine-pitch threads are cut on machines of relatively low power, and larger threads in harder materials require heavier machines with large power.

Thread Tapping Performance

Figure 5.14 summarizes the different factors that affect the performance measures of tapping in terms of quality, productivity, and cost. These include the following:

*WP characteristics.* The use of free-cutting metals is more recommended where better accuracy and surface finish at higher production rates and lower cost are achieved. General purpose HSS taps are used when the WP hardness is about 30 or 32 HRC; otherwise, highly alloyed HSS is recommended. The work material composition may affect the preparation of the hole before tapping. In this regard, reaming the hole improves the accuracy and finish in aluminum, although stainless and carbon steels do not require such a reaming process (7*Metals Handbook*, 1989). Tapping problems occur with WPs that are too weak to withstand tapping forces. Under such circumstances, a loss of dimensional accuracy, bad surface quality,

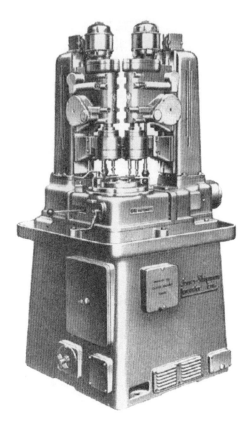

**FIGURE 5.13**   Jones and Shipman multiple-spindle automatic drilling and tapping machine.

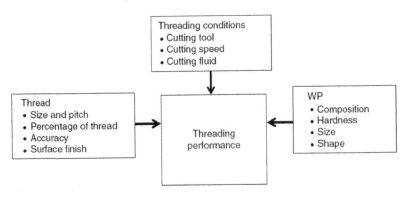

**FIGURE 5.14**   Factors affecting threading performance.

and WP damage may occur. For tapping blind holes, a clearance between the last full thread and the bottom of the hole should be compatible with the tap chamfer length. Such a clearance provides room for the produced chip to avoid tap breakage or hole damage by the compressed chip under the advancing tool.

*Thread features.* Thread size, pitch, and percentage of full depth to which the threads are cut determine the volume removed during the tapping operation. Larger volumes have a direct effect on the process efficiency and tool life. Conditions that cause dimensional variations in the tapped threads cause rough surface finish of threads. These include concentricity error between the tap holder and the spindle and the WP center. Worn tapes, chip entrapment in the tapped hole, and chip build-up on the cutting edges and flanks of the tool also cause dimensional variations and deterioration of the surface finish.

*Tapping conditions.* The WP material has the greatest effect on the tapping speed. The following recommendations should be followed (*Metals Handbook*, 1989):

- As the depth of the tapped hole increases, the speed should be reduced because of chip accumulation.
- In short holes, taps with short chamfers run faster than taps with long chamfers.
- As the pitch becomes finer, for a given hole, tapping speed can be increased.
- The amount of cutting fluid and the effectiveness of its application greatly influence the cutting speed.

During tapping, the teeth of the tap are more susceptible to damage by heat generated during threading and the chips that are more likely make the tape congested. Cutting fluids are, therefore, used in tapping all metals except CI. However, for tapping holes longer than twice the diameter or blind holes in CI, a cutting fluid or an air blast is recommended (*Metals Handbook*, 1989).

## 5.2.4   DIE THREADING

Die threading is a machining process that can be used for cutting external threads of 6.35–114 mm rapidly and economically in cylindrical or tapered surfaces using solid or self-opening dies. The process is faster than single-point threading on a lathe. Die threading of materials having hardness greater than 36 HRC causes excessive tool wear or breakage. Therefore, single-point threading or thread grinding is recommended for metals harder than 36 HRC. Die threading is capable of producing fine and coarse threads. The quality and accuracy of such threads are acceptable for most mass-produced articles. For a small shop, thread chasing may be less expensive than stocking a complete set of taps and dies.

### 5.2.4.1   Die-Threading Machines

1. *Drill press.* Easy to set and simple to operate. Threading can be cut manually or by using lead-control devices, which may require more rigid machine tools such as lathes.
2. *Manual turret lathes.* Used for threading small to medium quantities in parts that require other machining operations such as drilling, turning, reaming, and so on. Turret lathes can handle bar- and chucking-type work

and can thread larger parts that are difficult to manipulate in a drill press. Many turret lathes are equipped with lead-control devices.

3. *Automatic machines.* Include automatic turret lathes and single- or multiple-spindle automatic bar -or chucking-type machines, which are used for medium- or high-production lots because their setting time is long and their running cost is high. Threading is performed in addition to several other machining operations.

4. *Special threading machines.* Available only for die threading in either cylindrical- or irregular-shaped parts. WP loading and unloading can be manual, hopper-fed, or fully automatic. These machines usually incorporate lead-control devices. Bar-type machines with collets can handle long parts to thread rods, shafts, and pipes.

### 5.2.4.1.1 Solid Dies

Solid threading dies may be of one-piece construction with integral cutting edges or may have replaceable chasers. Nonadjustable, one-piece, solid dies (Figure 5.15) have all cutting edges in a rigidly fixed relationship and are available in standard sizes to fit various types of holders. Adjustable, one-piece, solid dies (Figure 5.16) have a slotted body, a spring, a relief hole, and an adjusting screw for small adjustments that compensate for tool wear and retain greater accuracy than is possible with nonadjusting dies. In Figure 5.17, the collet adjustability is provided by forcing the jaws inward as the outer nut is tightened. An inserted-chaser solid die consists of a holder and three or more chasers, which can be compensated for wear and can be

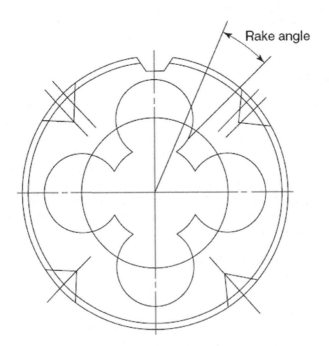

**FIGURE 5.15**   Solid nonadjustable die.

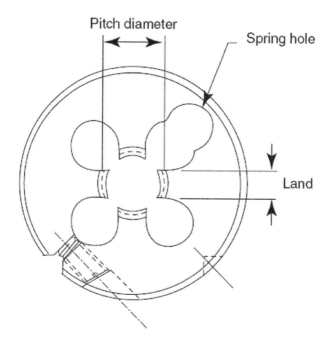

**FIGURE 5.16**   Solid screw-adjustable die.

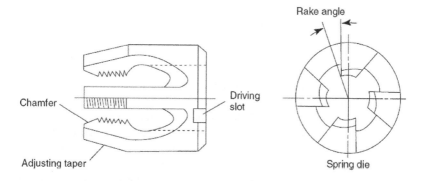

**FIGURE 5.17**   Spring-type collet-adjustable die and holder.

removed for resharpening operations. Like one-piece dies, it can be removed from the WP by being back-tracked over the cut thread. Circular chasers are used in solid dies in sets of five, where the die cuts better with less torque. Solid dies are not preferred for high production, because the spindle must be reversed for the die removal after the thread is cut, which is time-consuming and increases wear.

### 5.2.4.1.2   Self-Opening Die Heads

1. Revolving self-opening dies have a fixed WP and a rotating tool, as in the case of the drill press. The die is supported by a yoke that opens when it meets a stop at the end of the threading stroke, which retracts the chasers

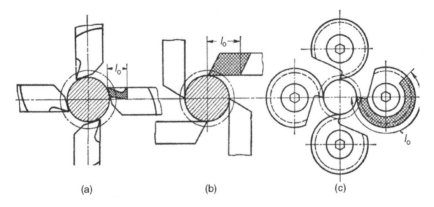

**FIGURE 5.18** Threading die heads: $l_o=$ maximum layer of stock available for sharpening: (a) radial chasers; (b) tangential chasers; (c) circular chasers. (From Arshinov, V. and Alekseev, G., *Metal Cutting Theory and Cutting Tool Design*, Mir Publishers, Moscow, 1970. With permission.)

from the WP. The die can then return to its starting position for the next threading cycle.

2. Stationary self-opening dies have a rotating WP and a fixed tool, as in the case of turret and capstan lathes. Similarly, the die opens and retracts to its starting position at the end of the threading stroke. Stationary dies may feed axially as the threading progresses.

The types of chasers used in self-opening dies are as follows:

- Radial chasers, which are restricted to soft and easy-to-cut materials, such as aluminum and free-cutting brass (Figure 5.18a).
- Tangential chasers (Figure 5.18b), which are especially suited for threading steel and other hard metals because of their long tool life. Repeated sharpening is permissible as long as a sufficient length of the chaser permits chaser holding for sharpening and securing in the die.
- Circular chasers (Figure 5.18c), which are made of sets of four or five with annular threaded form. They are normally used in high production for all metals that are threaded. They have a long lifespan, because they can be resharpened many times.

Ease of the die removal from the WP is the greatest advantage of self-opening dies. Dies that return in the open mode do not require spindle stopping regardless of whether either the tool or the WP is rotating. This improves machining productivity by more than 50% and reduces the possibility of thread damage by trapped chips. As the chasers make one trip over the WP, their wear is greatly reduced, and hence, their tool life is markedly increased. Figure 5.19 shows the general classification of threading die heads, and Figure 5.20 shows a self-opening die head with tangential chasers.

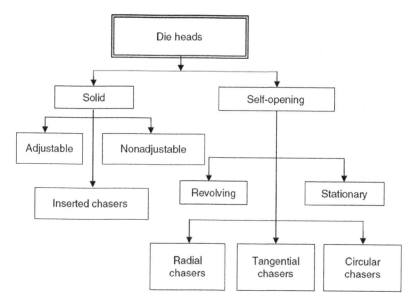

**FIGURE 5.19**   Classification of threading die heads.

**FIGURE 5.20**   Die head with tangential chasers. (From Alfred Herbert Ltd., Coventry, UK.)

### 5.2.4.2   Die-Threading Performance

Figure 5.14 summarizes the different factors that affect the performance measures of die threading in terms of quality, productivity, and cost. These include the following:

*WP characteristic.* The choice of free-cutting metals produces more accurate threads of better surface finish at higher production rate and low cost. Generally, the following recommendations are followed (*Metals Handbook*, 1989):

- For threading metals softer than 24 HRC, standard untreated HSS chasers are used.
- For metals of HRC 24–31, coated or surface-treated HSS chasers are recommended.

- For metals of HRC 36, more highly alloyed HSS chasers (M3, M4, M42, or T5) are used.
- Metals harder than 36 HRC are usually threaded by single-point tools.
- Harder metals require more power, rigid machine tools, and lead-control devices.
- Soft metals of non–free cutting grades form stringy chips that adhere to the chasers and cause dimensional variations and bad surface finishes.
- A WP threaded close to a shoulder should have a relief groove wide enough to admit the full chamfer of the chaser plus the first full thread and to provide extra clearance for over-travel without hitting the shoulder. Threading to a shoulder increases the tool cost and decreases the machining productivity.

*Thread features.* The diameter of the part being threaded has a significant effect on the threading procedure, production rate, and cost per threaded piece. The dimensional accuracy is mainly affected by the WP composition and hardness, the type and condition of the machine and cutting tools, and the type of cutting fluid.

*Die-threading conditions.* WP material has the greatest effect on the threading speed. In this regard, the following recommendations should be followed (*Metals Handbook*, 1989):

- Cuts that are made too slowly increase the threading time and the production cost.
- Quick threading results in excessive heat, short tool life, and poor threading accuracy.
- When threading without lead-control devices, the accuracy depends on the skill of the operator and the ability of the chasers to form a nut action.
- Manual control is satisfactory for threading diameters up to 6.4 mm.
- Sulfurized cutting oil is effective for most die-threading applications.

Threading using die heads has the following advantages:

1. Because the stopping and reversing of the spindle is eliminated, it is possible to save considerable time and increase the rate of production.
2. The use of die heads facilitates the withdrawal of any damaged chaser or the replacement of one chaser set by the other to suit the thread requirements so that several types of threads can be cut.
3. Thread manufacture with die heads is economical, because unskilled workers can operate the machines.
4. Thread accuracy is consistent.

However, die heads have the following limitations:

1. Square threads are difficult to cut using die heads.
2. Screw threads running up to the shoulder of the work cannot be cut.

### 5.2.5 Thread Milling

Thread milling is a machining process used for cutting screw threads with a single-form or multiple-form milling cutter (Figure 5.21). Threads having an accuracy of ±0.025 mm of pitch diameter and surface finish of 1.4 µm and spacing accuracy of multiple-start threads of ±0.01 mm can be cut. Thread milling makes smoother and more accurate threads than a tap or a die. It is more efficient than using a single–cutting point tool in a lathe. Thread milling is the most practical method for thread cutting near shoulders or other interfaces. Figure 5.22 shows thread-milled parts. This process is recommended for lot sizes greater than 20 units. Thread milling is used for cutting threads, usually of too large a diameter for die heads. As the milling cutter is held on a stub arbor, the length of the thread is limited to short threads. The cutter rotates at a cutting speed of 0.6 m/s, and the work rotates at the correct feeding speed. As the work rotates, the cutter is fed outward under the action of a master lead screw. Right-hand and left-hand threads can be machined by controlling the direction of tool feed and WP rotation. The disadvantage of the hob-type cutter is that it must revolve with a fixed relation to the work; this is not true for the cutter with annular teeth.

*Thread milling with a disk milling cutter.* This method is used for cutting long, coarse threads and threads with trapezoidal profiles. Sometimes the disk cutters are used to machine triangular threads, but they are not used for cutting square threads. During threading, the cutter rotates and provides a longitudinal feed by the pitch of the thread. The axis of the cutter arbor is set at the thread helix angle to the WP axis. When cutting multiple-start threads, the WP should be turned by $1/n$ of a revolution ($n$ is the number of starts), and the feed rate should be made equal to the lead of the thread.

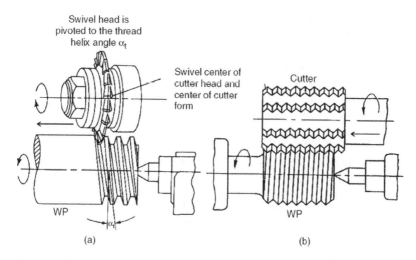

**FIGURE 5.21** Thread milling operations: (a) disk cutter and (b) multiple-thread cutter.

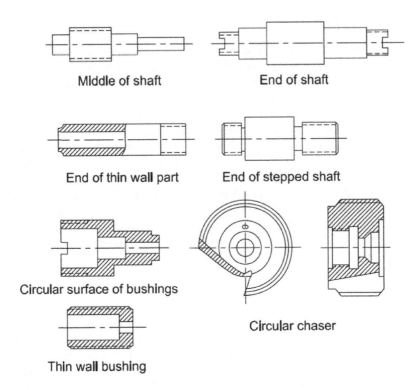

Middle of shaft

End of shaft

End of thin wall part

End of stepped shaft

Circular surface of bushings

Circular chaser

Thin wall bushing

**FIGURE 5.22** Thread-milled parts. (From Barbashov, F., *Thread Milling Practice*, Mir Publishers, Moscow, 1984. With permission.)

*Thread milling with multiple-thread milling cutter.* This method is used to produce short threads of 15–75 mm length and 3–6 mm pitch. The cutter should be 2–3 pitches longer than the thread being cut. Figure 5.23 shows milling straight threads with a multiple-thread milling cutter. External tapered threads can be cut using multiple-thread milling cutters having threads perpendicular to the axis of the cutter (Figure 5.24a). For internal tapered threads (Figure 5.24b), the cutter angle should be equal to the angle of the taper of the cut thread. Generally, the direction of feed is parallel to the generator of the thread surface.

### 5.2.5.1 Thread-Milling Machines

*Universal thread mills.* These machines have a lead screw and cut internal and external threads (with the exception of square threads). Change gears permit milling of threads with leads of 0.8–1520 mm. Pick-off gears in the cutter drive provide a wide range of speeds. The cutter head on the cross slide can be set at the proper angle for right-hand or left-hand thread helix angles. A single-form cutter must be set at such an angle and then allowed to traverse the full length of the thread.

*Planetary thread mills.* These machines are used to thread odd-shaped parts that are difficult to hold in a chuck. Consequently, the WP is held in a special fixture that does not rotate during thread cutting. The milling cutter rotates around its axis and

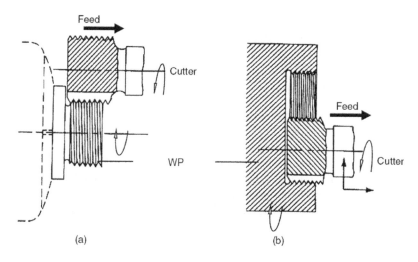

**FIGURE 5.23**   Milling straight (a) external and (b) internal threads.

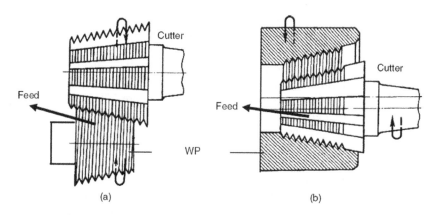

**FIGURE 5.24**   Milling (a) external and (b) internal tapered threads.

revolves around the work. Double heads can be used to cut both ends of the part, and external and internal threads can also be machined at the same time.

*NC machines.* These machines are used for thread milling together with other operations in a single WP operation. Long cutter life and high-quality threads are some of the advantages of these machines.

### 5.2.6   Thread Broaching

*Thread broaching* is a newly developed thread cutting process that has been employed in the automotive field. Typical parts include internal threads on steering-gear ball nuts and ball-race nuts for various circulating ball-type assemblies. The WPs are given one or two passes (rough and finishing cuts), heat-treated, and then finish ground on an internal thread grinder. The broaches used for the application have a special form and are guided by lead screws. Threads are cut by drawing the part and fixture against the revolving tool. Threading broaches are available in sizes up to 50 mm diameter and 750 mm length.

## 5.3 THREAD GRINDING

Thread grinding is the preferred method of threading when the WP hardness is greater than 36 HRC or less than 17 HRC, and when a high degree of accuracy is required. Threads are ground by the contact between a rotating WP and a rotating grinding wheel (GW) that has been shaped to the desired thread form. In addition to the rotation, there is a relative axial motion between the wheel and the WP to match the pitch of the thread being ground. The process can be used to produce either external or internal threads. Methods of thread grinding are classified in Figure 5.25.

### 5.3.1 CENTER-TYPE THREAD GRINDING

In this operation, the WP is held between centers or in the machine chuck. The material specifications and the form, length, and quality of the thread determine the number of passes required (from one to six passes). Depending on the design of the threading wheel, the following two basic methods can be identified:

1. *Single-rib-wheel traverse grinding.* The most versatile method, for which the highest accuracy can be obtained. The single-rib wheel is adaptable, by truing, to many different profile configurations (Figure 5.26a).
2. *Multi-rib-wheel grinding.* The wheels have two or more parallel grooves or ribs around the periphery of the wheel. Each rib is trued to the form of the thread to be ground. The thread form is imparted to the wheel by diamond or crush truing. Figures 5.26b and c show the different arrangements of multi-rib-wheel thread grinding, which include the following:
   A. *Traverse grinding.* More productive than single-rib-wheel grinding, because of the higher material removal rate per pass. However, the pitch

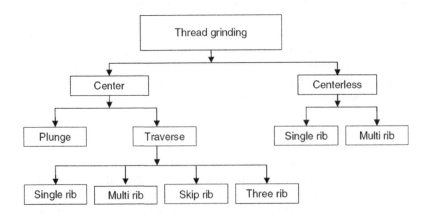

**FIGURE 5.25**  Thread-grinding methods.

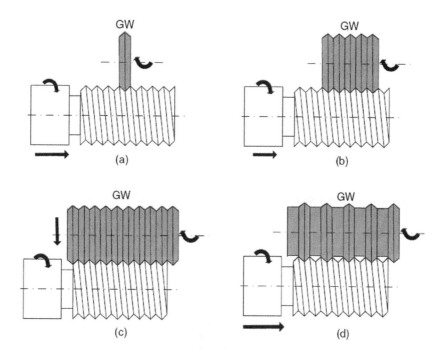

**FIGURE 5.26** Thread-grinding methods: (a) single-rib traverse grinding, (b) multi-rib traverse grinding, (c) multi-rib plunge grinding, and (d) skip-rib traverse grinding. (Adapted from *Metals Handbook*, Machining, Vol. 16, ASM International, Material Park, OH, 1984.)

should not exceed 1/8 of the wheel width, and threading against shoulders should be completely avoided (Figure 5.26b).

B. *Plunge grinding*. The most productive thread-grinding method and therefore used for the production of parts in substantial quantities. As shown in Figure 5.26c, the GW is advanced into the rotating WP.

C. *Skip-rib traverse grinding*. This process uses a wheel, which has a spacing that is twice that of the thread pitch, basically used for threading accurate and fine pitches. Threading is accomplished in two passes. In the first pass, the wheel grinds every other thread of the WP. In the second pass, the WP is advanced by a single pitch, and the untouched threads are then ground (Figure 5.26d).

D. *Three-rib traverse grinding*. As shown in Figure 5.27, the GW has a roughing rib (A) that removes two-thirds of the material and an intermediate rib (B) that takes the remainder of the material and leaves 0.13 mm for cleanup by the finishing rib (C). A flattened area (D) is used to finish the crest of the thread. The process produces more accurate threads than single-rib-wheel thread grinding.

## 5.3.2 Centerless Thread Grinding

Centerless thread grinding is the most productive method of grinding screw threads that uses either single-rib or multi-rib wheels. As shown in Figure 5.28, the regulating

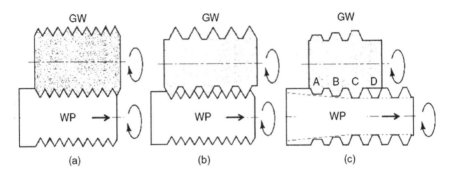

**FIGURE 5.27** (a) Conventional grinding, (b) skip-rib grinding, and (c) three-rib grinding. (Adapted from *Metals Handbook*, Machining, Vol. 16, ASM International, Material Park, OH, 1989.)

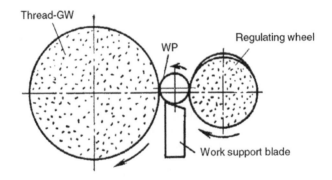

**FIGURE 5.28** Centerless thread grinding. (Adapted from *Metals Handbook*, Machining, Vol. 16, ASM International, Material Park, OH, 1989.)

wheel rotates in the same direction as the GW. Screw threads are cut by feeding the blanks between the grinding and the regulating wheels in a continuous stream as shown in Figure 5.29. Thread-grinding machines are classified as external, internal, or universal; the universal machines are capable of threading external and internal threads (Chernov, 1984; Acherkan, 1968). The machine structure depends on the following:

- The type of GW used (single-rib or multi-rib)
- The method of supporting the WP (centered or centerless)
- The method of restoring the contour of the GW (crushing or diamond dressing)

This process is similar to thread milling, in that it uses a GW having annular thread grooves formed around its periphery to cut a thread as the wheel and WP rotate to form and generate the thread. Internal or external threads can be finish-ground by means of a single- or multiple-edged GW. A vitrified bond is generally used with a fine grit of about 600 Mesh No. The process is carried on a special grinding machine

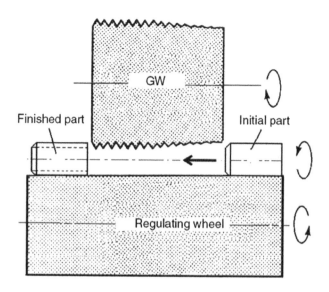

**FIGURE 5.29** Traverse centerless grinding of headless screws. (Adapted from *Metals Handbook*, Machining, Vol. 16, ASM International, Material Park, OH, 1989.)

having a master lead screw, change gears, and means of holding the work. The wheel rotates at a speed of 30 m/s, and work is rotated slowly. In the case of hardened stock, grinding is the only method of forming threads. The accuracy of thread grinding exceeds that of any other method, while the surface finish is exceeded only by a good thread rolling operation. Pitch diameters can be ground to an accuracy of ±0.002 mm/2.5 cm, and accuracy of lead may be maintained within ±0.007 mm in 50 cm of thread length. Distortions due to heat treatment may be eliminated by grinding. Parts that would be distorted by milling threads can be satisfactorily ground. Parts that demand high accuracies and freedom from distortion and stress cracks are usually made by this method. The GW has annular thread grooves around its periphery that can be produced either by crushing or by diamond dressing. The accuracy of thread profile is very important. In the case of crushing, a roller of hardened steel having the required thread on it is fed under pressure into the wheel face, while a voluminous supply of lubricant is applied as the wheel slowly rotates. Two basic methods of centerless thread grinding that can be identified are as follows:

A. *Plunge grinding.* In this arrangement (Figure 5.28), the wheel is plunged into the WP to the full depth. The WP then makes one revolution, while the wheel traverses one pitch. This method gives a uniform wheel wear but is used for short thread lengths.

B. *Traverse grinding.* The wheel is positioned at a full thread depth; then, the work is traversed past the wheel. The first thread form on the wheel removes the majority of metal and therefore is subjected to the most wear; the following threads affect the finishing. A single-rib wheel may be used for large threads (Figure 5.29).

## 5.4   REVIEW QUESTIONS

5.4.1   Mark true (T) or false (F).
   [ ] Thread cutting on a lathe can be performed using multiple-point threading tools.
   [ ] Self-opening dies reduce tool wear as well as the threading time.
   [ ] An acme thread is preferred over a square thread.
   [ ] Thread rolling is not a machining process.
   [ ] Change gears are not necessary when cutting threads on a lathe machine.
   [ ] Free cutting materials produce threads that are accurate and of good surface finish.
   [ ] Tapping blind holes is done more easily and faster than through holes.
   [ ] Thread grinding is recommended for WPs of 37 HRC.
   [ ] Traverse centerless thread grinding is faster than plunge centerless thread grinding.
   [ ] Plunge centerless thread grinding leads to more uniform wear than traverse grinding.

5.4.2   What are the main types of screw threads?

5.4.3   List the different methods of thread production.

5.4.4   List the different methods of thread cutting and grinding.

5.4.5   Show in a sketch how a thread is cut on a center lathe.

5.4.6   Calculate the suitable gear train when cutting the following threads on the lathe machine:
   • 3 mm pitch on a 6 mm lead screw
   • 13 tpi on a 4 tpi lead screw
   • 6 threads in 12 mm on a 6 mm lead screw
   • 2.5 mm pitch on a 6 tpi lead screw
   • 10 tpi on a lathe having a 6 mm pitch lead screw

5.4.7   Compare thread chasing and thread cutting on a lathe.

5.4.8   Compare thread milling using disk and multiple-thread cutters.

5.4.9   State the main advantages of self-opening die heads over solid dies.

5.4.10  Show the arrangement of thread chasers in threading die heads.

5.4.11  Show the arrangements of single-rib and multi-rib traverse grinding of threads.

5.4.12  Differentiate between skip-rib and three-rib thread grinding.

5.4.13  Show using line sketches how each of the following operations is performed:
   • Milling external tapered thread
   • Plunge centerless thread grinding
   • Traverse centerless thread grinding

## REFERENCES

Acherkan, N 1968, *Machine tool design*, 4 vols, Mir Publishers, Moscow.
Alfred Herbert Ltd, Machine Tool Manufacturers, Coventry, UK.

Arshinov, V & Alekseev, G 1970, *Metal cutting theory and cutting tool design*, Mir Publishers, Moscow.

ASM International. *Metals handbook* 1989, Machining, vol. 16, ASM International, Material Park, OH.

Barbashov, F 1984, *Thread milling practice*, Mir Publishers, Moscow.

Chapman, WAJ 1981, *Elementary workshop calculations*, Edward Arnold, London.

Chernov, N 1984, *Machine tools*, Mir Publishers, Moscow.

Rodin, P 1968, *Design and production of metal cutting tools*, Mir Publishers, Moscow.

# 6 Gear-Cutting Machines and Operations

## 6.1 INTRODUCTION

Gears are machine elements that transmit power and rotary motion from one shaft to another. An advantage they have over friction and belt drives is that they are positive in their action, a feature that most of the machine tools require, as exact speed ratios are sometimes essential. Thread cutting and indexing movements in gear cutting are typical examples, which require synchronized rotary and linear movements without any slip. As drive elements, gears are specifically used to

- Change the speed of rotation
- Change the direction of rotation
- Increase or reduce the magnitude of speed and torque
- Convert rotational movement into linear, or vice versa (rack and pinion drive)
- Change angular orientation (bevel gears)
- Offset the location of rotating movement (helical gears and worm gear sets)

Depending on the specific application, gears can be selected from the following types:

*Spur gears.* These are the most common type, which transmit power or motion between parallel shafts or between a shaft and a rack. They are simple in design and measurement. If noise is not a serious problem, spur gears can be used. For aircraft gas turbines, spur gears of extra high quality can operate at pitch-line speed above 2000 m/min. In general applications, spur gears are not allowed to work at speeds over 1200 m/min.

*Helical gears.* These are used to transmit motion between parallel or crossed shafts, or between a shaft and a rack by meshing teeth that lie along a helix at an angle to the shaft. Because of this angle, teeth mating occurs in such a way that more than one tooth of each gear is always in mesh. This condition permits smoother action than with spur gears. However, some axial thrust is inevitable in helical gears, causing loss of power and requiring thrust bearings. External helical gears are generally used when both high speed and high power are involved.

*Herringbone gears.* These are sometimes called double helical gears. These gears transmit motion between parallel shafts. They combine the principal advantages of spur and helical gears, because two or more teeth share the load at the same time. Because they have equal right-hand and left-hand helixes, axial thrust is eliminated. Herringbone gears can be operated at higher velocities than spur gears.

*Worm gear sets.* These are used where the ratio of the speed of the driving member (worm) to the speed of the driven member (worm wheel) is large and for a compact right-angle drive. They are frequently used in indexing heads of milling machines and in hobbing machines.

*Crossed-axes helical gears.* These operate with shafts that are nonparallel and nonintersecting (Figure 6.1a). The action between mating teeth has a wedging effect, which results in sliding on tooth flanks. Therefore, these gears have low load-carrying capacity but are useful where shafts must rotate at an angle to each other.

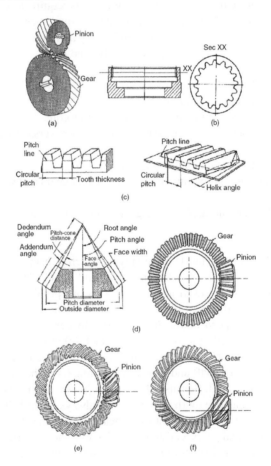

**FIGURE 6.1** Common gear types: (a) crossed-axes helical gears, (b) spur internal gears, (c) spur and helical racks, (d) straight bevel gear terminology and a pair in mesh, (e) spiral bevel gears in mesh, and (f) hypoid bevel gears in mesh. (Adapted from ASM International, *Metals Handbook*, Machining, ASM International, Material Park, OH, 1989.)

*Internal gears* (Figure 6.1b). They may be of spur or helical tooth form. Their main applications are as follows:

- Rear drives for heavy vehicles
- Planetary gears
- Toothed clutches
- Speed-reducing devices
- Compact design requirements

*Racks* (Figure 6.1c). A rack is a gear of infinite-pitch circular radius. The teeth may be at right angles to the edge of the rack and mesh with a spur gear, or at some other angle and engage a helical gear.

*Bevel gears.* These gears transmit rotary motion between two nonparallel shafts. Bevel gears are of the following types:

- Straight bevel gears (Figure 6.1d). The figure indicates the terminology of the bevel gear. It also shows a pair of bevel gears in mesh. The use of straight bevel gears is generally limited to low-speed drives and instances where noise is not important. These gears operate at high efficiencies of 98% or better and are used for nonparallel but intersecting shafts.
- Nonstraight bevel gears, which include spiral (Figure 6.1e), zerol, and hypoid gears. All these types are characterized by gradual and continuous engagement resulting in smooth running. Hypoid gears (Figure 6.1f) do not have as good efficiency as straight bevel gears but can transmit more power in the same space, provided that the speeds are not too high. They are used for nonparallel, nonintersecting shafts.

## 6.2  FORMING AND GENERATING METHODS IN GEAR CUTTING

Gears can be commercially produced by other methods such as sand casting, die casting, stamping, extrusion, and powder metallurgy. All these processes are used for gears of low wear resistance, low power transmission, and relatively low accuracy of transmitted motion. When the application involves higher values for one or more of these characteristics, cut or machined gears are used.

Gear cutting is a highly complex and specialized art, which is why most of the gear-cutting methods are single-purpose machines. Some of them are designed such that only a particular type of gear can be cut. Gear production by cutting involves two principal methods—forming and generating processes. Gear finishing involves four operations—shaving, grinding, lapping, and burnishing (Figure 6.2).

### 6.2.1  GEAR CUTTING BY FORMING

The tooth profile is obtained by using a form-cutting tool. This may be a multiple-toothed cutter used in milling, broaching machines, and shaping cutter heads, or a single-point tool form for use in a shaper and a bevel gear planer.

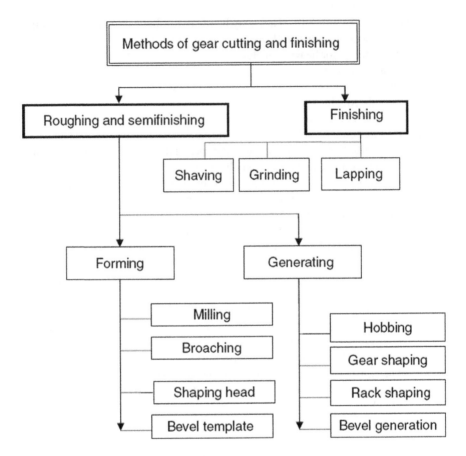

**FIGURE 6.2**   Methods of gear cutting and finishing.

### 6.2.1.1   Gear Milling

The usual practice in gear milling is to mill one tooth space at a time, after which the blank is indexed to the next cutting position. Figure 6.3 shows teeth in a spur gear cut by peripheral (horizontal) milling with a disk cutter. Similarly, end milling can also be used for cutting teeth in spur or helical gears and is often used for cutting coarse-pitch teeth in herringbone gears (Figure 6.4).

In practice, gear milling is usually confined to:

- One-of-a-kind replacement gears
- Small-lot production
- Roughing and finishing of coarse-pitch gears
- Finish milling of gears with special tooth forms

Although high-quality gears can be produced by milling, the accuracy of tool spacing on older milling machines was limited by the inherent accuracy of the indexing heads. Most indexing techniques used on modern machines incorporate numerical

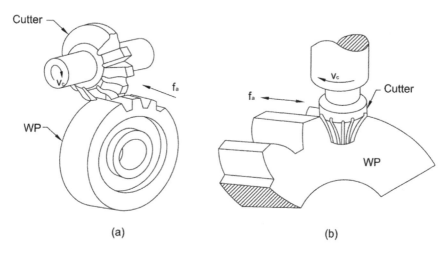

**FIGURE 6.3** Spur gear cutting on milling machines: (a) gear cutting on a horizontal milling and (b) gear cutting by end mill.

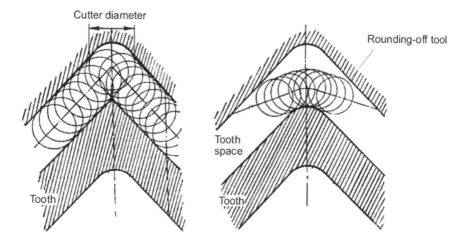

**FIGURE 6.4** Herring gear cutting and rounding off the vertex.

control or computer numerical control, and the accuracy can rival that of hobbing machines (*Metals Handbook*, 1989).

Moreover, as the tooth profile depends upon the module, pressure angle, and number of teeth, it is theoretically necessary to have a tool with a certain profile for each gear with a different number of teeth or module (module $m_g$ = pitch diameter/number of teeth). In actual practice, however, sets of gear tooth milling cutters, according to ASA B.9-1959, are used (eight cutters per set, or for more accurate gears 15, and less frequently 26 cutters for each module of gear. Each cutter in the set is designed for cutting a limited range of numbers of teeth (Table 6.1).

### TABLE 6.1
### Gear Cutter Sets for Milling

**8-Cutter Set for Spur Gears**

| Cutter Number | 1 | 2 | 3 | 4 | 5 | 6 | 7 | 8 |
|---|---|---|---|---|---|---|---|---|
| Number of teeth | 135-rack | $\dfrac{55}{134}$ | $\dfrac{35}{54}$ | $\dfrac{26}{34}$ | $\dfrac{21}{25}$ | $\dfrac{17}{20}$ | $\dfrac{14}{16}$ | $\dfrac{12}{13}$ |

**15-Cutter Set for Accurate Spur Gears**

| Cutter Number | 1 | $1\frac{1}{2}$ | 2 | $2\frac{1}{2}$ | 3 | $3\frac{1}{2}$ | 4 | $4\frac{1}{2}$ |
|---|---|---|---|---|---|---|---|---|
| Number of teeth | 135-rack | $\dfrac{80}{134}$ | $\dfrac{55}{79}$ | $\dfrac{42}{54}$ | $\dfrac{35}{41}$ | $\dfrac{30}{34}$ | $\dfrac{26}{29}$ | $\dfrac{23}{25}$ |

| Cutter Number | 5 | $5\frac{1}{2}$ | 6 | $6\frac{1}{2}$ | 7 | $7\frac{1}{2}$ | 8 | |
|---|---|---|---|---|---|---|---|---|
| Number of teeth | $\dfrac{21}{22}$ | $\dfrac{19}{20}$ | $\dfrac{17}{18}$ | $\dfrac{15}{16}$ | 14 | 13 | 12 | |

**According to ASA B.9-1959**

*Cutters for helical gears.* When cutting helical gears, the size of the cutter (cutter number), as obtained from Table 6.1, has to be modified due to the helix angle $\beta_g$.

$$\text{Equivalent teeth number } Z' = \frac{\text{number of teeth helical gear}}{\left(\cos\beta_g\right)^3}$$

Gear forming on milling machines (and shapers) has the following characteristics:
Advantages:

- General-purpose equipment and machines are used.
- Comparatively simple setup is needed.
- Simple and cheap cutting tools are used.
- It is suitable for piece and small-size production.

Drawbacks:

- It is an inaccurate process due to profile deviations and indexing errors.
- Production capacity is low due to the idle time lost in indexing, approaching, and withdrawal of the tool. However, productivity can be enhanced by multi-workpiece (WP) setup.

### Illustrative Example 1

The following helical gears are to be produced on a milling machine. Determine the cutter number for each case using both sets listed earlier:

1. Helical gear of helix angle $\beta_g = 10°$ and $Z = 22$ teeth
2. Helical gear of helix angle $\beta_g = 20°$ and $Z = 22$ teeth

## SOLUTION

1. $\beta_g = 10°$ and $Z = 22$ teeth
   8-cutter set

$$Z' = \frac{22}{(\cos 10)^3} = 23 \text{teeth}$$

Select cutter number 5.
15-cutter set

$$Z' = \frac{22}{(\cos 10)^3} = 23 \text{teeth}$$

Select cutter number $4\frac{1}{2}$.

2. $\beta_g = 20°$ and $Z = 22$ teeth
   8-cutter set
   Equivalent number of teeth $Z'$

$$Z' = \frac{22}{(\cos 20)^3} = 26.5 \text{ teeth}$$

Select cutter number 4.
15-cutter set

$$Z' = \frac{22}{(\cos 20)^3} = 26.5 \text{ teeth}$$

Select cutter number 4.

## Illustrative Example 2

Calculate the machining particulars for milling the helical gear.
$Z = 60 \text{teeth} \quad \beta_g = 45° \quad m_g = 5 \text{mm}$

The lead screw pitch of the milling machine table is 6 mm, and the indexing head is equipped with change gears of 24, 24, 28, 32, 40, 44, 48, 56, 64, 72, 86, and 100 teeth.

## SOLUTION

$$\tan \beta_g = \frac{\pi \cdot d_P}{L} \left( L = \text{lead of gear helix} \right)$$

Normal module:

$$m_n = m_g \cdot \cos \beta_g = 5 \times 0.707 = 3.5355 \text{ mm}$$

Then

---

Addendum:             $h_a = 3.5355$ mm

Dedendum:             $h_d = 1.25 \times m_n = 1.25 \times 3.5355 = 4.4193$ mm

Tooth height:         $h_t = 2.25 \times m_n = 2.25 \times 3.5355 = 7.9548$

Tooth thickness:      $s = 1.5708 \times m_n$

$= 1.5708 \times 3.5355 = 5.5528$ mm

Fillet radius:        $r = .4 m_n$

$= 0.4 \times 3.5355 = 1.4142$

Pitch diameter:       $d_p = \dfrac{Z \cdot m_n}{\cos \beta_g} = Z \cdot m_g$

$= 60 \times 5 = 300$ mm

Outside diameter:     $d_a = Z \cdot m_g + 2 m_n$

$= 60 \times 5 + 2 \times 3.5355 = 307.071$ mm

---

Indexing operation:

$$\text{Index crank movement} = \frac{40}{z} = \frac{40}{60} = \frac{2}{3}$$

The index crank is moved 2/3 rev every tooth from the 60 teeth.

Cutter selection:

Using the 15-cutter accurate set, the equivalent cutter, $Z'$, is selected by

$$Z' = \frac{Z}{\left(\cos \beta_g\right)^3} = \frac{60}{\left(\cos 45\right)^3} = 169.7 \approx 170 \text{teeth}$$

The cutter number 1 (135 – rack) is selected.

Table tilting:

The table is tilted by 45°.

Lead of the machine $L_m$ = pitch of machine lead screw × 40 = 6 × 40 = 240 mm.

Lead of gear $L_w$ is calculated from the following relation:

$$L_w = \frac{\pi d_p}{\tan \beta_g} = \frac{\pi \cdot Z \cdot m_g}{\tan \beta_g} = \frac{\pi \times 60 \times 5}{\tan 45} = 942.5 \text{ mm}$$

Therefore,

$$\frac{\text{Driver}}{\text{Driver}} = \frac{L_m}{L_w} = \frac{240}{942.5} \approx \frac{240}{960} = \frac{1}{4} = \frac{24}{48} \times \frac{32}{64}$$

Taking 960 instead of 942.5 will not change the helix angle much but enables standard change gears to be used.

Error in lead:

Lead produced = 960 mm

Lead desired = 942.5 mm

Error in lead = 960 – 942.5 = 17.5 mm

If this error is not acceptable, use other change gears.

### 6.2.1.2  Gear Broaching

Gear broaching is usually confined to cutting teeth in internal gears. However, not only internal but also external, spur, or helical gears can be broached. Figure 6.5a shows progressive broach steps in cutting an internal spur gear. The form of the space between gear teeth corresponds to the form of the broach teeth. The diameter of the broach increases progressively to the major diameter, which completes the tooth form on the WP. Figure 6.5b shows how an external spur gear is produced using a rotating broach. In such arrangements, the blank is withdrawn for indexing to cut another space between two teeth. Broaching is fast and accurate, and provides excellent surface quality. However, the cost of tooling is high; therefore, gear broaching is best suited to large production runs.

### 6.2.1.3  Gear Forming by a Multiple-Tool Shaping Head

This is a highly productive and accurate method of producing teeth in external and internal spur gears. This method is not applicable to helical gears. As in broaching of

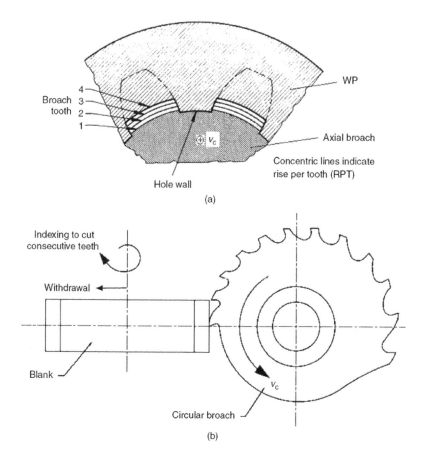

**FIGURE 6.5**  Gear broaching by forming: (a) broaching of an internal spur gear using an axial broach and (b) broaching of an internal spur gear using a rotating broach.

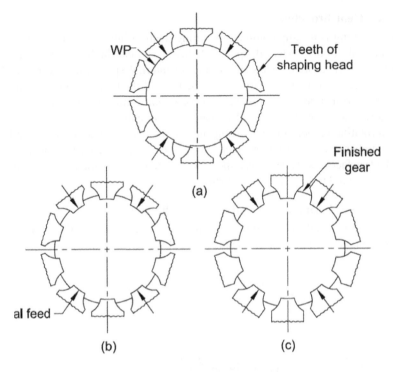

**FIGURE 6.6** Cutting with progressive gear shaping head: (a) starting of cut, (b) intermediate position, and (c) finished gear.

internal gears, all tool spaces are cut simultaneously and progressively (Figure 6.6). The cutter head has as many radially arranged form tools as the number of teeth on the gear being cut. The profile of the tool teeth has exactly the same shape as the gear tooth spaces. Prior to each cutting stroke, each tool is fed radially toward the blank by an amount equal to the prescribed infeed. All the tools are simultaneously retracted from the work on the return stroke to avoid rubbing the tool against the machined surfaces. The gear is finished when the tools reach the full depth of cut. Cutting speeds in this process are similar to those used for broaching the same work metal using the same tool material. Machines with shaping heads are available for cutting spur gears up to 500 mm in diameter with a face width up to 150 mm. For example, a machining time of not more than 1 min is required to produce a spur gear of 160 mm pitch diameter, face width of 30 mm, and a module of 4 mm; therefore, the process is best suited to large production runs. The drawbacks of the process are the comparatively complex shaping heads and the necessity of having a separate head for each gear size and module.

### 6.2.1.4 Straight Bevel Gear Forming Methods

Two methods available are as follows:

    a. *Straight bevel gear forming by milling.* This method is not widely used for two reasons:

- It is of very limited accuracy.
- The operation is time-consuming.

Sometimes, straight bevel gears are rough cut by milling and then finished by another method.

b. *Template machining.* This is a low-productive method used to cut large bevel gears of coarse pitch using a bevel gear planer (Figure 6.7); because the setup can be made with a minimum effort, template machining is useful when a wide variety of coarse pitch gears are required. Under these conditions, a high level of accuracy is possible. The setup utilizes two templates, one for each side of the gear tooth. Theoretically, a pair of templates would be required for each gear ratio, but in practice, a pair is designed for a small range of ratios. A set of 25 pairs of templates encompasses all 90° shaft angle ratios from 1:1 to 8:1 for either $14\frac{1}{2}$ or 20° pressure angles (*Metals Handbook*, 1989). After the roughing operations are performed by simple slotting tools, the templates are set up, and the teeth are finished by making two cuts on each side.

### 6.2.2 Gear Cutting by Generation

This technique is based on the fact that two involute gears of the same module and pitch mesh together—the WP blank and the cutter. So, this method makes it possible to use one cutting gear for machining gears of the same module with a varying number of teeth. Gear-generation methods are characterized by their higher accuracy and machining productivity than gear forming. They comprise hobbing and gear or rack shaping for the manufacture of spur and helical gears, worm and worm wheels, and bevel gear generation.

### 6.2.2.1 Gear Hobbing

Hobbing is a gear-generation method most widely used for cutting teeth in spur gears, helical gears, worms, worm wheels, and many special forms (Figure 6.8). Hobbing

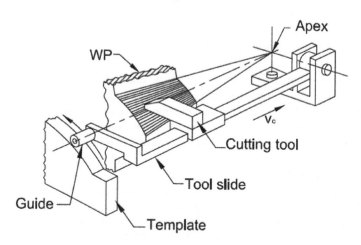

**FIGURE 6.7** Template machining using a bevel gear planer.

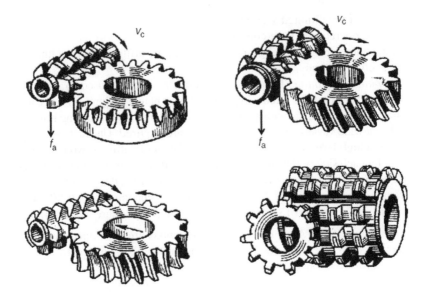

**FIGURE 6.8** Various products that can be hobbed.

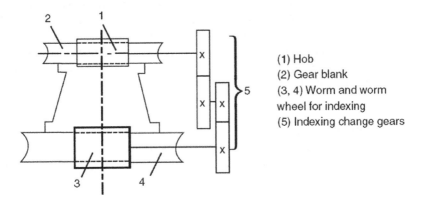

(1) Hob
(2) Gear blank
(3, 4) Worm and worm wheel for indexing
(5) Indexing change gears

**FIGURE 6.9** Elementary hobbing machine setup.

machines are not applicable to cutting bevel and internal gears. The tooling cost for hobbing is lower than for broaching and multiple-tool shaping heads. For this reason, hobbing is used in low-quantity production or even for a few pieces. Compared with milling, hobbing is fast, accurate, and therefore suitable for medium- and high-quantity productions. The hob is a fluted worm of helix angle $\alpha$ with form-relieved teeth that cut into the gear blank in succession. A simplified gear train of a hobbing machine is shown in Figure 6.9.

The use of hobbing is sometimes limited by the shape of the WP; for example, if the teeth to be cut are close to a shoulder or a flange, the axial distance is not large enough to allow the hob to over-travel at the end of the cut. This over-travel should be about one-half of the hob diameter plus an additional clearance to allow for the hob thread angle.

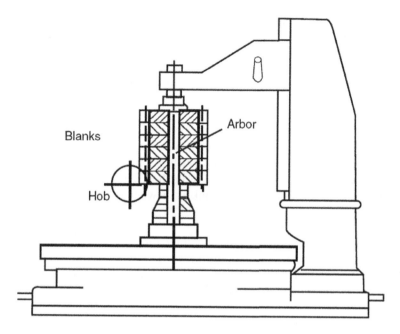

**FIGURE 6.10**   Clamping identical gears in one setup.

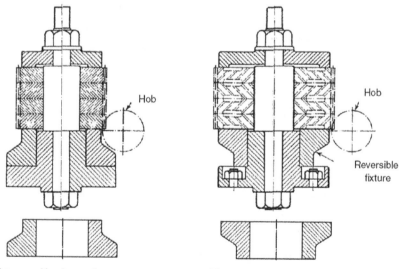

Fixture position for small gears          Fixture position for large gears

**FIGURE 6.11**   Interchangeable hobbing fixture to various size gears.

The ability to cut teeth in two or more identical gears in one setup can encourage the use of this method (Figure 6.10). Inexpensive fixturing is often utilized for cutting two or more gears at one time when the ratio of the face width to pitch diameter is small. A typical hobbing fixture is illustrated in Figure 6.11, which shows a common mandrel-type fixture for flat-face gears. Incorporated in the fixture is an interchangeable bottom plate to enable utilization of the same fixture for various sizes of gears. The clamp plate should be as large as possible and relieved to concentrate the clamping action near the outer edge of the blank. Figure 6.12 illustrates the cutting action used for different types of gears. The rotary motions imparted to the blank and hob are the same as those of worm wheel and worm gearing.

### 6.2.2.1.1 Hobbing of Spur Gears

The hob is set up so that the thread of the hob on the side facing the gear blank is directed vertically along the axis. This is done by setting the hob axis at an angle $\alpha_h$ to the horizontal equal to the helix angle of the hob. The hob attains a continuous feed motion along the axis of the gear blank as shown in Figure 6.12a.

### 6.2.2.1.2 Hobbing of Helical Gears

To cut helical gears, the hob is set up so that the thread of the hob facing the gear blank is directed at the helix angle of the teeth. This is done by setting the hob at an angle $\gamma = \beta_g \pm \alpha_h$, where $\beta_g$ is the helix angle of the helical gear being cut and $\alpha_h$ is the helix angle of the hob. If the hand of helical gear and that of the hob are different,

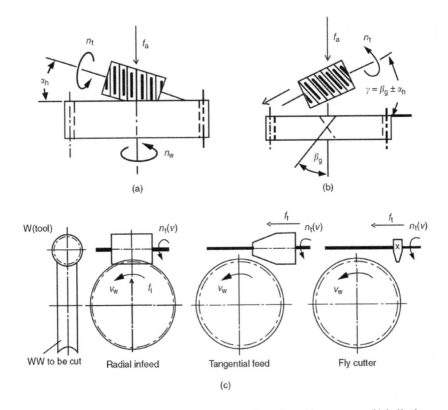

**FIGURE 6.12** Cutting action for different types of gearing: (a) spur gear, (b) helical gear, and (c) worm wheel.

the positive sign is considered; if the hand is the same, the negative sign should be used. Also, the hob attains a continuous feed motion along the axis of the gear blank (Figure 6.12b). In cutting helical gears, an incremental motion is imparted to the blank, with an angular velocity that would provide one full additional revolution of the blank during vertical feed of the hob through a distance equal to the lead of the helical teeth on the gear.

### 6.2.2.1.3  Hobbing of Worm Wheels

When cutting worm wheels, the axis of the hob is set perpendicular to the axis of rotation of the blank. The following principal motions are shown in Figure 6.12c:

1. Principal rotary cutting motion $v$ of the hob.
2. Continuous indexing rotary motion $v_w$ of the gear bank.
3. Feed motion of the hob, which may be either of the following:
   - Worm wheel hobbing through radial infeed $f_i$. The radial infeed ceases when the full depth of cut is reached.
   - Worm wheel hobbing through tangential feed $f_t$. The hob is set at the beginning to the full depth of cut and is fed tangentially into the blank.

The radial infeed method has a higher production capacity; however, a small part of the hob in the mid length is actually doing the cutting. As a result, the hob wears nonuniformly, which has an unfavorable effect on the tooth profile accuracy. If high gear accuracy is required, the tangential feed method is used. In this case, cylindrical hobs with a tapered start are used to perform the main cutting action by the tapered part and the sizing action by the cylindrical part (Figure 6.13). Fly cutters with tangential feed are used in piece production because they are considerably cheaper (Figure 6.12c).

### 6.2.2.1.4  Hobbing of Worms

Hobbing produces the highest-grade worm at the lowest machining cost, but it can only be used when production quantities are large enough to justify the high tooling cost. The number of flutes in a worm hob is increased to improve the surface finish.

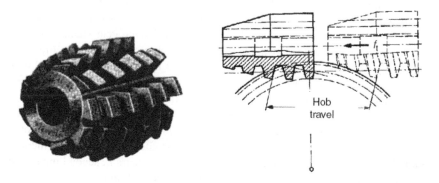

**FIGURE 6.13**  Cylindrical hob with tapered start for higher gear accuracy, as used in tangential feeding.

Gear hobbing is characterized by the following:

1. High accuracy.
2. Flexibility for any production volume.
3. Low cost.
4. Adaptability to cut metals with higher than average hardness.
5. Any external tooth that is uniformly spaced about the center can be hobbed using a suitable hob.
6. One hob of a particular module can be used to cut teeth of all involute spur and helical gears of any number of teeth of the same module and pressure angle. It is thus a versatile process.
7. The accuracy of hobbed gears depends upon:
   - Accuracy of the machine, blank, and tool
   - Care and accuracy of mounting work and hob
   - Feed method used
   - Machine rigidity
      A typical hobbing machine can produce gears of accumulated errors of tooth spacing not more than 20 μm.
8. The indexing is continuous, without an intermittent nature that can cause indexing errors.
9. Finish is dependent on the hob feed.
10. Hobbing cannot be used to cut the following:
   - Bevel gears
   - Internal gears
   - Gears having adjacent shoulders larger than the root diameter of gear and that are close enough to restrict the approach or run out of the hob

### 6.2.2.1.5  Kinematic Diagram and Gear Trains of Hobbing Machines

The hobbing machine is considered a model of versatile nature, capable of producing a wide spectrum of gear shapes. So, it is intentionally selected for investigation of its kinematic structure and gearing diagram. The whole kinematic structure of a hobbing machine is never simultaneously employed. The machine embraces different gear trains and change gears, which operate in accordance with the specific shapes of gears to be hobbed. The hobbing machine is ordinarily furnished with a complete selection of change gears to provide flexibility in producing gear shapes. A representative kinematic structure of a gear hobbing machine is given in Figure 6.14a. Accordingly, the same structure is also represented by the chain shown in Figure 6.14b. Accordingly, the motion is transmitted from the drive motor ($M_1$) through point (3) and speed change gears $i_v$. At point (4), the motion is branched into $1-R_1$ (hob rotation), a first input shaft (5) to differential $\Sigma_1$, point (6), indexing change gear $i_x$, point (2), point (9), and feed change gears $i_f$ to point (10). At point (10), the motion is branched to point (8), vertical feed screw $i_1$, a hob slide that imparts elementary motion $T_3$, and likewise branched to differential change gears $i_y$, second input shaft (7) of the differential $\Sigma_1$, output shaft (6) of the differential, indexing change gears $i_x$ and point (2), point (9), to the worktable to which the second elementary motion $R_4$ is imparted.

The representative structure shown in Figure 6.14 is applied in the most widely used gear hobbing machines produced in Europe and the United States. Figures 6.15a and 6.15b show in detail the different gear trains associated with a general-purpose hobbing machine. In the same figure, the different change gears ($i_v$, $i_x$, $i_f$, $i_y$) are calculated based on a given example. The machine setting to hob a helical gear is considered to be the most complex type of gear machining. For calculating the cutting speed change gears (Figure 6.15 a), the following is considered:

Assumed:

Hob diameter, $d_{hob}$ = 100 mm (one start)

Cutting speed $v$ = 60 m/min (high-speed steel [HSS] hob, mild steel blank)

Therefore,

$$n_t = \frac{1000v}{\pi d_{hob}} = \frac{1000 \times 60}{\pi \times 100} = 194 \text{ rpm} \tag{6.1}$$

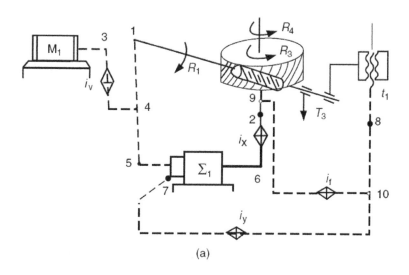

(a)

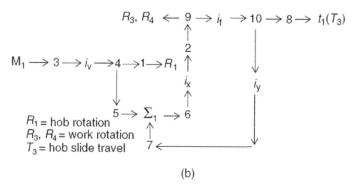

(b)

**FIGURE 6.14** Kinematic and chain structures of a gear hobbing machine: (a) Kinematic structure and (b) chain structure. (From Acherkan, N., *Machine Tool Design*, Mir Publishers, Moscow, 1968. With permission.)

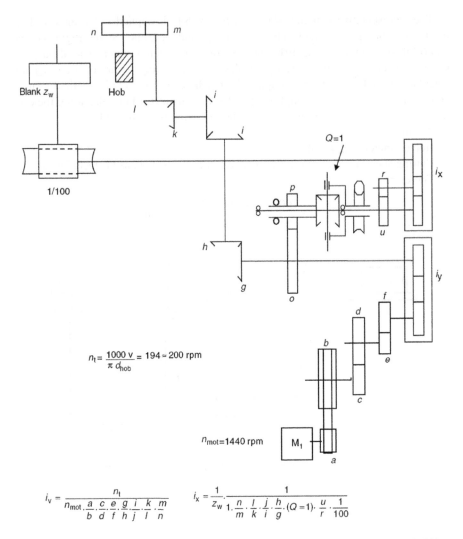

**FIGURE 6.15A** (a) speed and indexing gear trains $i_r$ and $i_x$ of general-purpose hobbing machine.

Assume $\underline{n}_t = 200$ rpm

Transmission ratio of cutting speed change gears is $i_v$

Consider the cutting speed gear train (Figure 6.15a):

$$n_{mot} \cdot \frac{a}{b} \cdot \frac{c}{d} \cdot i_v \cdot \frac{g}{h} \cdot \frac{i}{j} \cdot \frac{k}{l} \cdot \frac{m}{n} = n_t \qquad (6.2)$$

Then

$$i_v = \frac{n_t}{n_{mot} \cdot \dfrac{a}{b} \cdot \dfrac{c}{d} \cdot \dfrac{e}{f} \cdot \dfrac{g}{h} \cdot \dfrac{i}{j} \cdot \dfrac{k}{l} \cdot \dfrac{m}{n}} \qquad (6.3)$$

Transmission ratio of indexing change gears $i_x$ (1 rev of the blank gives $Z_w$ rev of hob, or $Z_w/k$ rev of hob, where $k$ = number of starts of the hob):

Therefore, 1 hob rev $= \dfrac{1}{Z_w}$ rev of blank.

$$1 \cdot \frac{n}{m} \cdot \frac{l}{k} \cdot \frac{j}{i} \cdot \frac{h}{g} \cdot (Q=1) \cdot \frac{u}{y} \cdot i_x \cdot \frac{1}{100} = \frac{1}{Z_w} \tag{6.4}$$

Then

$$i_x = \frac{1}{Z_w} \cdot \frac{1}{\dfrac{n}{m} \cdot \dfrac{l}{k} \cdot \dfrac{j}{i} \cdot \dfrac{h}{g} \cdot (Q=1) \cdot \dfrac{u}{r} \cdot i_x \cdot \dfrac{1}{100}} \tag{6.5}$$

Transmission ratio of feed change gears $i_f$:

Consider the feed gear train in Figure 6.15b

$$1 \cdot \frac{100}{1} \cdot \frac{2}{24} \cdot i_f \cdot \frac{s}{t} \cdot \frac{u}{v} \cdot \frac{w}{x} \cdot \frac{4}{20} \cdot \frac{5}{30} \cdot 10 = f \text{ mm/rev} \tag{6.6}$$

where $f$ is the axial or radial feed mm/rev of the WP.

Then

$$i_f = \frac{f}{1 \cdot \dfrac{100}{1} \cdot \dfrac{2}{24} \cdot \dfrac{s}{t} \cdot \dfrac{u}{v} \cdot \dfrac{w}{x} \cdot \dfrac{4}{20} \cdot \dfrac{5}{30} \cdot 10} \tag{6.7}$$

Transmission of differential change gears $i_y$:

Gear helix angle $= \beta_g$

$$\tan \beta_g = \frac{\pi m Z_w}{L} \quad L = \frac{\pi m Z_w}{\tan \beta_g}$$

Consider the differential gear train (one rev of blank provides $L$ [mm] longitudinal travel of hob):

$$1 \cdot \frac{100}{1} \cdot \frac{1}{i_x} \cdot \frac{r}{u} \cdot \frac{1}{2} \cdot \frac{30}{1} \cdot \frac{1}{i_y} \cdot \frac{s}{t} \cdot \frac{u}{v} \cdot \frac{w}{x} \cdot \frac{4}{20} \cdot \frac{5}{30} \cdot 10 = L \tag{6.8}$$

$C_1$ engaged, then $Q = \_1/2$

Then

$$i_y = \frac{\dfrac{100}{1} \cdot \dfrac{1}{i_x} \cdot \dfrac{r}{u} \cdot \dfrac{1}{2} \cdot \dfrac{30}{1} \cdot \dfrac{s}{t} \cdot \dfrac{u}{v} \cdot \dfrac{w}{x} \cdot \dfrac{4}{20} \cdot \dfrac{5}{30} \cdot 10}{\pi \cdot m \cdot Z_w / \tan \beta_g} \tag{6.9}$$

### 6.2.2.2  Gear Shaping with Pinion Cutter

This process is the most versatile of all gear-cutting processes. Although shaping is most commonly used for cutting teeth in spur and helical gears, this process is also applicable to cutting herringbone teeth, internal gears (or splines), chain sprockets, elliptical gears, face gears, worm gears, and racks. Shaping cannot be used to cut bevel gears.

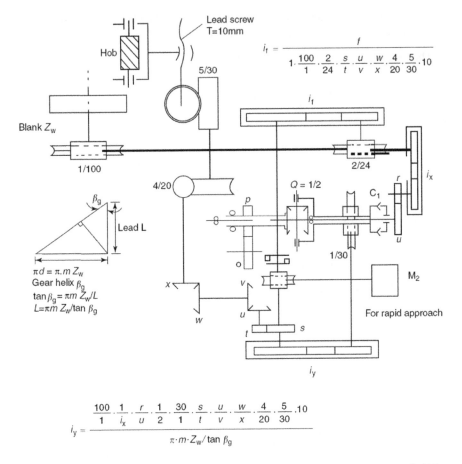

$$i_1 = \cfrac{f}{1 \cdot \cfrac{100}{1} \cdot \cfrac{2}{24} \cdot \cfrac{s}{t} \cdot \cfrac{u}{v} \cdot \cfrac{w}{x} \cdot \cfrac{4}{20} \cdot \cfrac{5}{30} \cdot 10}$$

$$i_y = \cfrac{\cfrac{100}{1} \cdot \cfrac{1}{i_x} \cdot \cfrac{r}{u} \cdot \cfrac{1}{2} \cdot \cfrac{30}{1} \cdot \cfrac{s}{t} \cdot \cfrac{u}{v} \cdot \cfrac{w}{x} \cdot \cfrac{4}{20} \cdot \cfrac{5}{30} \cdot 10}{\pi \cdot m \cdot Z_w / \tan \beta_g}$$

**FIGURE 6.15B** (b) feed and differential gear trains $i_f$ and $i_y$ of general-purpose hobbing machine.

Figure 6.16 shows the principle of gear shaping with a pinion cutter. In this process, the cutter is mounted on a spindle that reciprocates axially as it rotates. The WP spindle is synchronized with the cutter spindle and rotates slowly as the tool meshes and cuts while it is being fed into the work at the end of each return (upward) stroke. The downward movement of the tool represents the principal cutting motion. To prevent the flanks of the cutter teeth from scoring the blank as the cutter is returned upward, the blank (or the cutter) is withdrawn radially in the direction of arrow X.

Because tooling cost is relatively low, gear shaping is practical for any production volume. WP design often prevents the use of milling cutters or hobs (e.g., cluster gears), and shaping is the most practical method for such cases (Figure 6.17a). Shaping can also be applied in cutting a worm (Figure 6.17b) where the cutter involves no axial stroke. Figure 6.18 shows a simplified kinematic diagram and mechanical drives of a gear shaper. Table 6.2 illustrates some typical products produced on the Liebherr gear shaper WS1. The examples quoted are typical of the requirement of mass production. The table shows the product specifications, tooling, machining data, and the machining time, which ranges from 0.3 to 1.2 min.

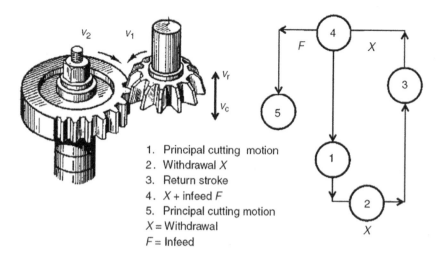

1. Principal cutting motion
2. Withdrawal $X$
3. Return stroke
4. $X$ + infeed $F$
5. Principal cutting motion
$X$ = Withdrawal
$F$ = Infeed

**FIGURE 6.16**   Principles of gear shaping.

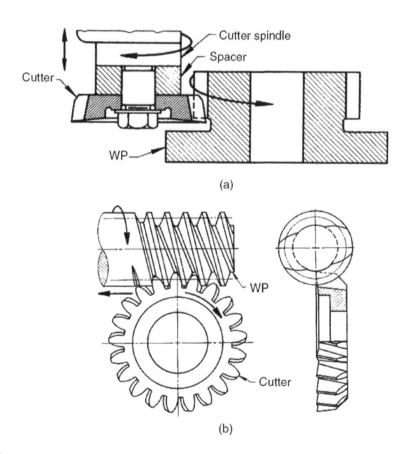

(a)

(b)

**FIGURE 6.17**   Shaping of (a) cluster gears and (b) a worm using gear shapers.

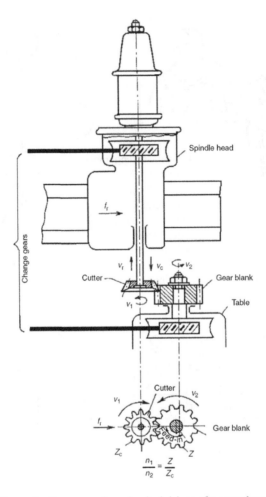

**FIGURE 6.18** Kinematic diagram and mechanical drives of a gear shaper.

Characteristics of gear shapers are as follows:

- They produce accurate gears.
- Both internal and external gears can be cut by this method.
- The production rate of gear shapers is lower than that of hobbers.
- Bevel and worm gears cannot be generated on gear shapers.

### 6.2.2.3   Gear Shaping with Rack Cutter

Gear shaping is performed by a rack cutter with three to six straight teeth (Figure 6.19). The cutters reciprocate parallel to the work axis when cutting spur gears and parallel to the helix angle when cutting helical gears. In addition to the reciprocating action of the cutter, there is synchronized rotation of the gear blank with each stroke of the cutter, with a corresponding advance of the cutter in a feed

**TABLE 6.2**

**Typical Products Machined on the Liebherr Gear Shaper Using HSS Cutters**

| Part Name | Product Specification | Machining Data | Cutter Teeth |
|---|---|---|---|
| Automotive gear 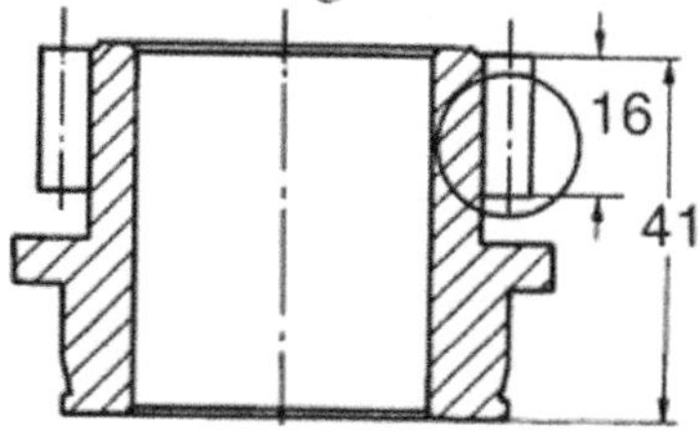 | Material: SAE3120<br>$Z = 17$ teeth<br>Module $m_g = 2.5$ mm<br>Helix $\beta_g = 31°$ | $n = 70$ stroke/min<br>$v = 52$ m/min<br>Number of cuts = 2<br>Rotary feed = 0.64 mm/stroke<br>$t_m = 1.2$ min | 46 |
| Lay shaft 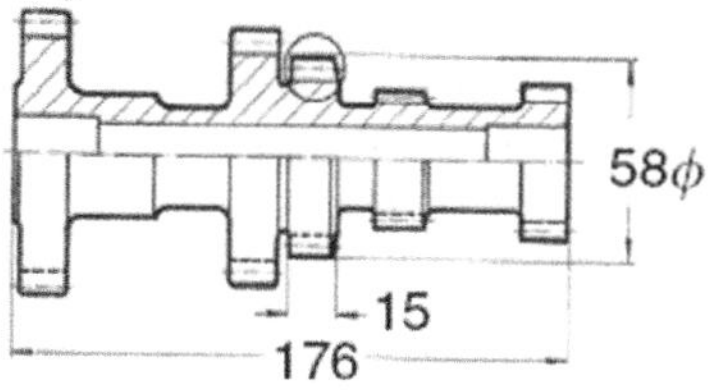 | Material: heat-treated steel<br>$Z = 16$ teeth<br>Module $m_g = 2.85$ mm<br>Pressure angle = 20°<br>Helix $\beta_g = 28°$ | $n = 900$ stroke/min<br>$v = 64$ m/min<br>Number of cuts = 3<br>Rotary feed = 0.65 mm/stroke<br>$t_m = 1.1$ min | 44 |
| Cluster gear 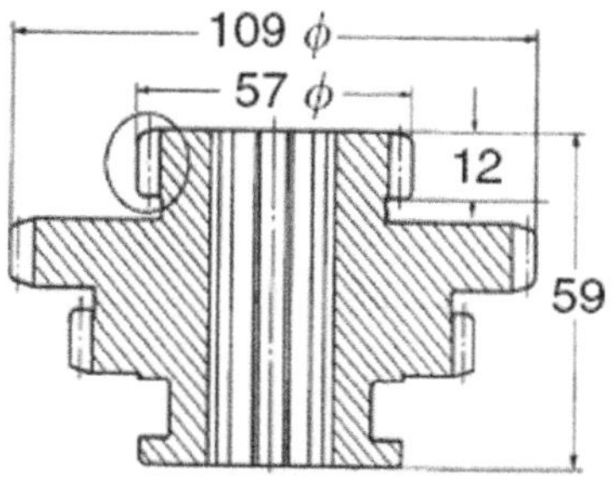 | Material: EC80<br>$Z = 26$ teeth<br>Module $m_g = 2$ mm<br>Pressure angle = 20° | $n = 1000$ stroke/min<br>$v = 56$ m/min<br>Number of cuts = 2<br>Rotary feed = 0.58 mm/stroke<br>$t_m = 1.05$ min | 38 |
| Starter pinion 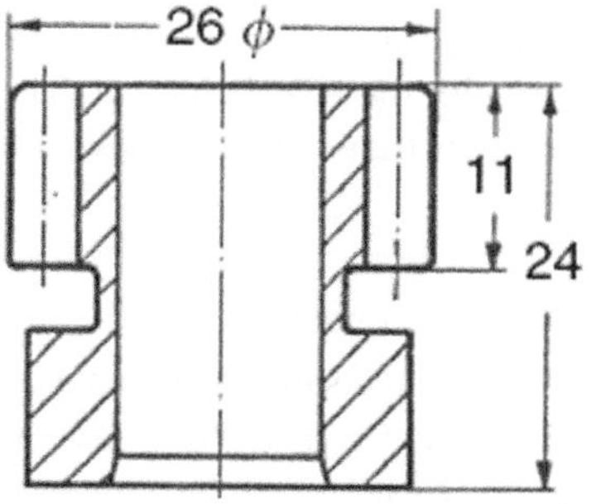 | Material: carbon steel<br>$Z = 9$ teeth<br>Module $m_g = 2.1$ mm<br>Pressure angle = 12° | $n = 1000$ stroke/min<br>$v = 50$ m/min<br>Number of cuts = 1<br>Rotary feed = 0.54 mm/stroke<br>$t_m = 0.3$ min | 64 |

*(Continued)*

**TABLE 6.2 (CONTINUED)**

**Typical Products Machined on the Liebherr Gear Shaper Using HSS Cutters**

| Part Name | Product Specification | Machining Data | Cutter Teeth |
|---|---|---|---|
| Clutch teeth | Material: 15CrNi6<br>$Z = 27$ teeth<br>Module $m_g = 5$ mm<br>Pressure angle = 20° | $n = 2000$ stroke/min<br>$v = 79$ m/min<br>Number of cuts = 1<br>Rotary feed = 1 mm/stroke<br>$t_m = 0.3$ min | 26 |
| Internal gear | Material: Bakelite<br>$Z = 60$ teeth<br>Module $m_g = 1$ mm | $n = 1250$ stroke/min<br>$v = 63$ m/min<br>Number of cuts = 1<br>Rotary feed = 0.34 mm/stroke<br>$t_m = 0.64$ min | 25 |

From High Production Gear Shaping Machine, WS1 Kaufbeurer Str. 141, Liebherr Verzahntechnik GmbH. D8960 Kempten, Germany. With permission.

movement. Rack cutters are less expensive than pinion cutters and hobs. A rack cutter is especially adapted for cutting of large gears of modules, typically of 5–10 mm.

### 6.2.2.4 Cutting Straight Bevel Gears by Generation

The generation principle of bevel gear cutting is based on reproducing the sides of the teeth on an imaginary crown gear in space by means of the cutting edges of rotating interlocking cutters or reciprocating two-tool generators. The profiles of the straight cutting edges coincide with the opposing sides of two teeth of the imaginary crown or generating gear that is in mesh with the gear being cut. The primary cutting motion, either rotation or reciprocation, is transmitted to these cutting edges.

#### 6.2.2.4.1 Interlocking Cutters (Completing or Konvoid Generators)

In this method, two interlocking disk-type cutters rotate at the same speed on axes inclined to the face of the mounting cradle, and both cut in the same tooth space. The gear blank is held in a work spindle that rotates in timed relation with the cradle on which the cutters are mounted (Figure 6.20). A simplified kinematic diagram of a Konvoid-type bevel gear generator is shown in Figure 6.21. A feed cam cycle begins with the work-head and blank moving into position for rough or finish cuts to provide

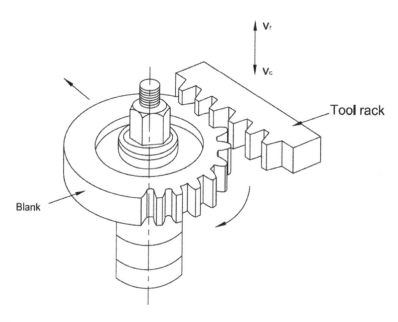

**FIGURE 6.19**   Principles of gear shaping using rack cutter.

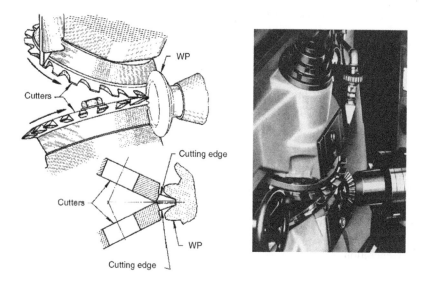

**FIGURE 6.20**   Bevel gear generating by interlocking cutters (Konvoid generators).

three different automatic programs (one plunge cutting program and two generating programs).

### 6.2.2.4.1.1   Plunge Cutting Program

This program is mainly used for roughing by machining of tooth space without generation (Figure 6.22a).

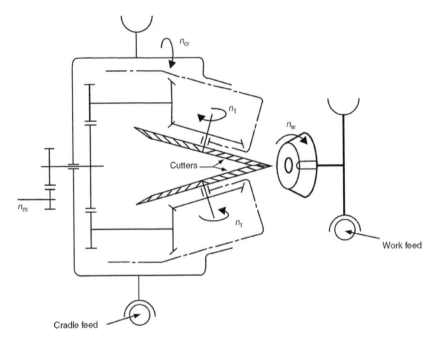

**FIGURE 6.21** Simplified kinematic diagram of Konvoid generators

The sequence of operation is as follows:

1. Plunge cutting
2. Withdrawal of WP for indexing
3. Return of WP to the clamping position after all tooth spaces have been milled

*6.2.2.4.1.2 Initial Generation Program* This program is used to cut spaces by generation (Figure 6.22b).
The sequence of operation is as follows:

1. Rolling return and indexing
2. Approaching the WP
3. Generation
4. Withdrawal
5. Return to the clamping position after all tooth spaces have been generated

*6.2.2.4.1.3 Infeed Generation* Program This program is used for finishing gears that have already been plunge-cut by the first program. It is similar to the second program, Figure 6.22c. The gear and pinion of the differential bevel gear, shown in Figure 6.23, have been produced on the bevel gear generating machine, model ZFTK 250x5, WMW. The corresponding machining data are listed in Table 6.3. If loading and unloading equipment is attached to the machine, it runs completely automatically, and the operation time to machine this differential set is reduced to 3.65 min.

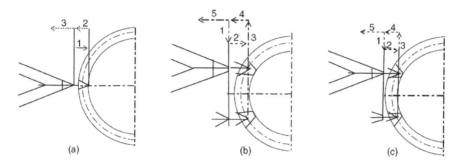

**FIGURE 6.22** Automatic programs of Konvoid generators: (a) plunge cutting program, (b) initial generation program, and (c) infeed generation program. (From WMW, Bevel Gear hobbing machine ZFTX 250x5, Technical Information, 108 Berlin, Mohrenstr, 61 WMW-Export.)

**FIGURE 6.23** Differential gear set, machined on bevel gear generating machine. (From WMW, Bevel Gear Hobbing Machine ZFTX 250x5, Technical Information, 108 Berlin, Mohrenstr, 61 WMW-Export. With permission.)

The tool life of the blades at the example machining conditions is about 13,000 teeth. The cutter blades can be resharpened 50 times, and with a single set of cutter blades, a total number of 650,000 teeth can be cut.

#### 6.2.2.4.2  Two-Tool Generators

These generators are also used to cut straight bevel gears but by means of two reciprocating tools that cut on opposite sides of a tooth (Figure 6.24). In a machine of this type, the gear blank (1) is rotated at $n_1$; also, the cradle (2) is rotated at $n_2$ with the reciprocating tools that represent kinematically the adjacent sides of a tooth in an imaginary crown gear. The slides (3) with tools (4) reciprocate at a speed $v_c$ along ways arranged on the face of the cradle (2) (Figures 6.24 and 6.25). The tools cut in their motion toward the gear apex. They do not cut on their return stroke, because they are withdrawn from the blank to avoid rubbing against the machined surfaces.

Figure 6.26 shows the successive positions of reciprocating tools and the gear blank during the generation process. For machining one of the side surfaces of the

**TABLE 6.3**

**Machining Data of the Differential Bevel Gear Set as Machined on the Bevel Gear Machine**

| Machining Data | Pinion | Gear |
|---|---|---|
| Material | 16 MnCr5 | |
| Cutting speed, $v$ (m/min) | 63 | |
| Module, $mg$ (mm) | 3.75 | |
| Number of teeth $Z$ | 9 | 22 |
| Machining time $t_m$ (min) | 1.5 | 3.45 |

From WMW, Bevel Gear Hobbing Machine ZFTX 250x5, Technical Information, 108 Berlin, Mohrenstr. 61 WMW-Export.

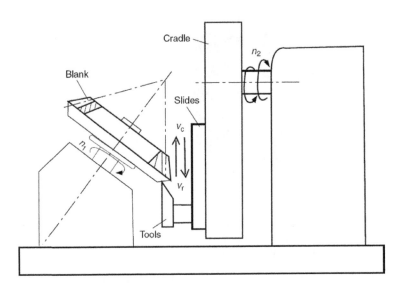

**FIGURE 6.24**   Operation of two-tool generators for the production of straight bevel gears.

tooth, the tool starts to cut into the blank (position a). Then, the second tool allocated to shape the other side of the tooth begins to cut (position b). At position c, both tools are in full engagement. Upon further rotation (roll) of the cradle, the tools run out of mesh with the gear blank (position d). At this stage, the first tooth has been generated. Then, the blank is automatically withdrawn from the engagement with the tools at the conclusion of each generating roll. The cradle work spindle rolls back to the starting position, where the blank is indexed to the next tooth. This procedure is repeated until all teeth are finished.

The two tools are not subjected to the same load, as one of them cuts into the blank for each tooth and wears faster than the other tool. To eliminate the effect of

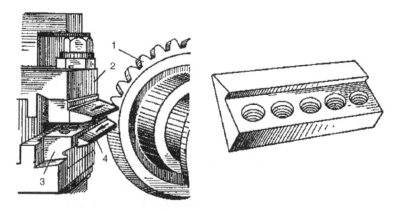

**FIGURE 6.25** Principles of two-tool generators.

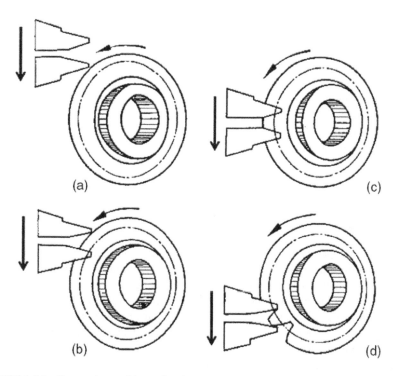

**FIGURE 6.26** Successive positions of reciprocating tools during straight bevel gear generation: (a) the tool starts to cut into the blank; (b) the second tool begins to cut; (c) both tools are in full engagement; (d) the tools run out of mesh with the gear

nonuniform wear on the profile accuracy, provision is made to make a finish cut after roughing, with most of the stock being removed in the roughing operation. The tooling cost of the two-tool generators is low, but production rates are lower than those of interlocking cutter generators, discussed previously.

Two-tool generators are usually used when:

- The bevel gears are beyond a practical size range (larger than 250 mm pitch diameter).
- Gears have integral hubs or flanges that project above the root line, thus preventing the use of other generators.
- Small production quantity or variety of gear sizes cannot be accommodated by other types of straight bevel gear generators.

## 6.3  SELECTION OF GEAR-CUTTING METHOD

Each gear-cutting method discussed so far has a field of application to which it is best adapted. These fields overlap, however, so many gears can be produced satisfactorily by more than one method. In such cases, the equipment availability often determines which machining method will be used. The type of gears to be cut is usually the main factor in the selection. However, one or more of the following factors should be considered in the final choice of the method:

- Size of the gear and its module
- Configuration of the WP to be machined
- Batch size
- Gear ratio
- Accuracy
- Cost related to the tool and the machine
- Cycle time and productivity

Figure 6.27 summarizes the possibilities of producing a certain gear type by cutting. The outcome of this layout leads to the conclusions displayed in Table 6.4.

## 6.4  GEAR FINISHING OPERATIONS

Gear finishing operations are distinguished from gear-cutting operations in that they are used for improving the accuracy, uniformity, and surface quality of the various gear tooth elements. The functional requirements of gears determine the degree of accuracy. Higher accuracy is necessary if the gears are required to operate quietly and at high speeds and to transmit heavy loads. Gear finishing methods include burnishing, shaving, lapping, and grinding. Unhardened teeth of gears are finished by shaving or burnishing, whereas hardened teeth are finished by grinding or lapping operations. Shaving is the main gear finishing process before hardening, whereas grinding is the main finishing process for hardened gears. A comparison between both processes with regard to machining time and machining allowance is presented in Figure 6.28 and Table 6.5. The effect of the gear module ($m_g$) on both the machining time and the machining allowance is clear. Both increase with increasing gear module. Figure 6.28 depicts how the time needed for grinding is about three times that needed for shaving. For the same module $m_g$, the machining time increases with the number of teeth for both shaving and grinding processes.

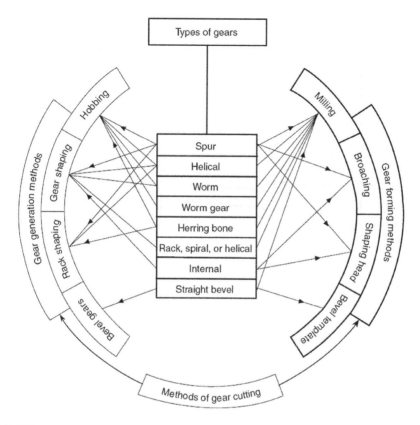

**FIGURE 6.27** Selection of gear cutting method.

**TABLE 6.4**

**Gear Cutting Methods and Their Capabilities to Produce Different Types of Gearing**

| Types of Gears | Gear Cutting Method | | | | | | | |
|---|---|---|---|---|---|---|---|---|
| | **Forming** | | | | | **Generation** | | |
| | **Milling** | **Broaching** | **Shaping Head** | **Bevel Template** | **Hobbing** | **Gear Shaping** | **Rack Shaping** | **Bevel Generators** |
| Spur | Spur | Spur external | | | Spur | Spur | Spur | |
| Helical | Helical | | | | Helical | Helical | | |
| Worm | Worm | | | | | | | |
| Worm wheel | Worm wheel | | | Straight bevel | Worm | Worm | Helical | Straight bevel |
| Herringbone | Herringbone | | | | Worm wheel | Rack | Herringbone | |
| Rack | Rack | Spur internal | | | | | | |
| Straight bevel | Straight bevel | | | | Herringbone | Internal | | |

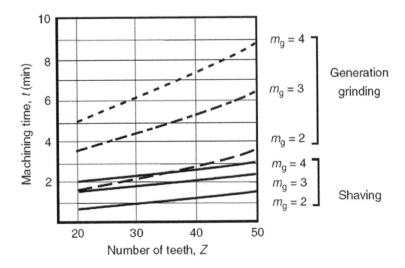

**FIGURE 6.28** Machining time for shaving and grinding. (From WMW, Gear Cutting Practice, Technical Information, Special Edition 12, 108 Berlin, Mohrenstr. 61 WMW-Export.)

**TABLE 6.5**
**Machining Allowances for Gear Shaving and Grinding**

| Gear Module (mm) | 1 | 2 | 3 | 4 | 5 |
|---|---|---|---|---|---|
| Shaving (μm) | 20 | 20 | 25 | 30 | 35 |
| Grinding (μm) | 50 | 60 | 80 | 100 | 120 |

From Düniβ, W., Neumann, M., and Schwartz, H., Trennen Spanen and Abtragen, VEB-Verlag Technik, Berlin, 1979.

### 6.4.1  FINISHING GEARS PRIOR TO HARDENING

### 6.4.1.1  Gear Shaving

Gear shaving is a finishing process based on consecutively removing thin layers of chips (2–10 μm thick) from the profiles of the teeth by a tool called a gear shaving cutter. Shaving is currently the most widely used method of finishing spur and helical gear teeth following the gear-cutting operation and prior to hardening the gear. It is not intended to salvage gears that have been carelessly cut, although it can correct small errors in areas such as tooth spacing, helix angle, tooth profile, and concentricity. Shaving reduces noise level and tooth-end load concentration and increases load-carrying capacity, surface quality, and accuracy.

#### 6.4.1.1.1  Principle of Operation

Shaving is performed with a cutter and gear at crossed axes; the value of the crossed-axes angle controls the finish produced to some extent. The smaller the

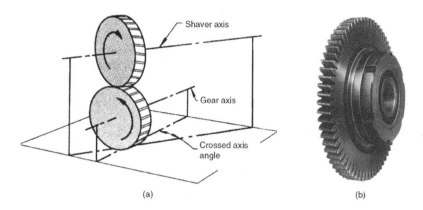

**FIGURE 6.29**   Gear shaving: (a) crossing of work gear with shaving cutter and (b) serrated shaving cutter.

angle, the finer the finish (Figure 6.29a). Angles ranging from 8° to 15° are generally ideal. In the shaving process, helical cutters of a helix angle 10–15° are generally used for spur gears, and vice versa. In some cases, helical gears are shaved by helical cutters. The action between gears and cutter is therefore a combination of rolling and sliding. Vertical serration (0.6–1 mm deep) in the cutter teeth (Figure 6.29b) takes thin hair-like chips from the profile of the gear teeth. Actually, one member of the pair is driven, and that causes the other to rotate. At the same time, a reciprocating axial feed movement is provided by the worktable. This movement ranges from 0.1 to 0.3 mm/rev of the work gear. After each stroke, the direction of cutter and work rotations is reversed to finish both sides of the teeth (Figure 6.30). Figure 6.31 shows the setup for machining spur and helical gears, respectively. The gear allowance increases from 10 to 130 μm for a corresponding increase of the gear module from 0.5 to 12.5 mm. During shaving, the tip of the cutter must not contact the root fillet; otherwise, uncontrolled, inaccurate involute profiles will result. The serration depth governs the total cutter life in terms of the number of sharpenings permitted. A shaving cutter is sharpened by regrinding the teeth profiles, thus reducing tooth thickness and consequently the general accuracy of shaved gears. Because the facilities necessary to produce high accuracy after resharpening of shaving cutters are not available in most gear manufacturing plants, cutters are ordinarily returned to the tool manufacturer for resharpening.

A rack-type shaving cutter can be used instead of the gear shaving cutter. In rack shaving, the rack is reciprocated under the gear to be shaved, and infeed takes place at the end of each stroke. Because racks longer than 500 mm are impractical, 150 mm is the maximum diameter of gear that can be shaved by the rack method.

Regardless of the previously mentioned limitation of rack cutters, shaving has been successfully used in the finishing of spur and helical gears of a very wide spectrum of sizes and modules, ranging from 6 to 5000 mm pitch diameter, and modules ranging from 0.15 to 12.5 mm. This process is ideal for finishing automotive and machine tool gearboxes after hobbing and before hardening.

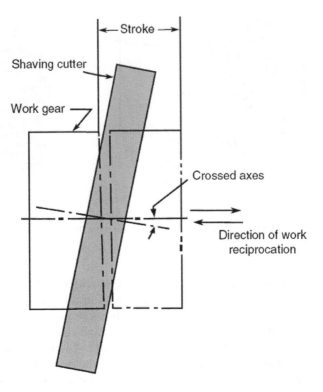

**FIGURE 6.30**    Axial traverse shaving.

#### *6.4.1.1.2 Comparison between Rotary and Rack Shaving*

A rotary cutter is much less expensive than a rack-type cutter, and its grinding cost is lower. Additional features include the following:

- Rotary cutters operate on simpler machines of smaller size.
- On rotary machines, internal gears can be shaved.
- A rotary cutter has a comparatively short tool life, and broken teeth cannot be replaced.
- Rotary cutters cannot remove excessive stock that impairs the final quality of shaved gears.

### 6.4.1.2   Gear Burnishing

Gear burnishing is another method of surface finishing for teeth of a gear, employed prior to heat treatment. It consists essentially of rolling the work gear with burnishing gears whose teeth are very hard, smooth, and accurate. The inaccuracies and asperities of the surface of the work gear are leveled by the kneading action of the material. Burnishing is of no use to gears that are to be subsequently heat-treated, as it may set up stresses that are released during heat treatment, hence leading to increased distortion, surface cracks, and peeling of the carburized and deformed surface layer.

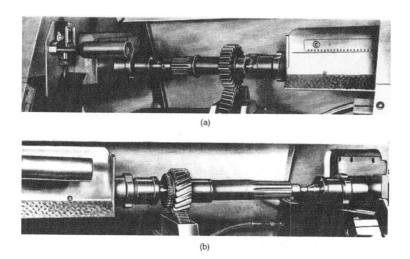

(a)

(b)

**FIGURE 6.31**  Gear shaving: (a) spur gear and (b) helical gear.

### 6.4.1.2.1  Principle of Operation

Three burnishing gears (spur or helical, depending on the type of burnished gear) are meshed with and spaced at 120° positions around the work gear. One of the burnishing gears is the driver, and the other two are idlers, which exert burnishing pressure against the work gear. The burnishing cycle starts by rotating the gears in one direction for the necessary period of time, followed by reversing the direction of rotation for an equal period of time (Figure 6.32). During burnishing, a lubricant is supplied to produce the desired surface quality and to prevent abrasion. Burnishing gears are used until worn beyond the usable accuracy and are then reground to restore the original accuracy. It is possible to regrind burnishing gears several times before discarding. It is advisable to use the largest permissible gears in order to obtain the longest usable tool life. As seen in Figure 6.32, the maximum limit of burnishing gears addendum diameter, $D_a$, is given by

$$D_a = \left(2\sqrt{3} + 3\right)d_a$$

where $d_a$ is the addendum diameter of work gear.

## 6.4.2  FINISHING GEARS AFTER HARDENING

### 6.4.2.1  Gear Grinding

Gear grinding is a specially adapted process to finish gears that have considerable stock to be removed after hardening in order to obtain the most accurate and the highest-quality gears. It is also frequently used in producing gear tools. The low rate of production and the high cost of gear grinding exclude the use of this method for mass production. As a rule, it is used only for finishing gears of precise machinery. Similarly to gear cutting, gear grinding also may be performed by forming or generation.

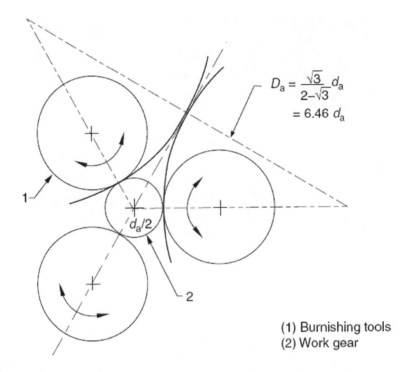

$$D_a = \frac{\sqrt{3}}{2-\sqrt{3}} d_a$$
$$= 6.46 \ d_a$$

**(1) Burnishing tools**
**(2) Work gear**

**FIGURE 6.32** Burnishing operation and maximum limit of addendum diameters of burnishing tools.

### 6.4.2.1.1 Formed Wheel Grinding

In this method, the grinding wheel has a profile corresponding to the tooth-space shape of the gear being ground and simultaneously machines the flanks of the two adjacent teeth.

The contour of the grinding wheel is profiled by a diamond dressing fixture (Figure 6.33a). The side diamonds are actuated by form templates and dress the tool profile on the grinding wheel. A variation of tooth forms can be produced by changing the contour of the templates by the dressing mechanism. Form grinding is performed by the wheel (1) that travels parallel to the axis of the work gear (2). After each full stroke of the wheel, the gear, mounted on an arbor, is automatically indexed by one or several teeth, and the cycle is repeated (Figure 6.33b). Grinding is completed by three or four passes of the wheel in each tooth space. Grinding allowance from 50 to 120 μm on each side may be removed. This gear-grinding method has a larger production capacity than generation grinding, but it is less accurate due to nonuniform wear of the wheel dressed to the tooth profile.

### 6.4.2.1.2 Generation Gear Grinding

This method is based on reproducing the mesh of the gear being ground with a rack whose tooth is represented by a form-grinding wheel or a pair of dish wheels. In Figure 6.34a, rotation (principal movement $v$) and reciprocating feed (movement in the direction of arrow $f$) are imparted to the grinding wheel. The gear is rotated

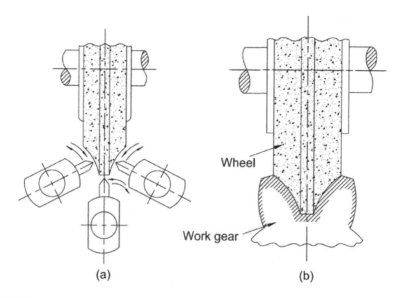

**FIGURE 6.33**    Gear grinding by forming: (a) diamond wheel dressing and (b) form grinding.

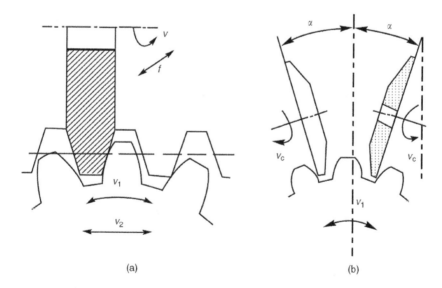

**FIGURE 6.34**    Gear grinding by generation: (a) rack type and (b) dish type.

about its axis at speed $v_1$ and is moved straight at speed $v_2$. These two movements ($v_1$ and $v_2$) are interrelated and form a complex generating roll movement. At this time, one tooth flank is ground. After reversal of the generating roll movement, the opposite flank in the same tooth space is machined. Upon the completion of the first tooth space, the grinding wheel is withdrawn, and the gear is indexed one tooth. Figure 6.34b illustrates grinding with two dish grinding wheels.

Disadvantages of the gear-grinding process include the following:

- The process is characterized by its low production capacity.
- The scratches or ridges formed increase both wear and noise. To eliminate this defect, ground gears are frequently lapped.
- Dimensional instability is an inherent characteristic of the method.
- The process requires complex and expensive gear-grinding machines to be tended by highly skilled operators.

### 6.4.3 GEAR LAPPING

Gear lapping is a microfinishing process performed on the gear after hardening. This method is based on the finishing of the gear teeth profiles using a lapping tool (called a lap) and fine-grained abrasive, with the purpose of imparting a high accuracy and fine surface finish to the gear teeth. It is, however, impossible to correct considerable errors (exceeding 30–50 μm) by lapping. Prolonged lapping associated with large allowance, besides being time-consuming, may distort the gear profile and impair the teeth accuracy. Usually, lapping is performed on special machines using three laps made of soft and fine-grained cast iron, where a lapping compound (oil and fine abrasives) is applied to the tools.

Figure 6.35 shows a setup for lapping a spur gear. The gear (3) meshes with the laps (1, 2, and 4), one of which is the driver lap. The axis of lap 2 is parallel to the axis of the work gear, whereas the axes of laps 1 and 4 cross with the gear axis at

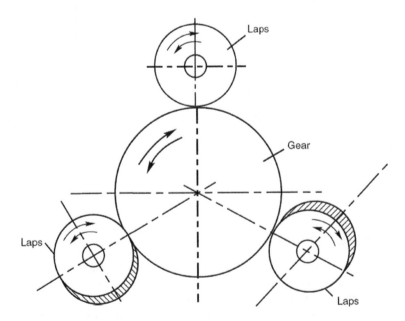

**FIGURE 6.35**   Gear lapping.

an angle of 3–5°. This setup increases the sliding of the abrasive grains across the surface of the tooth. Besides rotation of gear and laps, the work gear imparts an axial reciprocating movement to speed up the process and improve its quality. Gears with large errors are ground rather than lapped. Automotive gearbox gears that are finished before case hardening by shaving are usually finally lapped.

## 6.5 REVIEW QUESTIONS AND PROBLEMS

6.5.1 What are the disadvantages of milling a gear by a formed disk cutter?

6.5.2 Write down the relationships of the following in terms of module $m_g$ and the number of teeth $Z$ for a 20-gear tooth.
- Pitch circle diameter $d_p$
- Addendum $h_a$, dedendum $h_d$, clearance $(h_d - h_a)$
- Working depth $h_w$, tooth height $h_t$
- Outside diameter $d_a$
- Tooth thickness $s$, fillet radius $r$

6.5.3 What is the difference between hobbing and milling as gear-cutting processes? Discuss their fields of application.

6.5.4 Discuss the inaccuracies that may result from gear cutting by hobbing.

6.5.5 What are the advantages of gear generation by shaping?

6.5.6 Mention three gear types that may be produced on the following gear-cutting machines: hobbing, milling, gear shaping by rotary cutter.

6.5.7 Suggest only two types of gear-cutting machines to produce the following gear types: helical gears, worms, straight bevel gears, worm wheels.

6.5.8 Make the necessary setups for milling the following helical gears on a horizontal universal milling machine:
- 30° helix right-hand
- 30° helix left-hand
    Show the direction of table feed and work rotation for each case.
- Why is a heat treatment process not recommended after gear burnishing?
- Draw a sketch to illustrate the principle of gear lapping operation.
- Explain the main advantages and limitations when using a gear shaping head. Is it a forming or a generating gear production method?
- What are the advantages of helical gears over spur gears?
- What difficulty would be encountered in hobbing a herringbone gear? What modifications in design should be performed to permit it to be cut by hobbing?
- Can a helical gear be machined on a universal milling machine?
- Why is gear hobbing much more productive than gear shaping?
- Under what conditions can shaving not be used for finishing gears?
- Enumerate methods of finishing gears before and after hardening.
- An HSS hob of pitch diameter 70 mm is used to cut a spur gear of 48 teeth. A cutting speed of 30 m/min is used, and the gear has a face

width of 64 mm. The hob is fed axially at a rate of 2.1 mm/rev of the WP. What is the time required to achieve gear hobbing, provided that an approach and over-travel of 36 mm is assumed?

## REFERENCES

Acherkan N 1968, *Machine tool design* (four volumes). Mir Publishers, Moscow.

Düniβ, W, Neumann M & Schwartz H 1979, *Trennen spanen and abtragen*, VEB-Verlag Technik, Berlin.

High Production Gear Shaping Machine WS1 Kaufbeurer Str. 141, Liebherr Verzahntechnik GmbH. D8960 Kempten, Germany.

ASM International, *Metals handbook* 1989, Machining, vol. 16, ASM International, Material Park, OH.

WMW, Bevel Gear Hobbing Machine ZFTX 250x5, Technical Information, 108 Berlin, Mohrenstr, 61 WMW-Export.

WMW, Gear Cutting Practice, Technical Information, Special Edition 12, 108 Berlin, Mohrenstr, 61 WMW-Export.

# 7 Turret and Capstan Lathes

## 7.1 INTRODUCTION

Turret and capstan lathes are the natural development of the engine lathe, where the tailstock is replaced by an indexable multistation tool head called the capstan or the turret. This head carries a selection of standard tool holders and special attachments. A square turret is mounted on the cross slide in place of the usual compound rest in the engine lathe. Sometimes a fixed tool holder is also mounted on the back end of the cross slide. Dimensional control is effected by means of longitudinal (for lengths) and traversal (for diameters) adjustable stops.

Therefore, capstan and turret lathes bridge the gap between manual engine lathes and automated lathes and are most practical for batch and short-run production. In comparison with manual lathes, the chief distinguishing feature of capstan and turret lathes is the multiple tool holders that enable the setting up of all the tools necessary to produce a certain job. Except for sharpening, the tools need no further handling.

Considerable skill is required to set and adjust the tools on such machines properly. But once the machines are set, they can be operated by semiskilled operators. Eliminating the setup time between operations reduces the production time considerably. The development of this group of lathes has been enhanced to provide the level of accuracy required for interchangeable production.

The main advantages of turret and capstan lathes include the following:

1. Less skilled operators are needed as compared with center lathes.
2. There is no need to change tooling or move the work to another machine, as many operations can be performed without the need to change tooling layout.

## 7.2 DIFFERENCE BETWEEN CAPSTAN AND TURRET LATHES

The essential components and operating principles of capstan and turret lathes are illustrated schematically in Figure 7.1. Capstan lathes are mainly used for bar work, whereas turret lathes are applicable for large work in the form of castings and forgings.

In a capstan or ram-type lathe, the hexagon turret is mounted on a slide that moves longitudinally in a *stationary saddle* (Figure 7.2a). During setup of the machine, the saddle is positioned along the bed to give the shortest possible stroke for the job. The advantage of the capstan lathe is that the operator has less mass to move, resulting

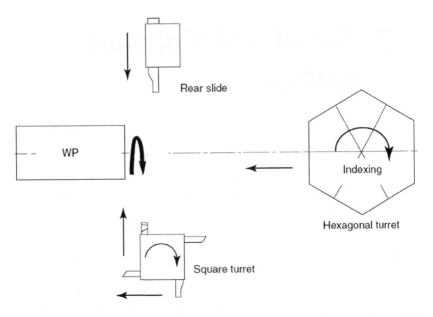

**FIGURE 7.1** Essential components and operating principles of capstan and turret lathes. (Adapted from ASM International, *Metals Handbook*, Machining, Vol. 16, ASM International, Materials Park, OH, 1989.)

in easier and faster handling. The disadvantage is that the hexagonal turret slide is fed forward such that the overhang is increased, resulting in the deflection of the ram slide, especially at the extreme of its position, which produces taper and reduces accuracy. In the turret- or saddle-type lathe, the turret is mounted directly upon a *movable saddle*, furnished with both hand and power longitudinal feed (Figure 7.2b). This machine is designed for machining chuck work in addition to bar work. Due to the volume of swarf produced, the guideways of the machine bed are flame-hardened and provided with covers that protect the sliding surfaces. The bed must be designed to allow free and rapid escape of swarf and coolant.

The advantages of the turret or saddle-type lathe include the following:

- It is more rigid and hence most suitable for heavier chucking work. Jobs up to 300 mm diameter can be machined on it.
- Its design eliminates the turret slide overhang problem inherent in the ram-type lathes.
- The power rapid traverse reduces the operator's handling effort.

Sometimes, saddle-type machines are built with a cross turret feeding on the saddle to meet the requirement of specific jobs. The eight-sided turret, while offering two additional tooling stations, has the disadvantage of increasing the interference between turret and cross-slide tools and limits the size of the tools that can be mounted on the turret stations.

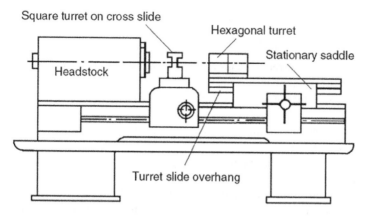

(a) Capstan

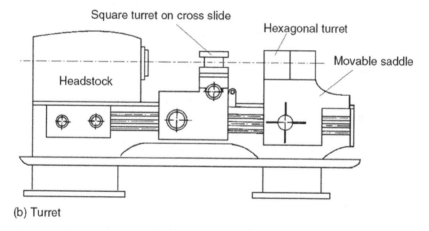

(b) Turret

**FIGURE 7.2** Difference between capstan and turret lathes: (a) capstan and (b) turret. (From Browne, J.W., *The Theory of Machine Tools*, Book 1, Cassell and Co. Ltd., London, 1965.)

## 7.3 SELECTION AND APPLICATION OF CAPSTAN AND TURRET LATHES

Machine selection is based on two factors: lot size and complexity of operation. A lot size of 10–1000 pieces is usually considered suitable for capstan and turret lathe work. For lot sizes under 10 pieces, these machines can compete with engine lathes strictly on a time basis but not on an economical basis. At the same time, it is impractical to use capstan lathes on very large lot sizes where the advantages of automatic equipment of turret-type machines can be economically utilized. A mathematical treatment should be developed for the determination of unit cost in terms of the lot size, taking into consideration many factors, such as machine cost, labor cost, machine-setter cost, and also the complexity of operations performed on the work.

Typically, turret and capstan lathe jobs contain multiple operations, such as turning, recessing, facing and boring, drilling, tapping, reaming, and so on. Jobs requiring simple operations should be done on simpler and less expensive center lathes. Once it is decided that a turret or a capstan lathe is the best suitable machine for the work, the size of the machine must be selected.

To finalize the selection process, the following aspects are to be considered:

1. Select a machine with sufficient power and rigidity to remove the metal at the most economical rate.
2. Choose the smallest machine that has ample swing and bed length for the job to be performed.
3. Choose between a ram- and a saddle-type machine. Long, accurate turning and boring operations dictate a saddle-type machine, while the ram type is preferred for ease of handling.
4. Determine whether a power feed or a manual feed machine is required.
   - Determine whether a cross-feeding hexagonal unit makes sense for the job.
   - Consider whether spindle speeds and carriage feeds lend themselves to the job.
   - Consider whether an automatic headstock control would be worthwhile.

A word of caution should be inserted here on the use of capstan lathes equipped with extra-large-capacity spindles: this machine is recommended only for light operations, in spite of its powerful spindle. If it is used for heavy work, excessive wear and ultimate breakdown result. Saddle-type lathes equipped with large-capacity spindles are recommended for heavy work and severe cuts.

## 7.4 PRINCIPAL ELEMENTS OF CAPSTAN AND TURRET LATHES

A ram or turret lathe has essentially the same elements as an engine lathe with additional elements like hexagonal turrets and front and rear cross slides. However, the controls used are more complex. The motor is more powerful to enable the machine to perform overlapped cuts. The elements of a standard turret and capstan lathe are described in the following sections.

### 7.4.1 HEADSTOCK AND SPINDLE ASSEMBLY

The headstock is heavier in construction than that of the engine lathe with a wider range of speeds. A typical layout from Heinemann Machine Tool Works-Schwarzwald is shown in Figure 7.3. The mounting of the free-running gears should be noted in addition to the use of roller bearings with a taper bore for the spindle. The multidisk clutch drive is widely used in conjunction with constant mesh gearing. The use of these clutches provides rapid acceleration and the ability to sustain high torque loads (Browne, 1965). In modern machines, pole-changing motors offer four speeds, which simplifies the design of the gearbox and limits its size. One of the chief characteristics of the turret headstock is the provision for rapid stopping and

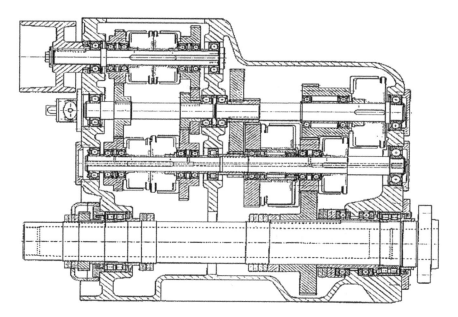

**FIGURE 7.3**   Typical headstock and spindle assembly of a turret lathe. (From Heinemann Machine Tool Works-Schwarzwald, Germany.)

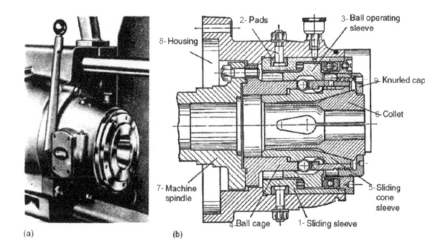

**FIGURE 7.4**   Hand-operated collet chuck: (a) general view and (b) sectional view.

starting, and for speed changing through speed preselectors. Through these measures, the minimum loss of time is realized. When components are turned from bar stock fed through the hollow spindle of the machine, a collect chuck is used. The bar is generally of round or hexagonal shape. Collect chucks may be pneumatically or hand operated. A sectional view of the hand-operated collet chuck is shown in Figure 7.4 (H. W. Ward and Co. Ltd.).

When the handle shown in Figure 7.4a is moved to the close position, the sliding sleeve 1 (Figure 7.4b) rotates and is therefore forced to move to the left, as the groove accommodating pads 2 are cut on a helix. Consequently, the sleeve forces the ball operating sleeve 3 to the left, which causes the right-hand (RH) ring of balls held in the ball cage 4 to move radially inward. This closes the sliding cone sleeve 5 and hence the collet 6. Moving the lever in the opposite direction reverses the action, and the left-hand (LH) ring of balls moves the sleeve to release the collet. In the position shown, the collet is closed. The machine spindle 7 and the housing 8 are bolted to the headstock. The knurled cap 9 adjusts the collet for variations of the machined bar size. By a slight modification, the design can be altered such that the sliding sleeve can be actuated pneumatically to reduce the operator's fatigue and reduce the chucking time.

### 7.4.2 Carriage/Cross-Slide Unit

The cross-slide unit on which the tools are mounted for facing, forming, recessing, knurling, and cutting off is made of four principal parts, namely, the cross slide, the square turret, the carriage, and the apron (Figure 7.5). The rear and front square turrets are mounted on the top of the cross slide. Each turret is capable of holding four tools ready for use. If additional tools are required, they are set up in sequence and can be quickly indexed and locked in the correct chucking position.

The slide is provided with a positive stop to control the depth of the cut. Dogs on the side of the cross slide engage these stops to regulate the cross-slide travel. The carriage has two hand wheels for manual longitudinal and cross feed. In some machines, besides hand feed, a power feed (rapid or slow) can be engaged by a lever.

### 7.4.3 Hexagonal Turret

The hexagonal turret is carried on a saddle and is intended for holding and bringing the tools in a forward feed movement. On the turret type, each face is provided with four tapped holes to accommodate screws for holding flanged holders and attachments in which tools are clamped. On capstan lathes, the turret may be circular; it

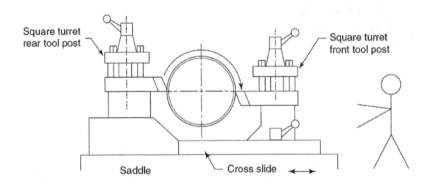

**FIGURE 7.5**  Cross slide and square turret tool posts.

has also six holes for accommodating shanked tool holders, which are normally used for small works that do not need to be held in a flat face. Two types of control are available, as described next.

### 7.4.3.1 Manually Controlled Machines

During the cycle of operations, it is necessary to bring each tool into a position relative to the work. The turret is located in each of six correct positions by some form of hand-operated arrangement in which the operator manually indexes the turret to the required position after releasing the clamp and locating plungers.

On the capstan lathe, means are provided whereby the turret is automatically indexed to the next position when it reaches the extreme end of its withdrawal movement from the previous position.

Various arrangements are adapted in this respect. Figure 7.6 illustrates diagrammatically one of the principles involved. An indexing plate (1) and a Geneva ring (2) are secured to the head (3). When the slide (4) is retracted, a spring loaded lever (5) contacts a projection (4) on the base slide (7). As the turret slide continues its retracting movement, the lever moves the locking bar (8) rapidly out of the slot (9) of the indexing plate. The slide moves further, and the pivoted finger (10) indexes the turret. Meanwhile, a lever passes over the projection prior to the end of the indexing motion. The locking bar moves rapidly and locks the turret (3) in position. A bevel gear (11) fixed in the underside of the turret meshes with a bevel pinion (12), the ratio being 5:1. The pinion shaft (13) carries a bush (14) in which six long screws (15) are filtered, one for each turret position. For one indexing movement, the bush rotates 5/4 of a revolution, providing the relevant screw, which moves to the dead stop fitted to the end of the base slide (Browne, 1965).

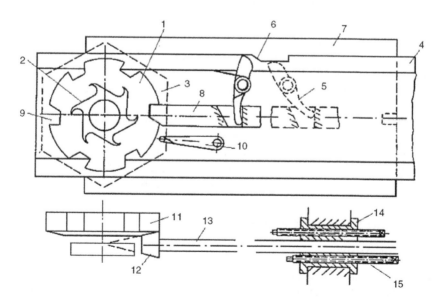

**FIGURE 7.6** Turret indexing mechanism. (From Browne, J.W., *The Theory of Machine Tools*, Book 1, Cassell and Co. Ltd., London, 1965.)

### 7.4.3.2 Automatically Controlled Headstock Turret Lathes

Automatic control of the headstock through the movement of the hexagonal turret results in considerable time savings on jobs where handling time constitutes a large part of the total floor-to-floor time (FFT). Starting, stopping, speed changing, and spindle reversing are all controlled by a unit actuated by the indexing of the turret head. The operator has to handle only the hexagonal turret, resulting in considerable time savings.

This control is best used on small machines where a large number of spindle changes take place in a short machining cycle. Plumbing fittings, aircraft fittings, small valve bodies, and other chucking work with short machining strokes are jobs ideally suited to the automatic controls. Bar work of the same cycle time and short strokes also shows time savings when the automatic spindle control is used instead of manually controlled machines.

The total percentage savings diminish as the machining time increases in relationship to handling time on longer-stroke jobs. For example, on a part requiring 0.5 min FFT, of which 0.15 min is machining time, the handling time will be 0.35 min. If the use of automatic control saves 0.15 min/piece, the total time is therefore reduced by 30%. On a longer-stroke job requiring 4.0 min FFT, in which 3.65 min is for machining and 0.35 min for handling time, if again 0.15 min is saved by the automatic control, the total time is reduced by only 3.7%. It can be readily seen through this comparison that a critical inspection should be made before deciding which type of control is to be used (*Tool Engineers Handbook*, 1959).

### 7.4.4 CROSS-SLIDING HEXAGONAL TURRET

Sometimes, the hexagonal turret has a cross-sliding ability to feed in four directions. This characteristic adds greatly to the versatility of the turret lathe on certain difficult types of work. This unit is used only on the larger-size turret lathe of the saddle-type construction. The mobility of this turret makes it especially adaptable to small-lot work, where multiple inner surfaces can be machined using a minimum of quickly set-up cutters (*Tool Engineers Handbook*, 1959).

Going beyond the small production lots, a cross-sliding turret offers other advantages for certain types of work. For example, it provides the possibility of machining large-diameter work, which prohibits the use of square turret cross slides. The graduated dial for the cross motion of the hexagonal turret enhances the accuracy and makes it the same as the square turret on the cross slides.

## 7.5 TURRET TOOLING SETUPS

### 7.5.1 JOB ANALYSIS

The jobs on a turret lathe are simply a series of basic machining operations such as turning, facing, drilling, boring, reaming, threading, and so on. The setup for any job consists of arranging these machining operations in their proper sequence. The best tooling setup considers the tolerances, the lot size, and the machining cost.

Once the turret lathe is properly tooled, an experienced operator is not required to operate the machine. However, skill is required in the proper selection and mounting of tools. The turret setup starts with the job analysis. The following guidelines should be considered in this analysis:

1. *Type of tooling.* In general, standard tools and holders should be used as much as possible, especially for small batches of work. Simple tool layout should be employed for small batches. Roller rest turning holders are used to support the bar work, whereas extended tooling is used for machining chucked parts such as castings and forgings.
2. *Machining operations.* Maximum productivity is achieved by a judicious combination of internal and external machining operations on both the square and the hexagonal turrets. As far as possible, cuts performed on square and hexagonal turrets should be combined in the tooling setup.
3. *Tool geometry and proper clamping.* Suitable cutting angles and edges should be ground on tools. Moreover, the tools should be set with minimum overhang and gripped firmly in their holders.
4. *Stops.* These should be set as accurately as part tolerances require. Hexagonal turret stops are usually set prior to cross-slide stops.
5. *Speeds and feeds.* Each cut should be performed using the highest speed and feed as cutter and job permit. The speeds and feeds recommended for various operations, and work and tool materials, are available in machining handbooks. Extracted and modified speeds and feeds are given in Table 7.1. The use of maximum speeds and feeds depends on the machine power available as well as the rigidity of tooling, the holding mechanism, and the workpiece (WP) itself.

    Multiple or combined cuts on different diameters of the WP may call for the use of different grades of carbides or high-speed steel (HSS) to get the proper surface speeds. The cutters for the larger diameters will be carbide, while HSS will give the best results on smaller diameters with correspondingly lower machining speeds.
6. *Production time.* If the production time $t_p$ consists of setup time $t_s$ and FFT $t_f$ (Figure 7.7), then

$$t_p = t_s + t_f \qquad (7.1)$$

The setup time $t_s$ is the time consumed in setting up the machine for a new job. It is evident that as the lot size decreases, the importance of setup time $t_s$ increases. Adding cutters to take multiple cuts decreases the FFT $t_f$. However, it usually increases the setup time $t_s$ and the tool cost. Simplifying the tooling setup by using fewer cutting tools reduces the setup time $t_s$ and the tool cost while increasing the FFT $t_f$. The FFT is the time that elapses between picking up a component to load for a machining operation and depositing it after machining.

The FFT is a combination of the machining time $t_m$ (productive time), and idle or auxiliary time $t_a$ (nonproductive time); therefore,

$$t_f = t_m + t_a \qquad (7.2)$$

**TABLE 7.1**

**Recommended Speeds and Feeds When Machining Different Materials on Capstan, Turret, and Automated Lathes Using HSS Tools**

| | | | | Material | | | |
|---|---|---|---|---|---|---|---|
| | | | | Structural, Case-Hardened, and Tempered Steels of $\sigma_u$ (kg/mm$^2$) | | | |
| Operation | Light Metals | Brass 70/30 | Free-Cutting Steels | Up to 50 | To 70 | To 85 | To 100 |
| *Cutting speeds*[a] (*m/min*) | | | | | | | |
| Turning, forming, cutting off[b] | 150–200 | 120–150 | 60–70 | 35–42 | 26–32 | 20–24 | 15–20 |
| Drilling | 80–120 | 70–120 | 40–50 | 30–35 | 20–26 | 16–20 | 12–15 |
| Threading | 30–50 | 30–60 | 5–9 | 5–7 | 4–6 | 2–4 | 1–3 |
| Tapping[c] | 10–20 | 8–16 | 5–8 | 3–7 | 3–5 | 2–3 | 1–2 |
| *Feeds* (*mm/rev*) | | | | | | | |
| Turning | 0.15–0.30 | 0.15–0.25 | 0.12–0.16 | 0.11–0.16 | 0.10–0.14 | 0.08–0.11 | 0.08–0.10 |
| Forming | 0.02–0.05 | 0.02–0.05 | 0.02–0.04 | 0.02–0.04 | 0.01–0.04 | 0.01–0.03 | 0.01–0.03 |
| Cutting off | 0.04–0.08 | 0.05–0.10 | 0.04–0.05 | 0.03–0.05 | 0.03–0.04 | 0.02–0.04 | 0.02–0.03 |
| Drilling[d] | 0.06–0.20 | 0.08–0.20 | 0.05–0.14 | 0.05–0.12 | 0.04–0.11 | 0.03–0.09 | 0.03–0.09 |
| Core drilling | 0.16–0.22 | 0.16–0.22 | 0.14–0.17 | 0.12–0.15 | 0.10–0.13 | 0.08–0.10 | 0.08–0.10 |

[a] Values are multiplied by a factor of 1.5–2.5 when carbide tools are used.
[b] Maximum values are used for turning; minimum values are applicable for forming and cutting-off operations.
[c] Lower values for small taps of 0.5 mm pitch, higher values for larger taps of 1.5 mm pitch. When cutting external threads using dies, the values here are reduced by 50%.
[d] Lower values for small drills ($\phi$2 mm), higher values for larger drills ($\phi$20 mm).
Reclassified and modified from Index-Werke, Esslingen, Germany Index 12-18-25 Berechnungsunterlagen.

The machining time $t_m$ is the time consumed in the actual cutting operation and is controlled by the use of proper cutting tools, feeds, and speeds. It can be saved by performing multiple cuts or by increasing chip removal rate.

The idle time $t_a$ is composed of the machine handling time $t_{mh}$ and the work handling time $t_{wh}$; therefore,

$$t_a = t_{mh} + t_{wh} \tag{7.3}$$

The machine handling time $t_{mh}$ is the time consumed in bringing the respective tools into the cutting positions. It can be reduced using multiple cuts, as several surfaces are machined with only one handling of the hexagonal turret unit, which is faster than a square turret.

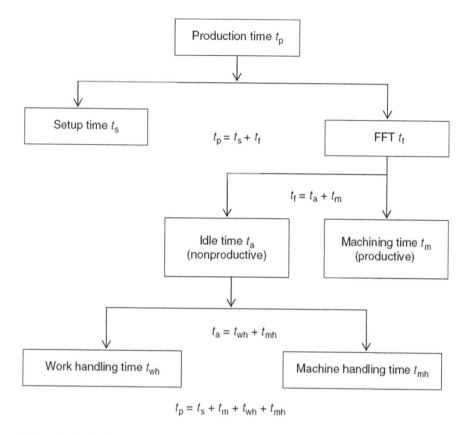

**FIGURE 7.7** Production time on a turret lathe.

Standard times (allowances) for machine handling time $t_{mh}$ on turret and capstan lathes are as follows:

Feed to bar stop = 0.04 min
Hexagonal turret indexing – 0.08 min
Square turret indexing = 0.20 min
Speed changing = 0.05 min
Feed changing = 0.05 min

The work handling time $t_{wh}$ is the time consumed in mounting or lifting the work. It is largely dependent upon the type of work-holding devices. In some jobs, the work handling time constitutes a fairly large part of the FFT. It is important to keep it to a minimum by the use of self-centering or pneumatic chucks. The four-jaw independent chucks, arbors, two-jaw box chucks, and special fixtures are necessary for some odd-shaped parts. However, such holding devices are slow and costly.

Fatigue allowance of 10–20% of various operating times should be added to total FFT.

## 7.5.2 TOOLING LAYOUT

Basically, the tools mounted on cross slides are used for turning, facing, necking, knurling, and parting off. Those mounted on the turret head are used for drilling, boring, reaming, threading, recessing, and so on. As previously stated, the accuracy and cost of the machined component mainly depends on the proper tool layout, which should be simple. For the preparation of tool layout, it is necessary to have the finished drawing, on which the machining allowances on different surfaces should be provided.

A preliminary operation sequence should be decided based on details such as:

- Tools, holders, and attachments required
- Tool layout drawn to the same scale of the component's final position
- Travels and clearances exactly checked before the final setup of the machine

The tooling layout differs according to the nature of the WP. In this regard, chucking or bar work is possible.

1. *Chucking work.* Figure 7.8 illustrates a standard tools setup for machining a cast iron ratchet wheel. In heavy cut stations (1 and 4), it is clear that the rigidity of the setup is enhanced by the use of pilot bar, fixed on the headstock and adapted in a socket in the turret head. Figure 7.9 represents the tooling setup for a cross-sliding hexagonal turret. Using such a basic setup with standard tools and tool holders, multiple cuts can be taken. The multiple turning heads (Figures 7.10 and 7.11) offer the possibility of multiple turning cuts at heavy metal removal (stations 1 and 4 in Figure 7.8) or for presetting tools for close-tolerance finish turning cuts (*Metals Handbook*, 1989). Figure 7.10 shows a knee-turning and boring attachment for chucking work; Figure 7.11 illustrates a combination tool holder for multiturning, chamfering, and boring also for chucking-type work.

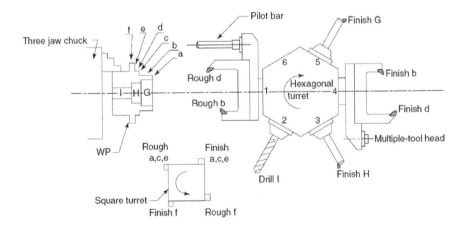

**FIGURE 7.8** Tooling layout for chucking-type work.

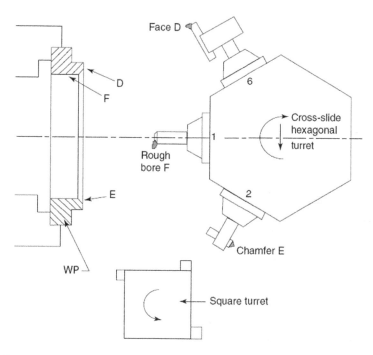

**FIGURE 7.9**   Cross-slide hexagon turrets for chucking work of large diameter.

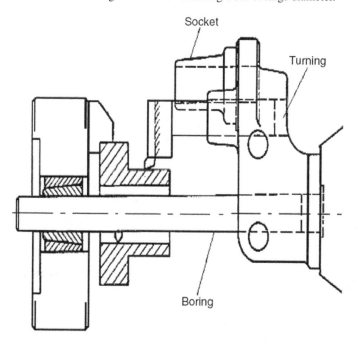

**FIGURE 7.10**   Knee-turning and boring attachment for chucking work. (From Herbert Machine Tools Ltd., U.K.)

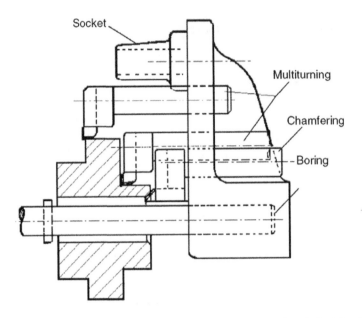

**FIGURE 7.11**   Combination tool holder (multiturning, boring, and chamfering) for chucking work. (From Herbert Machine Tools Ltd., U.K.)

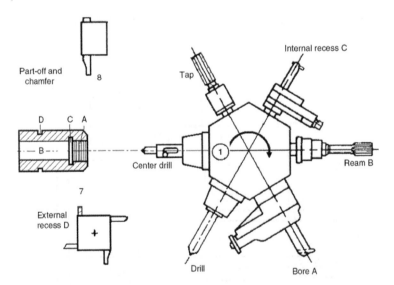

**FIGURE 7.12**   Tooling layout for producing a threaded adaptor. (Adapted from ASM International, *Metals Handbook*, Machining, Vol. 16, ASM International, Materials Park, OH, 1989.)

  2. *Bar work.* Figures 7.12 and 7.13 illustrate the tooling arrangement of a typical setup for bar work. Such an arrangement gives good production potential, realizing the minimum setup time. Figure 7.12 illustrates the multifunction capabilities of turret lathes in producing a thread adaptor shown in the same figure. Figure 7.13 illustrates the complex configuration of a

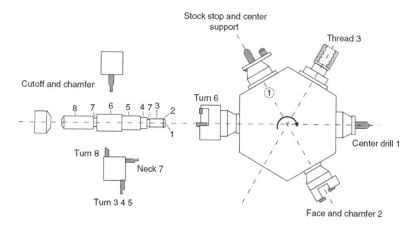

**FIGURE 7.13** Tooling layout for producing a steel shaft of complex configuration. (Adapted from ASM International, *Metals Handbook*, Machining, Vol. 16, ASM International, Materials Park, OH, 1989.)

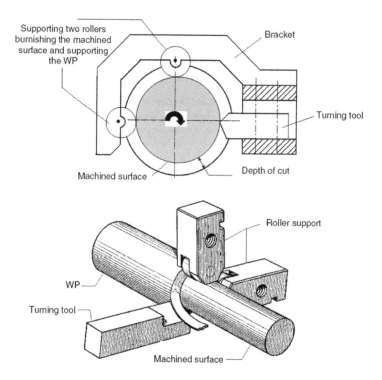

**FIGURE 7.14** Roller box turning attachment.

steel shaft machined on a turret lathe. The single-cutter holder in position turn 6 removes metal at a maximum rate, as the rolls support the work at the point of cut. Behind the cutting tools, the support rolls burnish the work to a fine surface finish and accurate size (Figure 7.14).

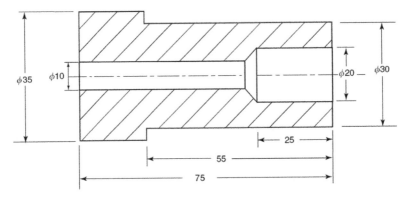

Dimensions (mm), Material: Brass 70/30

**FIGURE 7.15**   Sleeve to be machined on a turret lathe. (From Jain, R.K., *Production Technology*, 13th Edition, Khanna Publisher, Delhi, India, 1993.)

### Illustrative Example

Draw a tool layout for the component shown in Figure 7.15. Also, determine the FFT necessary for producing the component on a turret lathe.

### SOLUTION

Figure 7.16 shows the tools and standard holders required to produce the component. Table 7.2 lists the sequence of operations, speeds, feeds, and operation times for productive ($t_m$) and nonproductive ($t_i$) movements.

1. Selection of cutting speeds and feeds:

   Material: brass 70/30
   Turning tools: all carbides, K-type
   Twist drills: HSS
   Referring to Table 7.1, the following speeds and feeds are depicted:

| | |
|---|---|
| Turning | $v = 150–300$ m/min (select 150 m/min) |
| | $f = 0.15–0.25$ mm/rev (select 0.2 mm/rev and 0.1 mm/rev for parting off) |
| Drilling | $v = 70–120$ mm/min (select 70 m/min) |
| | $f = 0.08 – 0.2$ mm/rev, depending on diameter (select 0.2 mm/rev) |

2. Determination of spindle speeds:

| | |
|---|---|
| Turning | $n = \dfrac{1000v}{\pi D} = \dfrac{1000 \times 150}{\pi \times 40} = 1190\,\text{rpm}\left(\text{select }1000\,\text{rpm}\right)$ |
| Drilling | $n = \dfrac{1000v}{\pi D} = \dfrac{1000 \times 70}{\pi \times 20} = 1114\,\text{rpm}\left(\text{select }1000\,\text{rpm}\right)$ |
| | $= \dfrac{1000 \times 70}{\pi \times 10} = 228\,\text{rpm}\left(\text{select }2000\,\text{rpm}\right)$ |

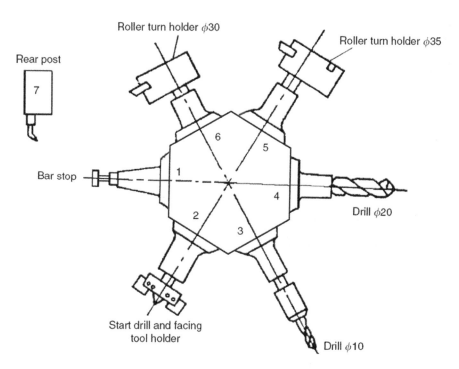

**FIGURE 7.16**  Tooling layout to produce the part in Figure 7.15. (From Jain, R.K., *Production Technology*, 13th Edition, Khanna Publisher, Delhi, 1993.)

Then the spindle operates at two speeds, 1000 and 2000 rpm.

3. Sample calculation of the productive time $t_m$:

Referring to Figure 7.16 and considering turret station 3, $\varphi 10$ mm × 75 mm, the spindle speed 2000 rpm and the turret feed 0.2 mm/rev, then

$$t_m = \frac{1}{n.f} = \frac{75}{2000 \times 0.2} = 0.19 \, \text{min}$$

4. Idle or nonproductive times $t_a$:

Use the previously suggested idle times,
0.08 min for turret indexing
0.05 min for switching over spindle speeds and feeds

5. Calculation of FFT ($t_f$):

All elementary times are added as shown in Table 7.2, and a fatigue allowance of 20% is considered:

$$t_f = 1.2 \left( t_m + t_a \right)$$

## TABLE 7.2
## Turret Work Sheet for the WP Illustrated in Figure 7.15

Work material: Brass 70/30
Bar size: φ40 mm
Tooling: turning, carbide K
  group; drilling, HSS

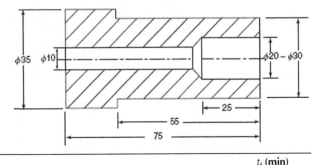

| Operation | Tooling Station | Sequence of Operation | Spindle Speed (m/min) | Feeds (mm/rev) | $t_m$ | $t_i$ |
|---|---|---|---|---|---|---|
| | | | | | \multicolumn{2}{c}{$t_f$ (min)} | |
| 1 | – | Index turret to position 1 | – | – | – | 0.08 |
| 2 | Turret 1 | Feed to bar stop | – | – | – | 0.08 |
| 3 | – | Index turret to position 2 | – | – | – | 0.08 |
| 4 | Turret 2 | Center drill and face | 1000 | – | 0.15 estimated | – |
| 5 | – | Select $f = 0.2$ mm/rev | – | – | – | 0.05 |
| 6 | – | Index turret to position 3 | – | – | – | 0.08 |
| 7 | – | Select $n = 2000$ rpm | – | – | – | 0.05 |
| 8 | Turret 3 | Drill φ10 × 75 mm | 2000 | 0.2 | 0.19 | – |
| 9 | – | Select $n = 1000$ rpm | – | – | – | 0.05 |
| 10 | – | Index turret to position 4 | – | – | – | 0.08 |
| 11 | Turret 4 | Drill φ20 × 25 mm | 1000 | 0.2 | 0.13 | – |
| 12 | – | Index turret to position 5 | – | – | – | 0.08 |
| 13 | Turret 5 | Turn φ35 × 77 mm | 1000 | 0.2 | 0.39 | – |
| 14 | – | Index turret to position 6 | – | – | – | 0.08 |
| 15 | Turret 4 | Turn φ30 × 55 mm | 1000 | 0.2 | 0.28 | – |
| 16 | – | Change to $f = 0.1$ mm/rev | – | – | – | 0.05 |
| 17 | Rear 7 | Parting off past center (19 mm) | 1000 | 0.1 | 0.19 | – |
| Determination of FFT ($t_f$) | | | | | 1.33 | 0.76 |

$t_f = t_m + t_i = 1.33 + 0.76 = 2.09$ min　　　　$t_f = 1.33 + 0.76$

  Considering a fatigue allowance of 20%, then $t_f = 2.5$ min

## 7.6　REVIEW QUESTIONS

7.6.1　Mark true or false.
  [ ] In turret lathes, the turret is mounted on a saddle.
  [ ] Capstan lathes are characterized by higher accuracy compared with turret lathes.
  [ ] Capstan lathes are ideal for heavy chucking work; therefore, they are equipped with powerful spindles.

[ ] Heavier cuts can be taken by automatic turrets rather than the capstan machines.

7.6.2    What is the difference between a turret lathe and a capstan lathe?

7.6.3    What is the difference between ram-type and saddle-type turret lathes? What are their advantages and disadvantages?

7.6.4    Why is the saddle-type lathe suited to repetitive manufacture of complex cylindrical parts?

7.6.5    Draw a sketch to show indexing and locking of a turret of a capstan lathe.

7.6.6    For what purpose is the cross-sliding turret used?

7.6.7    Define FFT.

## REFERENCES

Browne, JW 1965, *The theory of machine tools*, Book 1, Cassell and Co. Ltd., London.

Heinemann Machine Tool Works-Schwarzwald, Germany.

Herbert Machine Tools Ltd., UK.

Index-Werke, Esslingen, Germany Index 12-18-25 Berechnungsunterlagen.

Jain RK 1993, *Production technology*, 13th edn, Khanna Publisher, Delhi, India.

ASM International, *Metals handbook* 1989, Machining, vol. 16, ASM International, Materials Park, OH.

Wilson, FW and Harvey, PD, *Tool Engineers Handbook* 1959, ASTME, McGraw-Hill, New York.

H. W. Ward and Co. Ltd, Lionel Street, Birmingham, UK.

# 8 Automated Lathes

## 8.1 INTRODUCTION

Automated machine tools have played an important role in increasing production rates and enhancing product quality. Since they were introduced in industry, they have contributed to mass production of spare parts and machine components. Especially, automated lathes are high-speed machines, and therefore, the application of safety rules in their operation is obligatory for all attendants and servicing personnel. Moreover, machine tools of this type and their setup can be entrusted only to persons with comprehensive knowledge of their design and principles of operation. If all servicing instructions are strictly observed, the operation of automated lathes presents no hazard at all to the operator.

Fully automatic lathes are those machines on which workpiece (WP) handling and the cutting activities are performed automatically. Once the machine is set up, all movements related to the machining cycle, loading of blanks, and unloading of the machined parts are performed without the operator. In semiautomatic machines, the loading of blanks and unloading of the machined components are accomplished by an operator.

From early times, machine tools—especially lathes—have been fitted with devices to reduce manual labor. A considerable range of mechanical, hydraulic, and electrical devices, or a combination of these, have contributed to the development of automatic operation and control.

Lathes in which automation is achieved by mechanical means are productive and reliable in operation. Much time, however, is lost in switching over from one job to another. Therefore, automatics are used in mass production, while semiautomatics are used in lot and large-lot production. Machine tools and lathes that are automated by other than mechanical means (numerical control and computer numerical control, using numerical data to control their operating cycle) can be set up for new jobs much more rapidly and are therefore efficiently employed in lot, batch, and even single-piece production (Chapter 9). Automatic and semiautomatic lathes are designed to produce parts of complex shapes by machining the blank (or bar stock). They are designed to perform the following machine operations:

- Turning
- Centering
- Chamfering
- Tapering and form turning
- Drilling, reaming, spot facing, and counter boring
- Threading
- Boring
- Recessing
- Knurling
- Cutting off

Special attachments provide additional operations, such as slotting, milling, and cross drilling.

## 8.2  DEGREE OF AUTOMATION AND PRODUCTION CAPACITY

Generally, metal-cutting operations are classified into one of the following categories:

1. Processing or main operations in which actual cutting or chip removal takes place
2. Handling or auxiliary operations, which include loading and clamping of the work, releasing and unloading of the work, changing or indexing the tool, checking the size of the work, changing speeds and feeds, and switching the machine tool on and off

The operator of a nonautomated machine performs the handling or auxiliary operations. With automated machines, some or all of the auxiliary operations are performed by corresponding mechanisms of the machine. The faster the auxiliary operations are performed in the machine, the more WPs can be produced in the same period of time; that is, a higher production or automation rate is realized (Figure 8.1). The selection of a suitable degree of automation should be based on the feasibility of machining parts at the specified quality and desired rate of production. It should be emphasized that an increase in the number of spindles and the degree of automation leads to an increase in the time required for setting up the machine for a new lot of WPs, which calls for an increased lot size.

Each type of automatic lathe has an optimum range of lot size in which the cost per piece is minimal. Figure 8.2 shows that the physical and psychological effort exerted by the operator decreases with increasing degree of automation, and consequently,

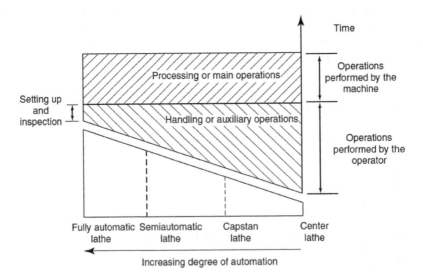

**FIGURE 8.1**   Degree of automation as affected by auxiliary operations.

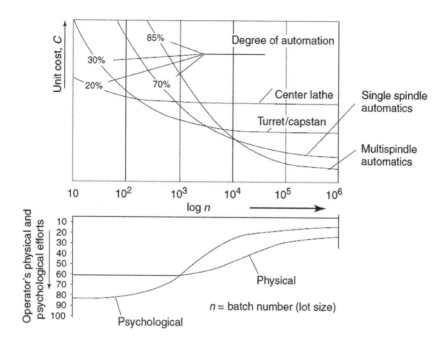

**FIGURE 8.2** Unit cost, as well as physical and psychological efforts against batch number, for different degrees of automation. (From Pittler Maschinenfabrik AG, Langen bei Frankfurt/M, Germany.)

the lot size also increases. Greater psychological effort is required with a lower degree of automation, whereas physical effort predominates with a higher level of automation. A higher degree of automation realizes the following advantages:

- Increases the production capacity of the machine.
- Ensures stable quality of WPs.
- Necessitates a lower number of machines in the workshop, thus achieving higher output per unit shop floor area.
- Reduces the physical effort required from the operator and releases him from tediously repeated movements and from monotonous nervous and physical stresses.
- Avoids direct participation of the operator and therefore enables him to operate several automatic machines at the same time.

To increase the production capacity of automated lathes, it is necessary to reduce the time required by the operating cycle of the machine through the following measures:

1. Concentrating cutting tools at each position or station. The concentration factor, $q$, represents the ratio of the total number of tools/number of stations. As a rule, automated lathes are multiple-tool machines.
2. Overlapping working travel motions.

3. Providing independent spindle speeds and feeds at each position or station.
4. Machining several WPs in parallel.
5. Employing throwaway tipped carbide cutting tools.

## 8.3   CLASSIFICATION OF AUTOMATED LATHES

The principal types of general-purpose automated lathes are visualized in Figure 8.3. They may be classified according to the following features:

1. *Spindle location (horizontal or vertical).* Vertical machines are heavier, more rigid, more powerful, and occupy less floor space. They are especially designed for machining large-diameter work of comparatively short length.
2. *Degree of automation (fully automatic or semiautomatic).* As mentioned earlier, the decision to choose between fully automatic and semiautomatic depends mainly on the lot size.

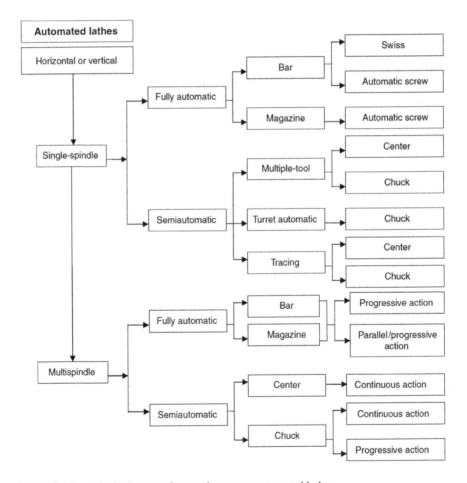

**FIGURE 8.3**   Principal types of general-purpose automated lathes.

3. *Number of spindles (single- or multispindle).*
   A. *Single-spindle automated lathe are classified as*
      - Fully automatic (Swiss-type and turret-screw automatics)
      - Semiautomatic
   B. *Multispindle automated lathes.* These machines have two to eight horizontal or vertical spindles. Their production capacity is higher than that of single-spindle machines, but their machining accuracy is somewhat lower. They are further classified as follows:
      - *Fully automatic.* These machines are suitable for both bar and magazine work. They are widely used for mass production and need a lot of setup work. Large multispindle automatics are equipped with an auxiliary small power motor, which serves to drive the camshaft when the machine is being set up. The rate of production of a multispindle automatic is less than that of a correspondingly sized single-spindle automatic. The production capacity of a four-spindle automatic, for example, is only 2.5–3 times (not 4 times) as large as that of a single-spindle automatic, assuming the same product size, shape, and material.
      - *Semiautomatic.* Semiautomatic multispindle machines are mostly of vertical type.
4. *Nature of WP stock (bar or magazine).* Automated lathes use either coiled wire stock (up to 6 mm in diameter), bar, pipe, or separate blanks. Bar stock is available in a great variety of shapes and sizes; however, it is considered poor practice to use bar stock over 50 mm in diameter, as the waste metal in the form of chips will be excessive. Separate blanks are frequently used in semiautomatics. The blanks should approach the shape and size of the finished product; otherwise, the cycle time increases, thus increasing the production cost.

Automated lathes are broadly classified according to the stock nature into the following main categories:

A. *Automatic bar machine.* These are used for machining WPs from bar or pipe stock.
B. *Magazine loaded machine.* These are used to machine WPs in the form of blanks, which have been properly machined to appropriate dimensions prior to feeding them into the machine.
   The introduction of any form of automatic feeding results in a higher degree of automation and economy, and makes it possible for one operator to observe a number of machines instead of being confined to one.
5. *WP size and geometry.* The size and geometry of the WP determine the suitable machine to be used. In this regard, long accurate parts of small diameters are produced on the Swiss-type automatics, whereas parts of complex external and internal surfaces are machined using turret-type automatic screw machines.

6. *Machining accuracy.* Generally, bar automatics are employed for machining high-quality fastenings (screws, nuts), bushings, shafts, rings, etc. The design configuration of the Swiss-type automatics makes them superior with respect to the production accuracy, especially when producing long, slender parts.

The machining accuracy of multispindle automatics is generally lower than that of single-spindle automatics due to the errors in indexing of spindles and the large number of spindle-head fittings.

## 8.4   SEMIAUTOMATIC LATHES

### 8.4.1   SINGLE-SPINDLE SEMIAUTOMATICS

The three main types of single-spindle semiautomatic lathes are as follows:

*   Multiple-tool semiautomatic lathes
*   Turret semiautomatic lathes
*   Hydraulic tracer–controlled semiautomatic lathes

All of these are equipped with either centers or chucks. WPs several times longer than their diameters are normally machined between centers, while short WPs with large diameters should be chucked.

1. *Multiple-tool semiautomatic lathes.* These machines operate on a semiautomatic cycle. The operator only sets up the work, starts the lathe, and removes the finished work. This feature allows one operator to handle several machine tools simultaneously (multiple machine tool handling). Figure 8.4 shows the multiple-tool machining of a stepped shaft, mounted

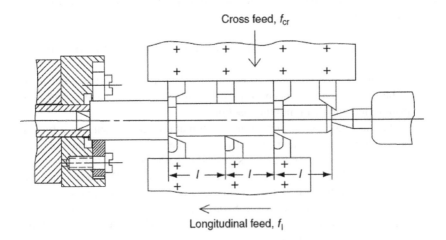

**FIGURE 8.4**   Multiple-tool machining of a stepped shaft.

between centers, using several tools mounted on the main and cross slides (cross and longitudinal feeds are designated by arrows). Tailstock centers are most often ball- or roller bearing–type to withstand heavy static and dynamic forces. These machines have found extensive applications in large-lot and mass production.

2. *Turret semiautomatic lathes.* Turret semiautomatics, commonly referred to as single-spindle chucking machines, are used basically for the same type of work carried out by the turret lathe. They generally require hand loading and unloading and complete the machining cycle automatically. These machines are used when production requirements are too high for hand turret lathes and too low for multispindle automatics to produce economically. The setup time is much shorter than that of multispindle automatics.

It is important to realize that during the automatic machining operation, the operator is free to operate another machine or to inspect the finished part without loss of time. The turrets normally consist of four or six tooling stations. Cross-slide tooling stations are also available in the front and rear slides.

The machine has a control unit that automatically selects speeds, feeds, length of cuts, and machine functions such as dwell, cycle stop, index, reverse, cross-slide actuation, and many others.

3. *Hydraulic tracer–controlled semiautomatic lathes.* These are intended to turn complex shaped and stepped shafts between centers by copying from a template or master WPs. Figure 8.5 represents the longitudinal and

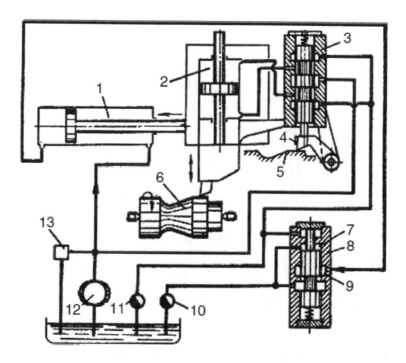

**FIGURE 8.5**   Hydraulic circuit diagram of a tracer control system.

cross-feed movements to produce the part (6). The casing (3) of the tracer valve is rigidly attached to the tracer slide, the valve spool being pressed by a spring to template (5) through a tracer stylus (4). Additionally, the feed pump (12) delivers oil into the right chamber of the cylinder (1). If a part of the template profile is parallel to the axis of the machine, the cylinder surface of the WP (6) is turned. Oil is exhausted from the left chamber of the cylinder (1) into the tank through the groove of an automatic regulator (8) and a throttle (10). As the stylus (4) moves downward or upward to the template profile, the speed and direction of the tracer control slide change. When the valve spool moves downward, the pump (12) delivers oil in the bottom chamber of the cylinder (2), the tracer-controlled slide also moves downward, and the oil from the upper chamber of the cylinder (2) is exhausted into the tank through the throttle (11). The valve (13) is a safety element for the system.

### 8.4.2 MULTISPINDLE SEMIAUTOMATICS

Multispindle semiautomatics may be of continuous or progressive action (Figure 8.6).

1. *Machines of continuous action.* This type is designed for holding the work either between centers or in a chuck. Its operation is shown diagrammatically in Figure 8.7.
   - The outer column (1), connected to the spindle (2), rotates continuously and slowly.

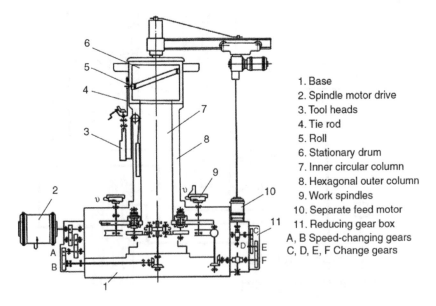

1. Base
2. Spindle motor drive
3. Tool heads
4. Tie rod
5. Roll
6. Stationary drum
7. Inner circular column
8. Hexagonal outer column
9. Work spindles
10. Separate feed motor
11. Reducing gear box
A, B Speed-changing gears
C, D, E, F Change gears

**FIGURE 8.6** Gearing diagram of a vertical multispindle semiautomatic. (From Acherkan, N., *Machine Tool Design,* Mir Publishers, Moscow, 1969. With permission.)

- The work is clamped in six chucks (Figure 8.7a) or between centers in six spindles (Figure 8.7b), and the longitudinal and cross-tool slides (4) are located on the outer column (1).
- The same machining operation is performed at each tooling station, except at the loading zone. Each slide is set up with the same tooling.
- In a definite zone, the work spindles cease to rotate, the finished work is removed from the chuck (3), and a new block (or bar) is loaded.

2. *Machines of progressive action.* This type is designed for chucking operations only. It is available with either six or eight spindles. Its operation is illustrated in Figure 8.8. Referring to Figure 8.8a:

- The carrier (1) is periodically indexed through 60°. Each spindle rotates at its own setup speed, independent of other spindles.
- A hexagonal column (5) carries only five tool slides (3 and 4). The WPs are clamped in chucks (2).

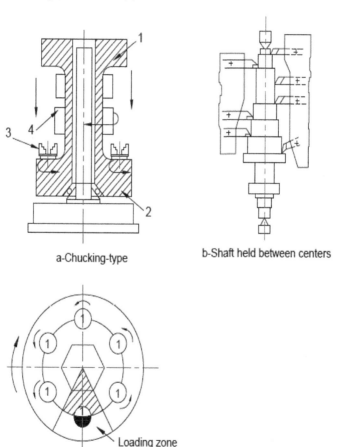

a-Chucking-type

b-Shaft held between centers

Loading zone

**FIGURE 8.7** Vertical multispindle semiautomatic machines of continuous action: (a) chucking type; (b) shaft held between centers. (From Maslov, D. et al., *Engineering Manufacturing Processes,* Mir Publishers, Moscow, 71970. With permission.)

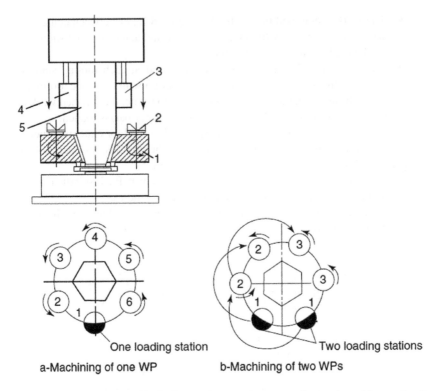

**FIGURE 8.8** Vertical multispindle semiautomatic-progressive action: (a) one loading station and (b) two loading stations. (From Maslov, D. et al., *Engineering Manufacturing Processes,* Mir Publishers, Moscow, 1970. With permission.)

- Work is loaded periodically in the loading station after the carrier (1) indexes through 60°, while the finished work is removed from the loading station.
- At the other five stations, the WPs are machined simultaneously; at each consecutive station, the work is machined by the tools set up at that station.
- The spindle speed is automatically changed to the setup value for each station, and each particular spindle stops rotating when it reaches the loading station.
- Referring to Figure 8.8b, the machine may be adjusted to perform the following machining duties:
  - Turning two different WPs in one operation cycle. This duty is applicable only if two tooling stations are sufficient to machine each WP on a six-spindle machine.
  - Turning both sides of the work consecutively.

For the last two machining duties, there should be two loading stations, and the spindle carrier should be indexed through 120° each time.

## 8.5  FULLY AUTOMATIC LATHES

These are mainly based on mechanical control systems and are characterized by a rigid linkage between the working and auxiliary operations. The two following mechanical systems are frequently employed:

1. A control system composed of a single camshaft, which provides all the working and auxiliary motions. This is the simplest arrangement and is further classified into two types:
   - *Systems in which the camshaft speed is set up and remains constant during the complete cycle*: These systems may be applied only for automatics with a short machining cycle (up to 20 s), such as Swiss-type automatics.
   - *Systems in which the camshaft speed is set up for the working movements*: Auxiliary movements are performed at a high constant speed of the camshaft that is independent of the setup. The mass of the camshaft with the cam drums and disks is comparatively large and has large inertia torques, which lead to impact loads at the moment of the camshaft switching over from high to low speeds or vice versa. This system is most extensively used for multispindle automatics and semiautomatics.
2. A control system composed of a main and an auxiliary camshaft. The machining cycle is completed in one revolution of the main camshaft. Trip dogs on the main camshaft perform the auxiliary operations through signals that link the operative devices for auxiliary movements to the auxiliary camshaft. This system is more complex than the preceding one due to the very large number of transmission elements and levers. It is used in automatic screw machines and vertical multispindle semiautomatic chucking machines.

### 8.5.1  SINGLE-SPINDLE AUTOMATIC

The range of work produced by these machines extends from pieces so small that thousands can be put in a houschold thimble to complex parts weighing several kilograms. Two distinct basic machining techniques involved are turret automatic screw machines and Swiss-type automatics.

### 8.5.1.1  Turret Automatic Screw Machine

This type is regarded as the final stage in the development of the capstan and turret lathes. Its main objective is to eliminate (as much as is possible) the operator's interference by extensive use of levers and cams. Although this machine was originally designed for producing screws, currently it is used extensively for producing other complex external and internal surfaces on WPs by using several parallel working tools. Figure 8.9 shows typical parts produced on turret automatic screw machines. A general view of a classical automatic screw machine is shown in Figure 8.10, along with its basic elements.

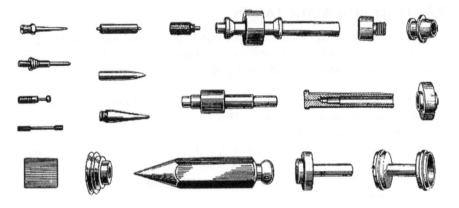

**FIGURE 8.9**  Typical parts produced on turret automatic screw machine. (From Acherkan, N., *Machine Tool Design*, Mir Publishers, Moscow, 1969. With permission.)

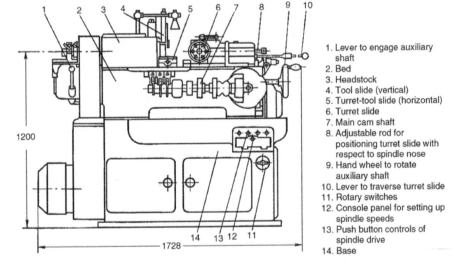

1. Lever to engage auxiliary shaft
2. Bed
3. Headstock
4. Tool slide (vertical)
5. Turret-tool slide (horizontal)
6. Turret slide
7. Main cam shaft
8. Adjustable rod for positioning turret slide with respect to spindle nose
9. Hand wheel to rotate auxiliary shaft
10. Lever to traverse turret slide
11. Rotary switches
12. Console panel for setting up spindle speeds
13. Push button controls of spindle drive
14. Base

**FIGURE 8.10**  General view of the automatic screw machine. (From Acherkan, N., *Machine Tool Design,* Mir Publishers, Moscow, 1969. With permission.)

The main specifications of turret automatic screw machines include the following:

a. Bar capacity
b. Maximum diameter of thread to be cut (in steel or in brass)
c. Maximum travel of turret
d. Maximum radial travel of cross slides
e. Maximum and minimum production times (maximum and minimum cycle times)
f. Range of spindle speeds (left and right)
g. Main motor power

h. Auxiliary motor power
i. Overall dimensions

An important feature of the turret automatic screw machine is the auxiliary shaft system, which will be described together with the main characteristics of this automatic.

### 8.5.1.1.1   Kinematic Diagram

A simplified kinematic diagram of a typical automatic screw machine is given in Figure 8.11. The main motor of 3.7 kW and 1440 rpm imparts the required motion to the following components:

1. *Main spindle.* For one pair (seven pairs existing) of pick-off gears, the motion is transmitted through the following gear train (Figure 8.11): main motor–24–46–20–50–back gear 56/37–pick-off gears *A/B*–sprockets 25/28 and 25/28. Accordingly, four spindle speeds are obtained (two forward *n* and two reverse $n_r$).

$$n = 1440 \times \frac{24}{46} \times \frac{20}{50} \times \frac{A}{B} \times \frac{28}{56} \times \frac{25}{28} \left( \text{slow forward} \right) \tag{8.1}$$

$$n_r = 1440 \times \frac{24}{46} \times \frac{20}{50} \times \frac{A}{B} \times \frac{25}{50} \times \left( \text{slow reverse} \right) \tag{8.2}$$

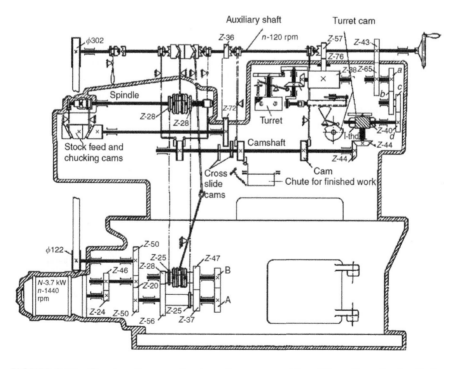

**FIGURE 8.11**   Gearing diagram of automatic screw machine. (From Boguslavsky, B. L., *Automatic and Semi-automatic Lathes,* Mir Publishers, Moscow, 1970. With permission.)

$$n = 1440 \times \frac{24}{46} \times \frac{20}{50} \times \frac{A}{B} \times \frac{47}{37} \times \frac{25}{28} \left(\text{fast forward}\right) \tag{8.3}$$

$$n_r = 1440 \times \frac{24}{46} \times \frac{20}{50} \times \frac{A}{B} \times \frac{47}{37} \times \frac{56}{28} \times \frac{25}{28} \left(\text{fast reverse}\right) \tag{8.4}$$

2. *Auxiliary and main camshafts.* Main motor–24–46–20–50–belt drive 122/302–auxiliary shaft–43–56–pick-off gears *a/b/c/d*–worm/worm wheel 1/40–turret camshaft–bevels 44/44–main camshaft.

Therefore, speed of auxiliary shaft:

$$n_{aux} = 1440 \times \frac{24}{46} \times \frac{20}{50} \times \frac{122}{302} \approx 120 \, \text{rpm} \tag{8.5}$$

and that of camshaft (main and turret camshaft):

$$n_{cam} = n_{aux} \times \frac{43}{65} \times \frac{a}{b} \times \frac{c}{d} \times \frac{1}{40} \, \text{rpm} \tag{8.6}$$

### 8.5.1.1.2  Working Features and Principle of Operation
All types of single-spindle turret automatics utilize a number of cross slides, with each one carrying a single tool, and some form of turret to manipulate another set of tools (Figure 8.12). All axial operations are performed by tools mounted in the turret, with only one turret station being in operation at once. Tools mounted on the four cross slides can perform consecutively or simultaneously to carry out operations such as turning, forming, grooving, recessing, cutting off, and knurling. The work is supplied as bar or tube stock, held firmly in the spindle by a collet chuck. After each piece has been completed, the bar is positioned for machining the next piece by being automatically moved forward and butted against a swing or turret stop. Provision is made to support bars extending out from the rear of the headstock to minimize whipping action, which causes excessive machine vibration.

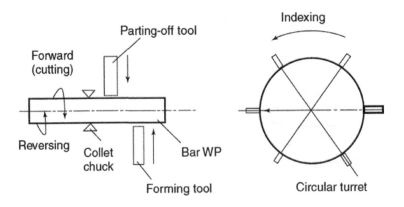

**FIGURE 8.12**  Essential components and operating principles of single-spindle automatic screw machine.

### 8.5.1.1.3   Spindle Assembly

A complete spindle assembly is shown in Figure 8.13b. A spring collet chuck arranged in spindles is commonly used in automatic bar lathes. Three widely used types are illustrated in Figure 8.13b through d:

a. Push-out type (Figure 8.13b), in which the collet ($a_1$) is pushed by the collet tube (s) to the right into the tapered seat of the spindle nose (b) for clamping.
b. Pull-in type (Figure 8.13c), in which the collet ($a_1$) is pulled by the collet tube (g) to the left into the tapered spindle nose (b) for clamping.
c. Immovable or dead-length type (Figure 8.13d), in which the shoulder of the chuck ($b_1$) bears against a nut (b) screwed on the spindle nose (Figure 8.13a). Hence, the axial movement is not exerted when the clamping sleeve (c) is actuated. The spring shown in Figure 8.13a shifts back tubes (c and g) to the left while releasing the bar. Figure 8.13a also shows the clamping levers (e) and clamping sleeve (d) actuated by the chucking cam drum. Nuts (i) are used for the fine adjustment of the ring (f) and the supporting levers (e) to adapt the collet to limited changes of the bar stock diameter.

The push-out and pull-in types have the disadvantage that during clamping, the bar (h) has an axial movement that affects the accuracy of axial bar positioning. Moreover, in the push-out type, the collet tube may be subjected to buckling if the tube is long and high clamping forces are exerted.

Spring collets locate the bar stock with high accuracy. Bars up to 12 mm in diameter will turn true on a length of 30 or 35 mm to within 20 or 30 µm, whereas bar stocks of 40 mm in diameter will run true within 50 µm on a length of 100 mm.

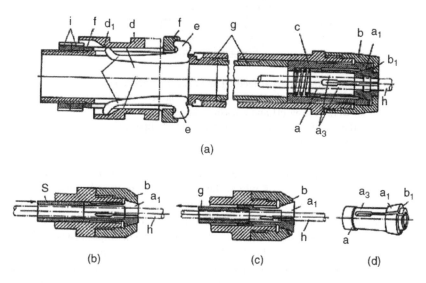

**FIGURE 8.13**   Spindle assembly and types of spring collets of automatic screw machines: (a) spindle assembly; (b) push-out type; (c) pull-in type; (d) dead-length type.

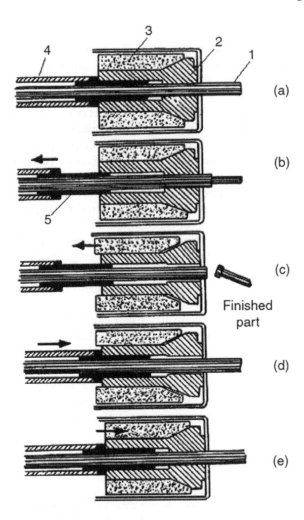

**FIGURE 8.14** Operation of the bar feeding and chucking mechanisms using a dead-length chuck: (a) bar feed, stock clamping, (b) retraction of the feeding finger that slides over the clamped WP, (c) work cutting off and releasing the collet, (d) feed of bar stock by feeding finger to stop of the first turret station, and (e) bar stock clamping. (From Boguslavsky, B. L., *Automatic and Semi-automatic Lathes,* Mir Publishers, Moscow, 1970. With permission.)

In the chucking arrangement, shown in Figure 8.14, the bar is clamped by the dead-length spring collet (2), on which the sleeve (3) is pushed by the collet tube, which closes the collet. The feeding tube (4) is arranged inside the collet tube and carries a spring feeding finger (5). The feeding finger (also called the bar stock pusher) is screwed in the frontal end of the feeding tube. The finger is a slitted spring bushing in which jaws are closed before hardening and tempering (Figure 8.15). The pressure of the slitted feeding finger is sufficient to move the bar through the open collet and slide over the bar when the collet closes. The bar stock feeding and chucking occur in the following order, shown in Figure 8.14.

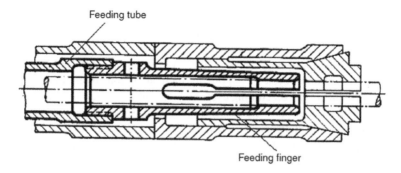

**FIGURE 8.15**  Feeding finger (stock pusher) in a pull-in collet chuck. (From Pittler Maschinenfabrik AG, Langen bei Frankfurt/M, Germany.)

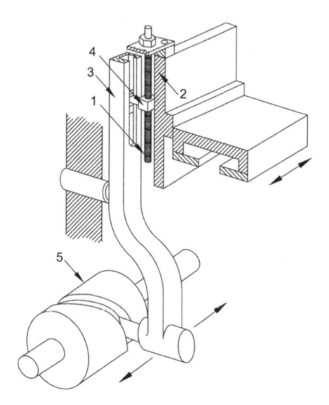

**FIGURE 8.16**  Mechanism of adjusting the travel of the feeding finger.

The stock feeding is actuated by the stock feeding cam, located under the spindle shown in Figure 8.11, which actuates the bar feeding mechanism. The amount of the stock feed movement can be adjusted by a screw (1 in Figure 8.16) through setting the sliding block (4) up or down in the slot of the rocker arm (3).

### 8.5.1.1.4   Control System

Figure 8.17 visualizes how the auxiliary shaft and the camshaft control the operation of a turret automatic screw machine. The auxiliary shaft rotates at a relatively high speed of 100–200 rpm. It helps mainly in bridging the idle (auxiliary) movements of the machining cycles by reducing their actuation times. In contrast, the cutting movements of the machining cycle are controlled by the camshaft, the speed of which is exactly equal to the production rate; that is, one WP produced per one revolution of the camshaft. The turret head has two types of movements: cutting and indexing movements. The cutting movement is actuated by the multicurve disk cam mounted on the camshaft, whereas the indexing movement is performed by the auxiliary shaft.

It should be emphasized that the control system of the automatic screw machine is based on the following:

1. All cutting movements performed by tools mounted on the cross slides and the turret head are controlled by the camshaft.
2. The idle or auxiliary movements are rapidly performed by an auxiliary shaft rotating at higher speeds.

The exact functions of both auxiliary shaft and camshaft are presented in Figure 8.17.

*8.5.1.1.4.1 Auxiliary Shaft*    An assembly of the auxiliary shaft is shown in Figure 8.18. The shaft carries three dog clutches (single-revolution clutches), which are operated through a lever system by trip dogs on drum cams that are mounted on the

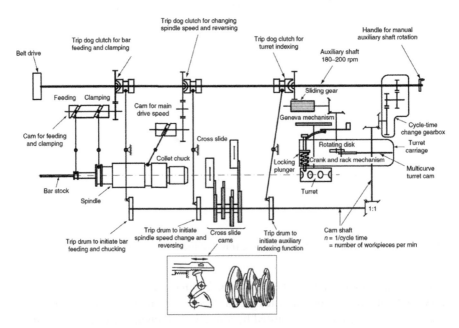

**FIGURE 8.17**   Control of automatic screw machine.

camshaft in the correct angular position (Figures 8.19 and 8.20). The dog clutches can be made to operate when required during the cycle of operations.

These three dog clutches, as arranged from left to right (Figure 8.18), when operated cause the following to occur:

1. Opening of the collect chuck, the bar feeding to the bar stop in turret, and closing the collet again for gripping the bar
2. Changing over of the spindle speed from fast to low or vice versa
3. Indexing of the turret head

Figure 8.21 shows a detailed drawing of the dog clutch (single-revolution). This type of clutch is not recommended for a rotational speed that exceeds 200 rpm. It operates in the following manner:

- The gear (1) is rotated only one revolution by the auxiliary shaft (2) and is then automatically disengaged.
- Long jaws are provided at the end of the gear (1) and clutch members (B), so that they do not disengage when a member is shifted to engage the jaws at the right end.

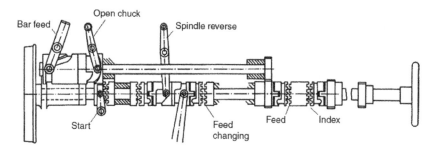

**FIGURE 8.18**   Auxiliary shaft assembly.

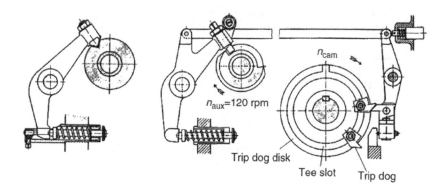

**FIGURE 8.19**   Trip levers controlling chucking and speed change over. (From Index-Werke AG, Esslingen/Neckar, Germany. With permission.)

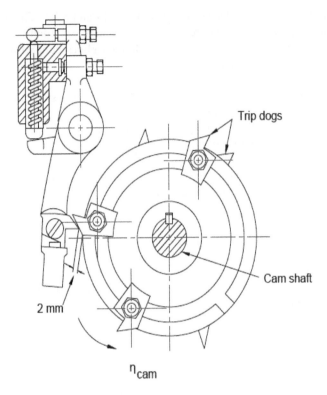

**FIGURE 8.20**  Turret indexing trips. (From Index-Werke AG, Esslingen/Neckar, Germany. With permission.)

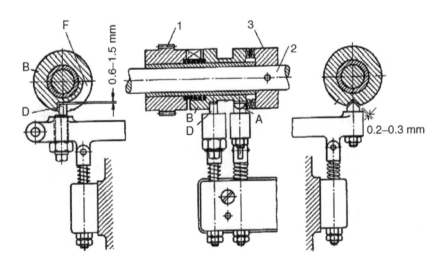

**FIGURE 8.21**  Details of dog clutch. (From Index-Werke AG, Esslingen/Neckar, Germany. With permission.)

- In the disengaged condition, the clutch member is retained by the lock pin (D).
- If the lock pin is withdrawn from its slot, the spring forces the clutch member to the right and engages the disk (3). The gear starts rotating through the clutch members.
- The pin slides along the external surface of a clutch member until it drops into the recess of a beveled surface, which forces the clutch member, upon further rotation, toward the left to disengage the gear from the rotating disk after one complete revolution.
- The lock pins (D and A) are backed by their springs.

*8.5.1.1.4.2 Camshaft* As previously mentioned, the auxiliary shaft transmits motion to the camshaft (Figure 8.11). As is also depicted in Figure 8.17, the camshaft consists of two parts, namely:

1. The front camshaft, which accommodates:
   - Disk cams of single lobe that control the cross-feed movement of the front, rear, and vertical tool slides
   - Three control or trip drums to initiate auxiliary functions by trip dogs that operate dog clutches, so that one revolution of the auxiliary shaft is imparted to perform chucking, speed changing or reversing, and indexing in the required times
2. A cross camshaft, which carries a multilobe cam to control the main movement of the turret head

Both camshafts are connected by bevels 44/44 (Figures 8.11 and 8.17) and rotate at rotational speed that equals the production rate of the machine. The rotational speeds of auxiliary shaft $n_{aux}$ and camshaft $n_{cam}$ are interrelated by Equation 8.7, which describes the cycle time $T_{cyc}$ in seconds (time/revolution of the camshaft):

$$T_{cyc} = \frac{60}{n_{cam}} = 60 \left( \frac{1}{n_{aux}} \times \frac{65}{43} \times \frac{a}{b} \times \frac{d}{c} \times 40 \right) s \tag{8.7}$$

If $n_{aux} = 120$ rpm, and for a calculated $T_{cyc}$, the pick-off gear ratio is given by

$$\frac{a}{b} \times \frac{d}{c} = \frac{30}{T_{cyc}} s \tag{8.8}$$

Many special devices, such as slot sawing attachments, cross drilling attachments, and milling attachments, are available to increase the productivity of automatic screw machines. These attachments require cams provided on the front camshaft to operate them.

*8.5.1.1.4.3 Turret Slide and Turret Indexing Mechanism* Figure 8.22 illustrates an isometric assembly of a turret slide during feeding travel as actuated by the multicurve cam (6). The basic elements of this assembly are also illustrated.

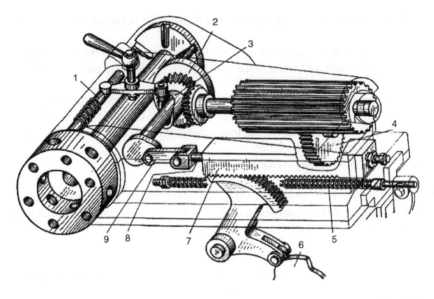

**FIGURE 8.22** Isometric assembly of a turret slide of automatic screw machine. (From Acherkan, N., *Machine Tool Design,* Mir Publishers, Moscow, 1969. With permission.)

The indexing cycle, actuated by auxiliary shaft, proceeds in the following manner (Figure 8.23):

a. *Beginning of indexing.* The roll of segment gear runs down along the drop curve of the turret cam. Under the action of the spring, the slide travels to the right. The single-revolution clutch on the auxiliary shaft is engaged. The crank begins to rotate.
b. *End of turret slide withdrawal to the stop.* The force of the spring acts against the stop.
c. *Beginning of turret rotation.* Upon further rotation of the crank, the rack travels forward to the left, so that the roll leaves the cam. The driver roll enters the slot of the Geneva wheel. The locking pin is fully withdrawn from the socket.
d. *End of turret rotation.* The crank passes the dead-center position, the rack begins to travel back to the right, the roll approaches the cam, the driver roll leaves the Geneva wheel slot, and indexing is completed. The locking pin reenters the socket.
e. *Beginning of turret slide approach by means of crank-gear mechanism.* Upon further crank rotation, the rack continues to travel to the right until the roll of gear segment lands on the cam. As the crank continues to rotate, the slide leaves the end stop and is rapidly advanced to the machining zone.
f. *End of indexing.* The slide approach is completed when the crank is in its rear dead-center position. At this moment, the single-revolution clutch is disengaged and locked, and with it, all the gears of the indexing mechanism and the crank gear mechanism are locked. At the end of turret indexing, the roll should be at the end of the drop curve.

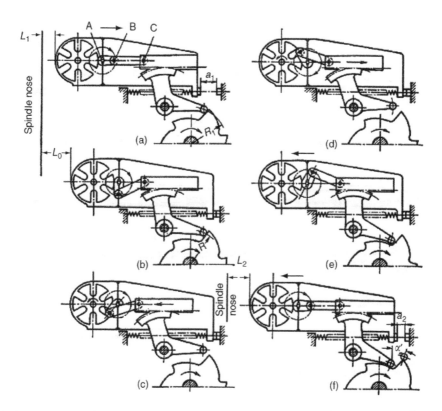

**FIGURE 8.23** Steps of turret indexing in an automatic screw machine: (a) beginning of indexing; (b) end of turret slide withdrawal to the stop; (c) beginning of turret rotation; (d) end of turret rotation; (e) beginning of turret slide approach; (f) end of indexing. (From Acherkan, N., *Machine Tool Design,* Mir Publishers, Moscow, 1969. With permission.)

### 8.5.1.2  Swiss-Type Automatic

A Swiss-type automatic is also called a long part, sliding headstock, or bush automatic. This type of automatic was originally developed by the watch-making industry of Switzerland to produce small parts of watches. It is now extensively used for the manufacture of long and slender precise and complex parts, as shown in Figure 8.24.

#### 8.5.1.2.1  Operation Features

A Swiss-type automatic has a distinct advantage over the conventional automatic screw in that it is capable of producing slender parts of extremely small diameters with a high degree of accuracy, concentricity, and surface finish. This is possible due to its different machining technique, which is based on the following exclusive features (Figure 8.25):

- The machining is performed by stationary or cross-fed single-point tools (at $f_2$) in conjunction with longitudinal working feed $f_1$ of the bar stock.

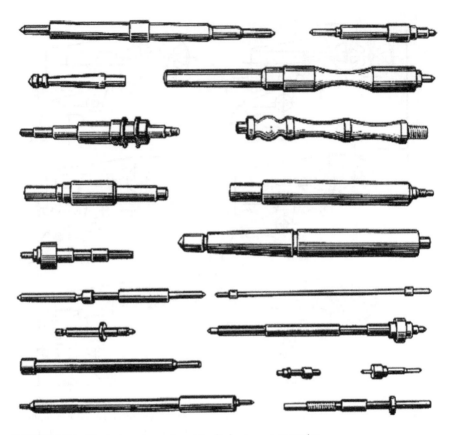

**FIGURE 8.24**   Typical parts produced by Swiss-type automatics.

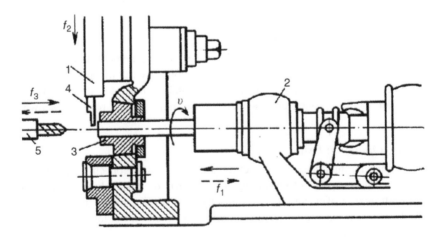

**FIGURE 8.25**   Operation of a Swiss-type automatic. (From Acherkan, N., *Machine Tool Design*, Mir Publishers, Moscow, 1969. With permission.)

- Longitudinal feed is obtained by the movement of the headstock or of a quill carrying the rotating work spindle.
- The end of the bar stock, projecting from the chuck, passes through a guide bushing, directly beyond which the cross-feeding tool (4) slides are arranged.
- Turning takes place directly at the guide bushing supporting the bar stock. The bushing then relieves the turned portion from tool load, which is almost entirely absorbed by the guide bushing. It is possible to turn a diameter as small as 60 μm.
- A wide variety of formed WP surfaces are obtained by coordinated, alternating, or simultaneous travel of headstock and the cross slides $f_2$.
- Holes and threads are machined by a multispindle end attachment carrying stationary or rotating tools performing axial feed $f_3$.

The clearance between bar stock and the guide bush is controlled to practically eliminate all radial movements. The best results are obtained by using centerless-ground bar stock as round as possible and of uniform diameter throughout the bar length. High machining accuracy is an important feature of the Swiss-type automatic. A tolerance of ±10 μm may be attained for diameters and ±20 to ±30 μm for lengths. When a wide tolerance is permitted and when the parts are not too long, the automatic screw machine is preferred for its higher productivity compared with the Swiss-type automatic. This is due to the reduced idle time of the automatic screw, whose control is based on the auxiliary shaft system.

### 8.5.1.2.2  *Machine Layout and Typical Transmission*

The Swiss-type automatic bears a slight resemblance to a center lathe. Figure 8.26 shows a general layout of this machine. The bar is fed by a sliding headstock and

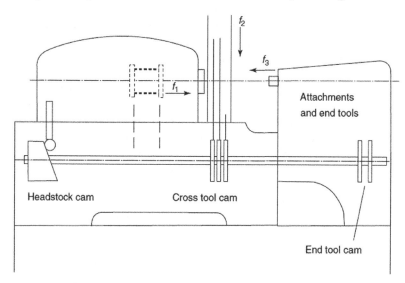

**FIGURE 8.26**  General layout of a Swiss-type automatic. (From Browne, J. W., *The Theory of Machine Tools*, Cassel and Co. Ltd., 1965.)

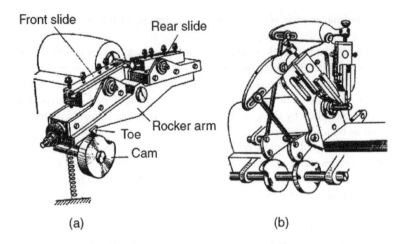

**FIGURE 8.27** Radial feed of slides in Swiss-type automatics: (a) rocker arm and (b) overhead tool slides. (From Boguslavsky, B. L., *Automatic and Semi-automatic Lathes,* Mir Publishers, Moscow, 1970. With permission.)

held in a collet chuck. The movement of the headstock is controlled by a bell or disk cam designed to suit each component. The tool-slide block carries four or five radial tool slides; the radial movement of each slide is controlled by a cam (Figure 8.27). A precise stationary bush is inserted in the tool block to guide round bars (Figure 8.25). A running bush must be used for a hexagonal bar. The bar is moved past the radially acting tools by the headstock to provide the longitudinal feed. In addition, it can move backward or pause during cutting operation as dictated by the operational layout.

A simplified line diagram of a typical transmission is shown in Figure 8.28. The machine is equipped with two motors. The main motor drives a back shaft at relatively high speed and imparts rotation to the spindle. The motion is then transmitted from the backshaft to the main camshaft through: worm and worm wheel—change gears, belt drive—worm, and worm wheel—main camshaft. On the main camshaft are various cams to control machine movements. The end attachment has an independent motor drive. The three spindles carrying end tools can be either stationary, all running, or a combination of both. The attachment can be shifted laterally according to the required sequence by a cam mounted on the camshaft. One revolution of the camshaft presents the cycle time and produces one component. The change gears in Figure 8.28 are selected according to the required cycle time.

### 8.5.1.2.3 General Guidelines When Operating Swiss-Type Automatics

These guidelines are to be followed when operating Swiss-type automatics:

1. A parting-off tool is used as a bar stop at the end of the machining cycle.
2. Recessing is accomplished by cross-feeding tools, as the WP is stationary.
3. Low feeds should be used when turning far from the guide push.
4. Using a wide parting-off tool initiates vibrations and inaccurate and rough WPs.

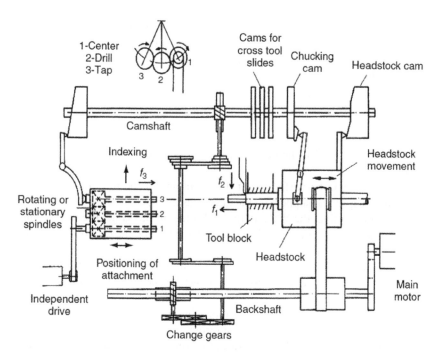

**FIGURE 8.28**  Simplified transmission diagram of a typical Swiss-type automatic. (From Browne, J. W., *The Theory of Machine Tools,* Cassell and Co. Ltd., 1965.)

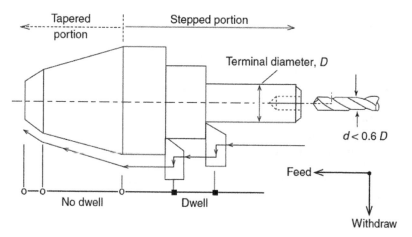

**FIGURE 8.29**  Dwells when operating Swiss-type automatics.

5. When turning stepped WPs, one tool is used, starting with the smallest diameter; then there is a stop in feeding (or a *dwell*), before the tool is moved to the next larger diameter; then it is fed for turning, and so on (Figure 8.29).

6. When turning from taper to cylinder and vice versa or from taper to taper, dwells should be avoided (Figure 8.29).

7. When positioning the tool from neutral for taper turning, a slight dwell must be allowed; otherwise, incorrect taper is produced.
8. Wide forms and long tapers are produced at lower accuracy and lower surface quality.
9. Centering must be performed before drilling deep holes.
10. Usually, a dwell is allowed after each productive motion, followed by a nonproductive one.
11. When drilling a hole of large diameter ($d = 0.6$–$0.7$ of the terminal diameter $D$), perform drilling in two stages to avoid causing the bar to recede by the axial cutting force (Figure 8.29).
12. Drilling should be performed at the beginning of the machining cycle to ensure proper WP support by the guide bush.
13. During threading, the WP and the tool rotate in the same direction (counterclockwise); accordingly, the cutting speed will be the difference of both speeds.
14. When drilling, it is preferable to use a rotating twist drill in the direction opposite to WP spindle rotation to increase the cutting speed and to shorten the machining time.
15. It is not advisable to cut with two tools simultaneously, at an angle greater than 90° to each other, to avoid chattering.
16. Tapping must be finished before the cutting-off tool reaches a diameter equal to that being threaded to avoid breakage of the WP due to the threading torque.

## 8.5.2 Horizontal Multispindle Bar and Chucking Automatics

The principal advantage of the multispindle automatics over the single-spindle automatics is the reduction of cycle time. In contrast to single-spindle automatics, where the turret face is working on one spindle at a time, in multispindle automatics, all turret faces of the main tool slide are working on all spindles at the same time.

Multispindle automatics are designed for mass production of parts from a bar stock or separate blanks. The distinguishing characteristic is that several WPs are machined at the same time from bars or blanks. According to the type of stock material of the WP, they are classified as bar- or chucking (magazine)-type automatics. Chucking machines have the same design of bar automatics, with the exception of stock feeding mechanisms. Typical parts produced by these machines are illustrated in Figure 8.30.

### 8.5.2.1 Special Features of Multispindle Automatics

Multispindle automatics have the following distinctive features:

- They may be of parallel or progressive action.
- Simultaneous cutting by a number of tools is possible.
- A nonindexable central main tool slide has a tooling position for each spindle, serving the same function as the turret of a single-screw automatic. It provides one or more cutting tools for each spindle and imparts axial feed to these tools (Figure 8.31).

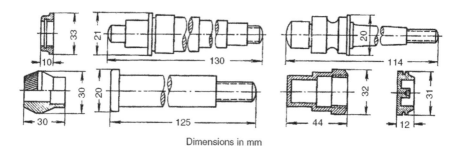

Dimensions in mm

**FIGURE 8.30** Typical parts produced on multispindle automatics. (From Acherkan, N., *Machine Tool Design,* Mir Publishers, Moscow, 1969. With permission.)

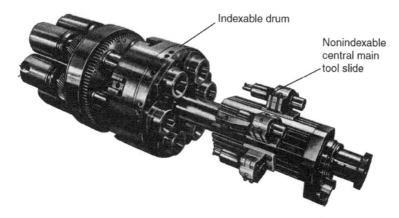

**FIGURE 8.31** Nonindexable central main tool slide of a multispindle automatic. (From Pittler Maschinenfabrik AG, Langen bei Frankfurt/M, Germany.)

- Progressive-action automatics are available in four, five, six, or eight spindles. Six-spindle automatics are the most common. The equispaced work spindles are carried by a rotating drum (headstock) that indexes consecutively to bring each spindle into a different working position (Figure 8.31).
- Parallel-action automatics are simpler than progressive-action automatics in construction, as no indexing is required.
- Multispindle automatics (parallel or progressive) have a nonindexable cross slide at each position, so that an additional tool can be fed crosswise (Figure 8.32).
- Cams are used to control the motions of the cross slides and the main tool slide. They are either specially designed or selected from a standard range at some sacrifice of optimum output.

The advantages and limitations of multispindle automatics are given here.
Advantages:

- More tooling positions and resultant higher productivity
- Greater variety of work that can be produced

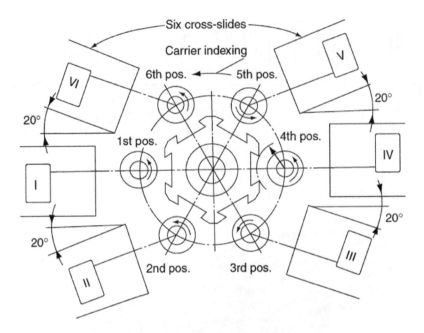

**FIGURE 8.32**   Cross slides of a progressive-action multispindle automatic. (From Acherkan, N., *Machine Tool Design*, Mir Publishers, Moscow, 1969. With permission.)

- Possibility of producing two pieces per cycle
- More economical use of floor space when continuous high output is required
- Simplicity of cams controlling the movement of different parts of the machine

Limitations:

- Higher loss when the machine is not running, according to its higher capital cost.
- Setup of the machine is rather tedious and requires a long time, as the space of the head stock and main tool slide is crowded by many tools and attachments.

### 8.5.2.2 Characteristics of Parallel- and Progressive-Action Multispindle Automatic

Figure 8.33 illustrates a multispindle automatic with parallel and progressive action. A parallel-action multispindle automatic is characterized by the following:

1. Its spindles are arranged vertically (Figure 8.33a), and it is usually a four-spindle machine.
2. The same operation is performed simultaneously in all spindles.
3. During one operating cycle, as many WPs are completed as the number of spindles.

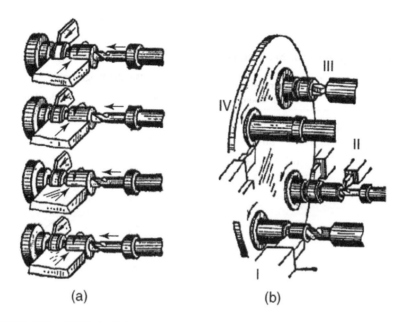

**FIGURE 8.33** Multispindle bar automatics: (a) parallel-action and (b) progressive-action. (From Chernov, N., *Machine Tools,* Mir Publishers, Moscow, 1975. With permission.)

4. Each spindle has usually two cross slides. The first is used for forming or chamfering and the other for cutting off the stock.
5. The machine can be equipped with multiple-tool spindles for drilling, boring, or threading operations.
6. Such a machine produces comparatively short parts of simple shape from bar stock. This type is also known as a straight four-spindle bar automatic.

A progressive-action multispindle automatic is characterized by the following:

1. The arrangement of spindles is radial about the axis of the spindle drum.
2. Four, five, six, or eight spindles are mounted in the spindle drum, which indexes periodically through an angle equal to the central angle between two adjacent spindles.
3. Only one machining stage is performed at each spindle position, and each WP passes consecutively through all positions according to the sequence of operations established in the setup (Figure 8.33b).
4. The setup is designed so that the WP is completely machined in one full revolution of the spindle drum, one part being completed at each indexing.
5. One of the positions is the loading or feeding position. In the bar-type automatic, the finished WP is cut off in this position, and the bar is fed out to the stop and then clamped by the collet chuck. In the chucking type, the finished part is released by the chuck; in the loading position, a new blank is loaded into the chuck from the magazine and clamped.

**TABLE 8.1**

**Switching Sequence for Six- and Eight-Spindle Bar and Chucking Automatics**

| Automatic | Progressive Bar/Chucking | Parallel/Progressive | |
|---|---|---|---|
| | | Bar | Chucking |
| Loading and Feeding Stations | (6) | (3)–(6) | (1)–(2) |
| | (8) | (4)–(8) | (1)–(2) |
| Six-spindle | 1–2–3–4–5–(6) | (spindle diagram) | (spindle diagram) |
| Eight-spindle | 1–2–3–4–5–6–7–(8) | (spindle diagram) | (spindle diagram) |

Six-spindle, Bar arrangement (circles):

```
  3   4
2       5
  1   6
```
1–2–(3)4–5–(6)

Six-spindle, Chucking arrangement (circles):

```
  6   5
1       4
  2   3
```
3–5–(1)4–6–(2)

Eight-spindle, Bar arrangement (circles):

```
  4   5
3       6
2       7
  1   8
```
1–2–3–(4)5–6–7–(8)

Eight-spindle, Chucking arrangement (circles):

```
  8   7
1       6
2       5
  3   4
```
3–5–7–(1)4–6–8–(2)

From Technical Data, Mehrspindel-Drehautomaten, 538-70083. GVD Pittler Maschinenfabrik AG, Langen bei Frankfurt/M, Germany.

A parallel- or progressive-action multispindle automatic is characterized by the following:

1. Sometimes, provision may be made in the design of six- or eight-spindle machines for two loading (or feeding) positions, usually diametrically opposed in the case of bar automatics and adjacent in the case of chucking automatics. Table 8.1 illustrates the switching sequence for six- and eight-spindle bar and chucking automatics.
2. Single indexing is required in the case of a bar automatic; double indexing is required in the case of a chucking automatic (Table 8.1).
3. Two WPs are completely machined during one full revolution of the spindle drum in a bar automatic, whereas two revolutions of the spindle drum are needed to produce two WPs in the case of a chucking automatic.
4. Parallel/progressive action is applicable for machining parts of simple shape at a high rate.

### 8.5.2.3    Operation Principles and Constructional Features of a Progressive Multispindle Automatic

The multispindle automatic has a rigid frame base construction, in which the top brace connects the headstock and the gearbox mounted at the right side of the heavy base. The base also serves as a reservoir for cutting fluid and lubricating oil. The headstock has a central bore for the spindle drum with the work spindles.

The gearing diagram of the spindles of a horizontal four-spindle automatic is shown in Figure 8.34. The power is transmitted from an electric motor (7 kW, 1470 rpm) through a belt drive, change gears $Z_1/Z_2$, continuously meshing gears $Z_3$ and $Z_4$, a long central shaft, central gear $Z_5$, and a gear ($Z_6$) to impart rotational motion to the spindles. The long central shaft should be hollow and strong to have sufficient torsional rigidity. It is evident that all spindles rotate in the same direction at the same speed. Both bar and chucking multispindle automatics are made in a considerable range of sizes. The sizes are mainly determined by the diameter of stock that can be accommodated in the spindles. The following are the main specifications of multispindle bar automatic DAM 6 × 40:

| | |
|---|---|
| **Number of work spindles** | **6** |
| Maximum bar diameter (mm) | Ø42, Hex. 36, Sq. 30 |
| Maximum bar length (mm) | 4000 |
| Maximum length of stock feed (mm) | 200 |
| Maximum turning length (mm) | 180 |
| Maximum traverses | |
| Bottom and top slides (mm) | 80 |
| Side slide (mm) | 80 |
| Height of centers over main slide (mm) | 63 |
| Speed range, normal (rpm) | 100–560 |
| Speed range, rapid (rpm) | 400–2240 |
| Progressive ratio (rpm) | 1.12 |
| Range of machining time per piece, normal (s) | 8.9–821 |
| Range of machining time per piece, rapid (s) | 5.5–206 |
| Rated power of drive motor (kW) | 17 |
| Overall dimensions (L × W × H) (mm) | 6000 × 1400 × 2280 |
| Weight (kg) | 11,000 |

A brief description of the machine elements is as follows:

*Spindle-drum (carrier) and indexing mechanism.* The spindle drum (2) is supported by and indexes in the frame of the headstock (I). It is indexed by the Geneva mechanism (3) through index arm (4) Figure 8.35a), which revolves on the main camshaft (5). The indexing motion is geared to the drum. During the working position of the machine cycle, the spindle drum is locked rigidly in position by a locking pin (6), which is withdrawn only for indexing (Figure 8.35b). A Geneva cross of five parts is preferred to index the drum of four-, six-, and eight-spindle automatics. The

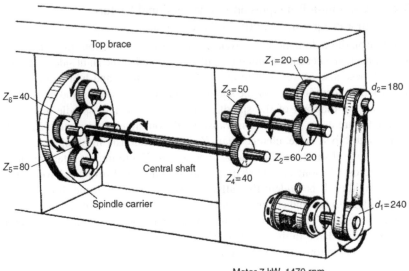

**FIGURE 8.34** Gearing diagram of a four-spindle automatic. (From Boguslavsky, B. L., *Automatic and Semi-automatic Lathes,* Mir Publishers, Moscow, 1970. With permission.)

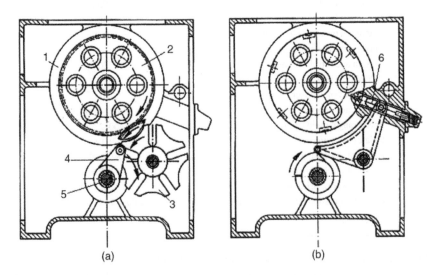

**FIGURE 8.35** Drum indexing and locking of a six-spindle automatic: (a) indexing and (b) locking.

division into five parts renders a favorable transmission of acceleration and power, thus granting a light and smooth indexing of the spindle drum.

*Spindle assembly.* Figure 8.36 shows a section through a typical assembly of one of the machine spindles. The collet opening and closing unit is similar to that of the single-spindle automatic. The spindle is mounted on fixed front bearings and a

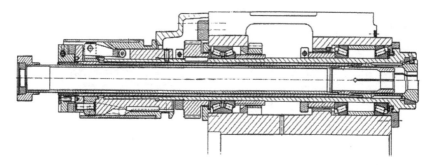

**FIGURE 8.36**  Spindle assembly of a six-spindle automatic. (From Pittler Maschinenfabrik AG, Langen bei Frankfurt/M, Germany.)

**FIGURE 8.37**  High-speed drilling attachment. (From VEB-Drehmaschinenwerk/Leibzig, Pittlerstr, 26, Germany, Technical Information Prospectus Number 1556/e/67.)

floating rear double raw tapered roller bearing; thus, differential thermal expansion between spindle and housing is allowed. The spindle expands only backward, so that its running accuracy is not affected.

### 8.5.2.3.1  Tool Slides

1. The main tool slide (end working slide) is a central block that traverses upon a round slide on an extension to the spindle drum to provide accurate alignment of the slide with the spindles. The main slide is advanced and retracted (Figure 8.37). The end tools are mounted directly on the main slide by means of T-slots or dovetails. Every tool mounted upon the slide must have the same feed and stroke. These tools are intended for plain turning, drilling, and reaming operations. Special attachments and holders for independent feed tool spindles are used when the feed of any cutting tool must differ from that of the main slide. These attachments and holders are actuated by drum cams. Figure 8.37 shows a holder carrying a high-speed drilling attachment, whereas Figure 8.38 shows an independent feed, high-spindle speed attachment. The drive mechanism of the end tool slide is shown in Figure 8.39 and performs the following steps:
   - Rapid approach of the tool slide may be either 75 or 120 mm, while the working feed may be adjusted in a range from 20 to 80 mm. The rapid approach is effected by the advance of the carriage (1) with the feed

**FIGURE 8.38** Independent feed/high-speed drilling attachment. (From VEB-Drehmaschinenwerk/Leibzig, Pittlerstr, 26, Germany, Technical Information Prospectus Number 1556/e/67.)

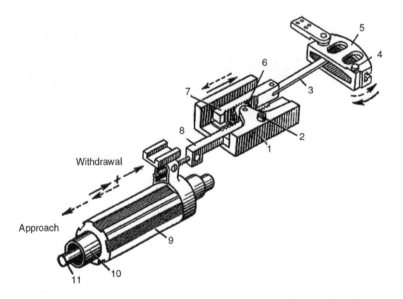

**FIGURE 8.39** Drive mechanism of the end tool slide. (From Chernov, N., *Machine Tools*, Mir Publishers, Moscow, 1975. With permission)

    lever (5) held stationary. The carriage is traversed by a corresponding cam of the main slide through the roll (2) (Figure 8.39).

- Rapid approach proceeds until the carriage runs against a stop screw (not shown in the figure). The gear (6) travels together with the carriage. This gear meshes simultaneously with the rack (7) of the tie rod (3) and the rack of the main slide (8). As rack (3) is stationary, the rack (8) and correspondingly, the main tool slide travels a distance twice that of the carriage.
- At the end of rapid approach, the carriage stops and is held stationary by the stop screw and the carriage driving cam.
- Immediately after this, another cam mounted on the camshaft actuates the feed lever (5) (Figure 8.39) through the roll (4). The rack (7) moves

the gear (6), which imparts the movement to the rack (8) and to the end tool slide.

- The length of the main slide working travel is set up by positioning a link (3) in the slot of the lever (5) with the aid of the scale located on the lever.
- Rapid withdrawal is engaged at the end of working feed. In this case, both the carriage (1) and lever (5) return to their initial positions at the same time.

2. Cross slides are intended for plunge-type cutting operations such as facing, grooving, recessing, knurling, chamfering, and cutting. They are directly mounted on the headstock of the machine and move radially to the center line of the work. Figure 8.40a and b show the cross-slide arrangement for bar and chucking six-spindle automatics, respectively. The cross slides are cammed individually; each is driven by its own cam drum. Therefore, the feed rate can be different for each side tool. The side tools feed slowly into the work to perform their cutting operations and then return to clear out spindles for indexing. In general, two slides are allocated for making heavy roughing and forming cuts. The other slides are used to complete subsequent finishing operations to the required accuracy. Except for a stock feed stop at one position, the tools on the main tool slide move forward and make the cut essentially simultaneously. At the same time, the tools in cross slides move inward and make their plunge cuts (Figure 8.41).

*Camming and cyclogram.* The main camshaft, either directly or indirectly, controls the cam movements. Hence, cams of various machining operations must be selected from a range of standard cams according to rise and feeds required. The cams for the idle motions, such as stock feeding, chucking, indexing, and so on, are standard cams and are not changed. Multispindle cams are generally composed of specially shaped

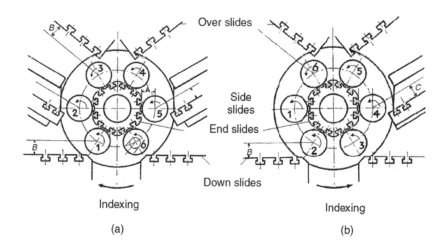

**FIGURE 8.40**  Cross-slide arrangement for (a) bar automatic and (b) chucking automatic.

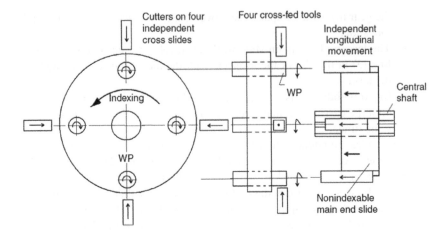

**FIGURE 8.41** Simultaneous working movements of main end slide and cross slides of a four-spindle automatic.

**FIGURE 8.42** Standard cams of cross overslides. (From Pittler Maschinenfabrik AG, Langen bei Frankfurt/M, Germany.)

segments that are bolted onto a drum to control motions. Figure 8.42 shows the cams of the cross overslides, which reduce the need for special cams.

Figures 8.43 and 8.44 show the developments of the cross overslide and the main tool-slide cam drums of a six-spindle automatic DAM 6 × 40. The working feeds of both drums occupy about 105°, and the auxiliary activities occupy 255° of the total cycle time 360°. Figure 8.45 shows the complete cyclogram of the working and auxiliary cams of a four-spindle bar automatic (on the basis of camshaft rotation angle). The cyclogram shows the sequence of events in the production of a single piece during one complete revolution of the main tool-slide cam or cross-slide cams. Prior to stock feeding, there is a rapid rise or jump toward the work, and at the end of the cut, an equal and rapid withdrawal or drawback is followed by a dwell while indexing and stock feeding. The dwell is denoted by a horizontal line, while a rising or a falling line denotes movement. Cyclograms may be of circular or developed types. The

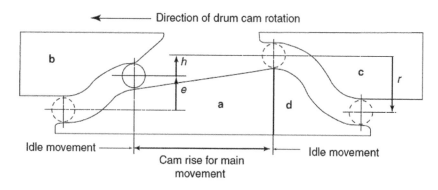

**FIGURE 8.43**   Development of the cross overslide in the direction of drum cam rotation. (From Pittler Maschinenfabrik AG, Langen bei Frankfurt/M, Germany.)

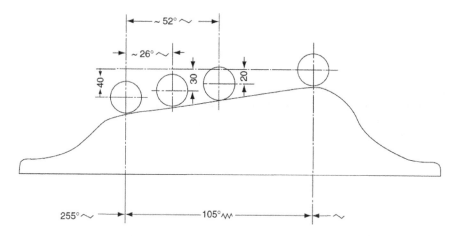

**FIGURE 8.44**   Development of the main tool-slide cam. (From Pittler Maschinenfabrik AG, Langen bei Frankfurt/M, Germany.)

developed cyclograms are more easily read. Chucking events occur during the rapid drawback of the slide (Browne, 1965).

*Setting time and accuracy of multispindle automatics.* Setting the multispindle automatic for a given job requires 2–20 h, while a piece can be often completed every 10 s. The precision of multispindle chucking or bar automatics is good but seldom as good as that of single-spindle automatics. Tolerances of ±13 to ±25 μm on the diameter are common (*Metals Handbook,* 1989), and the maximum out-of-roundness may reach 15 μm.

## 8.6   DESIGN AND LAYOUT OF CAMS FOR FULLY AUTOMATICS

The production of a WP on an automatic machine represents a symphonic master work in which different instruments (cams and tools) contribute in harmony to compose or produce the work in a predetermined playing or cycle time. The machine is

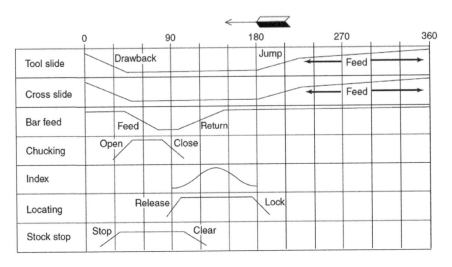

**FIGURE 8.45** Developed cyclogram of working and auxiliary cams of a four-spindle bar automatic. (From Browne, J. W., *The Theory of Machine Tools,* Cassell and Co. Ltd., 1965.)

the orchestra; the contributors (owners) of the work are the process engineer, the cam designer, and the machine setter. Setting up an automatic involves all the preparatory work required to manufacture a WP in accordance with the part drawing and specifications.

Setting up includes the following steps:

1. Planning the sequence of operation
2. Working out the calculation sheet for the setup
3. Manufacturing the necessary cams and tooling
4. Setting up the kinematic trains to obtain the required speeds and feeds
5. Installing and adjusting cams and tools on the machine

### 8.6.1  Tooling Layout and Establishing a Sequence of Operation

The sequence of operations is worked out on the basis of the specifications of the automatic as given in the machine manuals and the specifications of the WP as obtained from the working drawing.

There are general rules for developing the tool layout of a general-purpose automatic. These are based on the following machine processing features:

1. Determine the quickest and best operation sequence before designing the cams.
2. Use the highest spindle speeds recommended for the material being machined, provided the tools will stand such conditions.
3. Begin finishing only after rough cutting is completed.
4. Wherever possible, overlap the working operations and try to increase the number of the tools operating simultaneously at each position.

5. Overlap idle operations with one another and with the working operations.

6. Do not permit substantial reduction of the WP rigidity before rough cuts have been completed.

7. Obtain accurate dimensions along the WP length by cross-slide tools and not turret tools.

8. Speed up cutting-off operations, which require much time, especially in solid stock. The feed may be decreased near the end of the cut, where the piece is separated from the bar.

9. Wherever possible, break down form-turning operations into a rough cut and a finish cut.

10. Provide a dwell at the end of the cross-slide travel for clearing up the surface, removing the out-of-roundness, and improving the surface finish.

11. When deep holes are drilled, it is sometimes necessary to withdraw the drill a number of times. This facilitates chip removal and permits drill cooling.

12. When drilling a hole, first spot (center) drill the work using a short drill of a larger diameter.

13. When drilling a stepped hole, first drill the largest diameter and then smaller diameters in succession. This approach reduces the total working travel by all drills.

14. If strict concentricity and alignment are required between external and internal surfaces or stepped cylindrical surfaces, finish such surfaces in a single turret position.

15. Do not combine thread cutting with other operations. The calculated length of working travel should be increased by two or three pitches in comparison with thread length specified in the part drawing. Moreover, the actual length of travel is reduced by 10–15% of the calculated length by correspondingly reducing the radius of the cam at the end of working travel movement. Thus, the slide feed lags behind the tap or die movement along the thread being cut, excluding any possibility of stripping the thread due to incorrect feed of the slide by the cam. The tap or the die should have a certain amount of axial freedom in its holder.

16. If only two or three positions of the turret are occupied:
    - Index the turret through every other position and use a swing stop.
    - Machine two WPs every cycle.

17. To obtain equal machining times at all positions of multispindle automatics, divide the length to be turned into equal parts, or increase the feed or cutting speeds at positions where a surface of longer length is to be turned.

18. In all cases, use standard tools and attachments whenever possible.

*Tooling layout.* The tooling layout for all operations consists of sketches drawn to a convenient scale of WP, tools, and holders in the relative position that they occupy at the end of the working travel movement. The lengths of travel should be indicated. These sketches are checked against the setup characteristics of the working members and are also used to check whether the tools and slides interfere with one another during operation. The tooling layout serves as initial data for working out the operation sheet and the cam design sheet.

## 8.6.2 CAM DESIGN

The cam design for automatics is tedious work including definite steps depending on the type of automatic machine. However, the following main guidelines are to be generally observed in the design process:

1. Determine the number of spindle revolutions required for each operation and idle movements.
2. Overlap those operations and idle movements that can take place simultaneously.
3. Proportion the balance of spindle revolutions on the surface of the cam so that the total of these revolutions equals the full circumference of the cam.

Within the spaces reserved for turret operations, in the case of automatic screw machines, lobes are developed to feed the tools on the work. The radial height (throw) of these lobes equals the length a tool will travel on the work, and the gradient of the cam lobe governs the rate of tool feed. The lobes are connected by drops or rises, and in these spaces, idle movements take place. The cross-slide cams revolve at the same rate as the turret cam, and operations performed by cross-slide tools are laid out on these cams.

The cam blank surface is divided into 100 equal divisions (Figure 8.46). The radius $R$ is equal to the distance from the cam follower center to the fulcrum of the

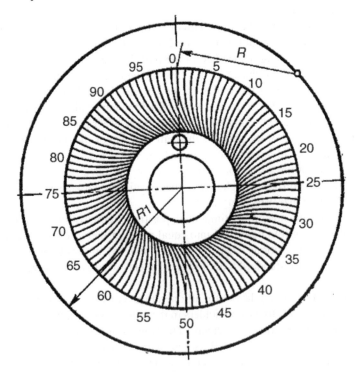

**FIGURE 8.46** Blank of disk cam. (From Chernov, N., *Machine Tools*, Mir Publishers, Moscow, 1975. With permission.)

lever carrying the roll; the arc centers are located on the lever fulcrum circle of the radius $R_1$, given in the machine manual. The cams are drawn on a full scale. Marking out the cam begins from a zero arc and proceeds clockwise, provided that the turret-slide cam is watched from the rear side of the machine and the cross-slide cams from the turret side. Feeding and clamping the bar begins from zero arc. Whenever a tool has not been moved, the corresponding cam outline is formed by a circle arc drawn from the cam center. To construct the withdrawal and approach curves of the turret-slide cam and the cross-slide cams, a special drawing template is provided as a supplement to the machine's service manual (Figure 8.47).

The detailed procedure of cam layout for different automatics can be carried out in the following sequences.

1. Single-spindle screw automatic:
   • Determine the machine size and specifications.
   • Determine the operational sequence.
   • Determine the tool geometry and tool material to be employed.
   • Select permissible cutting speeds for the material to be machined according to the operations to be performed.
   • Calculate the spindle revolutions per minute by

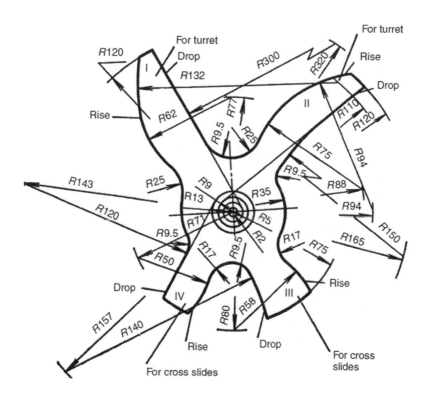

**FIGURE 8.47**  Cam template for drawing curves for idle travel movements of an automatic screw machine.

$$n = \frac{1000v}{\pi D} \text{rpm}$$

where

    $v$ = cutting speed in meters per minute

    $D$ = work diameter in millimeters

- Determine the travel of each tool, considering an approach of 0.5 mm to avoid damage to tools.
- Select the feed rate (mm/rev) for each tool depending on the material to be machined, the type of operation to be performed, and the tool material.
- Calculate the number of spindle revolutions for each cutting operation and the idle movements.
- Determine the cycle time expressed in number of spindle revolutions required to complete one component and determine the corresponding ratio of the pick-off gears $(a/b) \times (c/d)$, as previously mentioned.
- Calculate the hundredths of cam surface needed for both cutting and idle operation by converting revolutions into hundredths.
- Establish the operation and cam design sheets.

2. Swiss-type automatic:
- Reproduce the component accurately to a suitable scale, showing various dimensions.
- Determine the operation sequence.
- Select the cutting tools.
- Determine the travel of all tools, as well as feeds and cutting speeds, required spindle revolutions for each operation element, lobes of the plate cams, and so on.
- Compile the cam layout sheet.
  - Determine the sequence of operations.
  - Determine the rises and falls on the cams and the time (expressed in degrees) required to perform the movements.

3. Multispindle automatic:
- Determine the machine to be employed.
- Determine tool geometry and tool material.
- Select permissible cutting speed for cutting and threading if necessary.
- Calculate the number of spindle revolutions for each operation.
- Determine the throw of the main tool slide.
- Determine the feed rate of the main tool slide.
- Determine the cutting time and establish the idle time.
- Determine the throw of each cross-slide cam required.
- Find the time relationship between the parting-off tool and the threading operation to make sure that the threading operation is completed before the component is finally cut off.
- Establish timing chart (cyclogram) so that the cycle time can also be determined.
- Draw the tool layout and record data.

### 8.6.3 ILLUSTRATIVE EXAMPLES ON CAM DESIGN AND TOOLING LAYOUT FOR SINGLE- AND MULTI-SPINDLE AUTOMATICS

The following examples illustrate the cam layout procedure as applied to different types of automatic.

#### Illustrative Example 1

The component shown in Figure 8.48 is to be produced by mass production on a turret automatic screw. It is made of brass 85, and the bar stock is of 28 mm diameter. To increase productivity, the machine is supplemented by a milling attachment to take care of milling the product after the cutting-off operation by gripping and moving it far from the machining area while the next product is in operation.

Design and draw a set of plate cams required for the production of this component. Illustrate the tool layout without considering the cam layout necessary for the milling, attached for simplicity.

#### SOLUTION

After the machine is specified, the operational sequence is determined as indicated in Figure 8.49.

1. The cutting speed for brass in the case of using HSS tools:

Turning: $v$ = 132 m/min
Threading: $v$ = 42.5 m/min

Determination of spindle rotational speed:

$$\text{Turning}: n_s = \frac{1000v}{\pi D} = \frac{1000 \times 132}{\pi \times 28} = 1500\,\text{rpm}$$

$$\text{Threading}: n_s = \frac{1000 \times 42.5}{\pi \times 18} = 750\,\text{rpm}$$

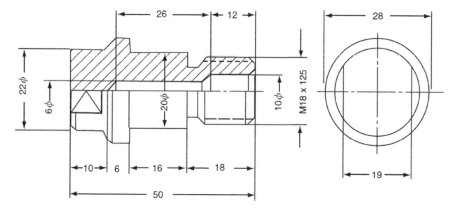

**FIGURE 8.48** Product to be produced on an automatic screw machine. (From Index-Werke AG, Esslingen/Neckar, Germany. With permission.)

Turret stations 1-6

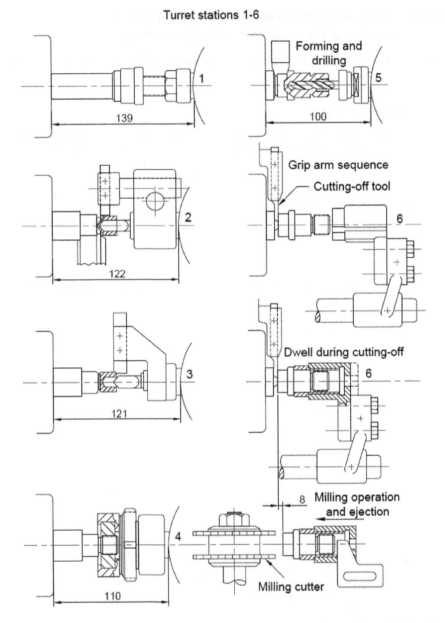

**FIGURE 8.49** Operational sequence for the selected part. (From Index-Werke AG, Esslingen/Neckar, Germany. With permission.)

2. The sequence of operation is illustrated in Table 8.2, and the tool layout is given in Figure 8.49.
3. The throw or travel of each tool is determined as in Table 8.2 column (b). Add up to 0.4 mm approach to avoid tool damage.

## TABLE 8.2
## Operation Sheet of the Part Produced on Automatic Screw Machine Index 24

Part Drawing (see Figure 8.48)  Machine: Index 24  Work Material: Brass Ms 58, 28Ø

| | | | | | | Spindle Speed (rpm) | Turning 1500 / Thread Cutting 750 |
| | | | | | | Cutting Speed (m/min) | Turning 132 / Thread Cutting 42.5 |

| Tool Station | Operation Sequence (a) | Throw (mm) (b) | Feed/Rev (mm) (c) | Revolutions Per Operation (d) | Number (e) | Cam From (f) | Cam To (g) | Main (In rev) (h) | Auxiliary (%) (i) | Cam Drawing From (k) | Cam Drawing To (l) |
|---|---|---|---|---|---|---|---|---|---|---|---|
| Turret Slide | | | | | | | | | | | |
| 1 | Feed stock to stop (1 s) | | | | | | | | 5 | 0 | 5 |
| | Index turret (2/3 s) | | | | | | | | 3 | 5 | 8 |
| 2 | Turn for thread/drill 10Ø | 17 | 0.133 | 128 | 23 | – | – | 128 | | 8 | 31 |
| | Index turret | | | | | | | | 3 | 31 | 34 |
| 3 | Chamfer for thread or break hole edge | 1.5 | 0.14 | 11 | 2 | | | 11 | | 34 | 36 |
| | Index turret | | | | | | | | 4 | 36 | 40 |
| 4 | Thread cutting   on 1:2   off 1:1 | 16th | 1.25 | 32 | 6 | | | 32 | | 40 | 46 |
| | Index turret | 16th | | 16 | 3 | | | 16 | | 46 | 49 |
| 5 | Drilling 6Ø with high speed drill 1600 rpm | 25.5 | 0.11 / 0.23 | 110 | 20 | | | 110 | | 49 | 53 |
| | Index turret (half) 1/3 s | | | | | 73 | 77 | | 2 | 53 | 73 |
| 6 | Vacant hole to clear grip arm | | | | 4 | 77 | 96 | | | 73 | 75 |
| | Index turret (half) 1/3 s | | | | 4 | 96 | 0 | | 2 | 98 | 0 |

(Continued)

**TABLE 8.2 (CONTINUED)**
**Operation Sheet of the Part Produced on Automatic Screw Machine Index 24**

Part Drawing (see Figure 8.48)  Work Material: Brass Ms 58, 28Ø  Machine: Index 24

| | Turning | Thread Cutting |
|---|---|---|
| Spindle Speed (rpm) | 1500 | 750 |
| Cutting Speed (m/min) | 132 | 42.5 |

| Tool Station | Operation Sequence | Throw (mm) | Feed/Rev (mm) | Revolutions Per Operation | % of Cam Circumference | | | Figures for Calculating Cycle Time | | | |
|---|---|---|---|---|---|---|---|---|---|---|---|
| | | | | | | Cam | | | | Cam Drawing | |
| | | | | | Number | From | To | Main (In rev) | Auxiliary (%) | From | To |
| | a | b | c | d | e | f | g | h | i | k | l |
| Grip Arm | Swing down (half) | 6 | | | 6 | 72 | 78 | | 3 | 75 | 78 |
| | Dwell | | | | – | – | – | | 1 | 78 | 79 |
| | Advance for picking up | | | | | | | | 6 | 79 | 85 |
| | Dwell while cutting off | | | | 4 | 85 | 89 | | – | | |
| | Dwell after cutting off | | | | 2 | 90 | 92 | | 1 | 89 | 90 |
| | Relief arm from stop | | | | – | – | – | | – | | |
| | Withdrawn | | | | 10 | 92 | 102 | | 3 | 90 | 93 |
| | Swing up (half) | | | | 27 | 7 | 34 | | 5 | 93 | 98 |
| Cross Slides | Front: 28–16Ø | 6 | 0.04 | 150 | | | | | | | |
| | Back: 28–20Ø | 4 | 0.04 | 100 | 18 | 55 | 73 | | | | |
| | Cutting off 28–3Ø | 12.5 | 0.065 | 192 | 35 | 50 | 85 | 22 | | 85 | 89 |
| | 3-center | 1.5 | 0.065 | 22 | 4 | – | – | | | | |
| | Past center | 1 | 0.1 | 10 | 2 | 89 | 91 | | | | |
| | | | | | | | | 319 | 42 | | |

$$\text{Revolutions/four pieces} = \frac{319}{58} \times 100 = 550 \text{ rev}$$

$$\text{Cycle time} = \frac{550}{1500} \times 60 = 22 \text{ s}$$

4. Selected feeds per revolution are illustrated in Table 8.2 column (c). The following notes are especially important:
   - Feeds for forming and cutting off are much smaller than those for drilling and turning.
   - In high-speed drilling (turret station 5), the feed for a stationary drill equals 0.11 mm/spindle revolution. If the drill is rotating in the opposite direction from the spindle at 1600 rpm, then the equivalent feed should equal 0.11 × (1500 + 1600)/1600 = 0.23 mm/rev.
   - In threading, the feed equals the pitch of the thread. To avoid the possibility of stripping the thread, the slide feed should lag behind the die movement by about 10% of its throw (Figure 8.50); see turret station 4. Also it is evident from Figure 8.49, at turret station 5, that threading is performed alone and not combined with other operations.

5. Calculation of the number of work spindle revolutions for different operations, from Table 8.2 column (d), according to

$$n^* = \frac{\text{Throw}}{\text{Feed}}$$

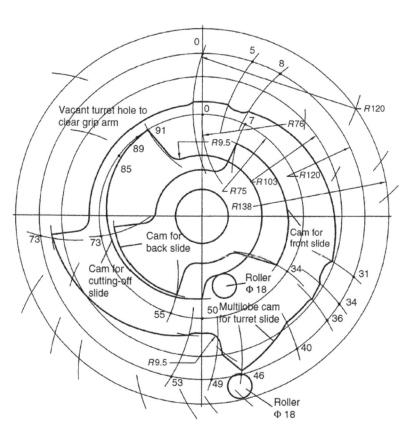

**FIGURE 8.50** Cam layouts for the selected part. (From Index-Werke AG, Esslingen/Neckar, Germany. With permission.)

For example, at turret station 2:

$$n^* = \frac{17}{0.133} = 128 \, \text{rev and so on.}$$

6. Determination of the cutting or working time expressed in spindle revolutions without considering time of overlapped operations (Table 8.2), column (h).

Sum of spindle revolutions: $n = 319$ rev

7. Determination of idle movements in hundredths of cam circumference (Table 8.2), column (i), assuming the following allowances.

| | |
|---|---|
| Feeding stock to stop and clamping (1 s) | = 5% |
| Turret indexing (2/3 s) | = 3–4% |
| Turret half indexing (1/3 s) | = 2% |
| Grip arm allowances | |
| Swing down (half) | = 3% |
| Dwell | = 1% |
| Advance for picking up | = 6% |
| Dwell after cutting off | = 1% |
| Withdraw | = 3% |
| Swing up (half) | = 5% |

Columns (e), (f), (g), (k), and (i) (Table 8.2) could now be completed.

8. Calculation of cycle time in seconds, referring to Table 8.2.

$$\text{Total revolutions / pieces} = \frac{319}{58} \times 100 = 550 \, \text{rev}$$

$$\text{Cycle time,} \, T_{cyc} = \frac{550}{1500} \times 60 = 22 \, \text{s}$$

Therefore, the cam work sheet (Table 8.3) for the required component can be constructed, and the cam layout is shown in Figure 8.50.

## Illustrative Example 2

It is required to produce in mass production the long part (brass) shown in Figure 8.51 on the Swiss-type automatic, Model 1 π16 Stankoimport (spindle speed: 400–5600 rpm, bar capacity: 16 mm, and rated power: 3 kW).

Suggest an operational sequence and tooling layout. Establish a cam design sheet and calculate the product cycle time.

### SOLUTION

Figure 8.52 illustrates the proposed tooling layout. The operational sequence is shown in Table 8.4. Three tools are sufficient to perform the work. Tools I and II are mounted on the rocker arm, and tool III is mounted on the overhead slide.

**TABLE 8.3**

**Cam Work Sheet of the Part Produced on Automatic Screw Machine Index 24**

| Operation | Hundredths | Overlapped | Range |
|---|---|---|---|
| Feed to stop | 5 | | 0–5 |
| Index turret | 3 | | 3–8 |
| Turn for thread/drill 10Ø | 23 | | 8–31 |
| Front slide forming 28–16Ø | 27 | 7–34 | |
| Index turret | 3 | | 31–34 |
| Chamfer/countersink | 2 | | 34–36 |
| Index turret | 4 | | 36–40 |
| Threading | | | |
| ON | 6 | | 40–46 |
| OFF | 3 | | 46–49 |
| Index turret | 4 | | 49–53 |
| Cutting off 28/3Ø | 35 | 50–85 | |
| Drill 6Ø | 20 | | 53–73 |
| Back slide, forming 28–20Ø | 18 | 55–73 | |
| Index turret (half) | 2 | | 73–75 |
| Vacant turret hole to clear grip arm | | 77–96 | |
| Swing down grip arm (half) | 3 | | 75–78 |
| Dwell of grip arm | 1 | | 78–79 |
| Advance for picking up | 6 | | 79–85 |
| Cutting off 3Ø-center | 4 | | 85–89 |
| Dwell while cutting off | 4 | 85–89 | |
| Dwell after cutting off | 1 | | 89–90 |
| Cutting off paste center | 2 | 89–91 | |
| Relief arm from stop | 2 | 90–92 | |
| Withdrawn | 3 | | 90–93 |
| Swing up (half) | 5 | | 93–98 |
| Index turret (half) | 2 | | 98–100 |

The procedure is carried out according to the following sequence:

1. Determination of the spindle speed:

$$v = 100\,\text{m}/\text{min}\left(\text{WP}:\text{brass}, \text{Tool}:\text{HSS}\right)$$

$$n = \frac{1000v}{\pi D} = \frac{1000 \times 100}{\pi \times 14} = 2274\,\text{rpm}$$

The spindle speed $n$ is selected to be 2240 rpm.

2. The tool travel (throw) and the selected feeds are listed for each operation in columns (c) and (d) of Table 8.4. Accordingly, the number of spindle revolutions column (e) can be calculated. For example, for operation 6 (Table 8.4):

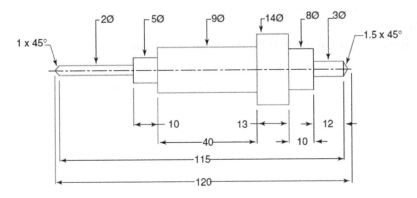

**FIGURE 8.51**  Part to be produced on a typical Swiss-type automatic.

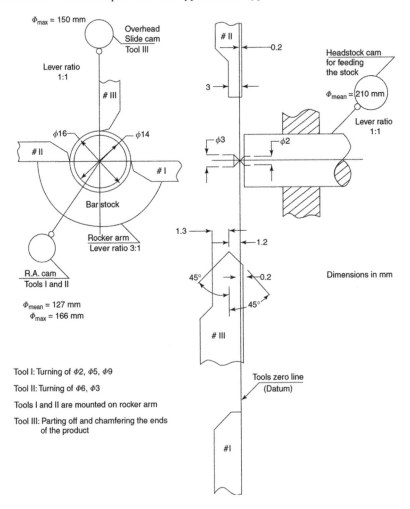

**FIGURE 8.52**  Proposed tooling layout.

| Tool travel | = 15.1 mm |
| Feed | = 0.05 mm/rev |
| Number of revolutions | = 15.1/0.05 = 302 rev |

From Table 8.4, the total number of revolutions to perform the main (productive) operations = 996.

3. Allowances for idle (nonproductive) activities are assumed in degrees; column (g).
   Total of idle activities = 123°.
   Therefore, total of main activities = 237°.

4. Determination of the time $T_{cyc}$:

$$\text{Expressed in revolutions} = 996 \times \frac{360°}{237°} = 1513 \text{ rev}$$

$$\text{Expressed in seconds} = \frac{1513}{2240} \times 60 = 40.5 \text{ s}$$

5. The main (productive) activities, as expressed in degrees instead of revolutions, are listed in column (f) of Table 8.4.

6. Rises and drops on different cams are calculated, column (i), by considering the lever ratio of each slide, column (h).

7. The cam layout data are completed:
   - Degrees on cam circumference, column (j).
   - Lobe radii (mm), column (k).

8. Use data in columns (j) and (k) to draw the cams.

## Illustrative Example 3

A batch size of 50,000 pieces is to be produced on a six-spindle bar-type automatic. The part (Figure 8.53) is made of steel 20 ($\sigma_u$ = 40–50 kg/mm$^2$). The bar size is 27 mm in diameter. Provide a tooling layout and calculate the cycle time.

### SOLUTION

*Tooling layout.* The sequence of operation and tooling layout should be written in advance, and the best one should be chosen with regard to lower tooling cost. In multispindle automatics, a proper sequence of operation necessitates distributing the machining operation so that all operations have approximately the same machining time. Elements of operation that require much time are sometimes divided between two or even three positions in order to increase the production capacity.

The tooling setup is shown in Figure 8.54. It may be written as follows:

First station: Turn length 40 mm and spot drill before drilling by the central main slide. Rough front forming by cross slide.

Second station: (Drilling needs too much time, so it is divided into three parts, performed in second, third, and fourth stations.) Turn remainder (40 mm) and drill first part (30 mm) by the central main slide.

## TABLE 8.4
## Cam Design Sheet for the Long Part Produced on a Swiss-Type Automatic

Part Shape (see Figure 8.51) Material: Brass 70 Bar Stock of 14 mm Diameter

Machine: 1 π16—Swiss-type, Stankoimport Cutting Speed, $v$ = 100 m/min Spindle Speed, $n$ = 2240 rpm Cycle Time, $T_{cyc}$ = 40.5 s

| Operation No. | Sequence of Operation | Cam Name | Tool Travel (Throw) (mm) | Feed mm/rev | Rev No. of Revolutions of Spindle | Degrees Productive (Main) | Degrees Nonproductive (Idle) | Lever Ratio | Rise or Drop on Cam (mm) | Cam Layout Data Degrees (Range) | Lobe Radius (mm) (Range) |
|---|---|---|---|---|---|---|---|---|---|---|---|
| | a | b | c | d | e | f | g | h | i | j | k |
| 1. | Open chuck | HS | | | | | 10 | | | 0–10 | 105–105 |
| 2. | Back movement of HS | HS | 60.1 | | | | 30 | 1:1 | 60.1 | 10–40 | 105–39.9 |
| 3. | Close chuck | HS | | | | | 15 | | | 40–55 | 39.9–39.9 |
| 4. | Exit tool III | III | 8.1 | | | | (4) | 1:1 | 8.1 | (55)–(59) | 75–66.9 |
| 5. | Enter tool I | I | 7.0 | | | | 10 | 3:1 | 21.0 | 55–65 | 63.5–42.5 |
| 6. | Turn φ2 mm | HS | 15.1 | 0.05 | 302 | 72 | | | | 65–137 | 39.9–55 |
| 7. | Dwell tool I to clean up | HS | | | | | 2 | | | 137–139 | 55–55 |
| 8. | Exit I to φ5 mm | I | 1.5 | | | | 4 | 3:1 | 4.5 | 139–143 | 42.5–47 |
| 9. | Dwell tool I | I | | | | | 2 | | | 143–145 | 47–47 |
| 10. | Turn φ5 mm (I) | HS | 5.0 | 0.07 | 71 | 17 | | 1:1 | 5.0 | 145–162 | 55–60 |
| 11. | Dwell tool I | HS | | | | | 2 | | | 162–164 | 60–60 |
| 12. | Exit I to φ9 mm | I | 2.0 | | | | 6 | 3:1 | 6.0 | 164–170 | 47–53 |
| 13. | Dwell tool I | I | | | | | 2 | | | 170–172 | 53–53 |
| 14. | Dwell φ9 mm (I) | HS | 25.0 | 0.08 | 313 | 75 | | 1:1 | 25.0 | 172–247 | 60–85 |
| 15. | Dwell tool I | HS | | | | | 2 | | | 247–249 | 85–85 |
| 16. | Exit I to φ16 mm | I | 3.5 | | | | 10 | 3:1 | 10.5 | 249–259 | 53–63.5 |
| 17. | Stock feeding | HS | 9.3 | | | | 10 | 1:1 | 9.3 | 259–269 | 85–94.4 |

(Continued)

# TABLE 8.4 (CONTINUED)
## Cam Design Sheet for the Long Part Produced on a Swiss-Type Automatic

Part Shape (see Figure 8.51) Material: Brass 70 Bar Stock of 14 mm Diameter

Machine: 1 π16—Swiss-type, Stankoimport
Cutting Speed, $v$ = 100 m/min Spindle Speed, $n$ = 2240 rpm Cycle Time, $T_{cyc}$ = 40.5 s

| | | | | Rev | Degrees | | | | Cam Layout Data | |
| | | | | | | | | | | |
| Operation No. | Sequence of Operation | Cam Name | Tool Travel (Throw) (mm) | Feed mm/rev | No. of Revolutions of Spindle | Productive (Main) | Nonproductive (Idle) | Lever Ratio | Rise or Drop on Cam (mm) | Degrees (Range) | Lobe Radius (mm) (Range) |
| | a | b | c | d | e | f | g | h | i | j | k |
|---|---|---|---|---|---|---|---|---|---|---|---|
| 18. | Enter II to φ14.2 mm | II | 0.9 | | | | (3) | 3:1 | 2.7 | (269)–(272) | 63.5–66.2 |
| 19. | Feeding II to φ8 mm | II | 3.1 | 0.05 | 62 | 14 | | 3:1 | 9.3 | 269–283 | 66.2–75.5 |
| 20. | Dwell II | II | | | | | 2 | | | 283–285 | 75.5–75.5 |
| 21. | Turn φ8 mm (II) | HS | 5.0 | 0.08 | 63 | 15 | | 1:1 | 5.0 | 285–300 | 94.3–99.3 |
| 22. | Dwell tool II | II | | | | | 2 | | | 300–302 | 75.5–75.5 |
| 23. | Enter II to φ3 mm | II | 2.5 | 0.04 | 63 | 15 | | 3:1 | 7.5 | 302–317 | 75.5–83 |
| 24. | Dwell tool II | II | | | | | 2 | | | 317–319 | 83–83 |
| 25. | Turn φ3 mm (II) | HS | 5.7 | 0.06 | 78 | 19 | | 1:1 | 5.7 | 319–338 | 99.3–105 |
| 26. | Dwell tool II | HS | | | | | 2 | | | 388–340 | 105–105 |
| 27. | Exit II to φ16 mm | II | 6.6 | | | | (8) | 3:1 | 19.5 | (340)–(348) | (83)–63.5 |
| 28. | Enter III to φ3.2 mm | III | 6.4 | | | | 8 | 1:1 | 6.4 | 340–348 | 66.9–73.3 |
| 29. | Feed III for parting off | III | 1.7 | 0.04 | 44 | 10 | | 1:1 | 1.7 | 348–358 | 73.3–75 |
| 30. | Dwell | III | | | | | 2 | | | 358–360 | 75–75 |
| | | | | | 996 | 237 | 123 | | | | |

Cycle time in
revolutions = $\dfrac{996}{237} \times 360 = 1513$ rev

$T_{cyc} = \dfrac{1513}{2240} \times 60 = 40.5$ s

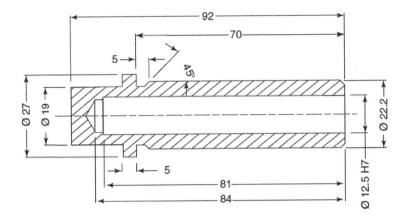

**FIGURE 8.53**   Part produced on a six-spindle automatic.

Third station: Rough face, and form by cross slide, support, and drill second
    part (30 mm) by the central main slide.
Fourth station: Finish rear forming by cross slide, support, and drill third part
    by the central main slide.
Fifth station: (Reaming is not recommended with cutting-off station.) Fine
    face, form, chamfer, and size the outer flange by cross slide, support, and
    ream by the central main slide.
Sixth station: Cut off by cross slide.

## DETERMINATION OF SPINDLE SPEEDS AND TOOL FEEDS

The material of the WP: steel 20
    The bar stock: 27 mm diameter
    HSS is selected as tool material for all tooling. The recommended cutting speeds
and feeds are listed in Table 8.5.

### Spindle Speeds

The cutting speeds for the different operations are calculated as follows.

$$\text{Turning}: n = \frac{1000v}{\pi \times D} = \frac{1000 \times 45}{\pi \times 27} = 530\,\text{rpm}$$

$$\text{Forming}: n = \frac{1000 \times 30}{\pi \times 27} = 353\,\text{rpm}$$

$$\text{Cutting off}: n = \frac{1000 \times 30}{\pi \times 19} = 502\,\text{rpm}$$

$$\text{Drilling}: n = \frac{1000 \times 40}{\pi \times 12.5} = 1018\,\text{rpm}$$

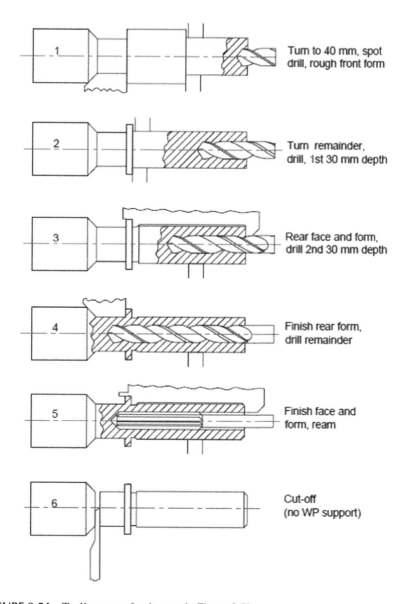

**FIGURE 8.54**    Tooling setup for the part in Figure 8.53.

$$\text{Reaming}: n = \frac{1000 \times 15}{\pi \times 12.5} = 382\,\text{rpm}$$

The smallest spindle speed of 353 is selected to suit the most severe operation of rough forming at the first station. All spindles run at the same and the lowest speed. For this reason, multispindle automatics generally operate at a smaller spindle speed as compared with single-spindle automatics. The next lower spindle speed,

$n_s$ = 350 rpm, is selected from those available and listed in the machine service manual. Pick-off gears are used to provide this speed.

If thread cutting is to be performed, the cutting speed is selected according to the material to be cut and the pitch of the thread. The relatively low threading speed is achieved by rotating the threading tool in the same direction of the spindle rotation such that the difference realizes the required threading speed.

## TOOL FEEDS AND CALCULATION OF THE MACHINING TIME FOR EACH OPERATION

The machining time $t_m$ (s) is calculated in terms of tool travel $L_t$ (mm), spindle speed $n_s$ (rpm), and tool feed rate $f$ (mm/rev):

$$t_m = \frac{L_t \times 60}{n_s \times f} s \qquad (8.9)$$

The operational machining time ($t_m$) is calculated for the different stations. Consider again the sequence of operation and tool layout.

First station:

Turn, $L$ = 40 mm, $f$ = 0.1 mm/rev, $n$ = 350 rpm, $t_m$ = 69 s
Rough form, $L$ = 4.5 mm, $f$ = 0.1 mm/rev, $n$ = 350 rpm, $t_m$ = 8 s, overlap

Second station:

Turn, $L$ = 40 mm, $f$ = 0.1 mm/rev, $n$ = 350 rpm, $t_m$ = 69 s
Drill, $L$ = 30 mm, $f$ = 0.1 mm/rev, $n$ = 350 rpm, $t_m$ = 52 s, overlap

Third station:

Drill, $L$ = 30 mm, $f$ = 0.1 mm/rev, $n$ = 350 rpm, $t_m$ = 52 s
Rough form, $L$ = 4.5 mm, $f$ = 0.1 mm/rev, $n$ = 350 rpm, $t_m$ = 8 s, overlap

Fourth station:

Drill, $L$ = 30 mm, $f$ = 0.1 mm/rev, $n$ = 350 rpm, $t_m$ = 52 s

## TABLE 8.5
## Recommended Speeds and Feeds for Different Operations, Tool HSS and WP = Steel 20

| Operation | Speed (m/min) | Feed (mm/rev) |
|---|---|---|
| Turning | 45–55 | 0.05–0.18 |
| Forming and cutting off | 30–40 | 0.02–0.05 |
| Drilling | 40–50 | 0.04–0.12 |
| Reaming | 10–15 | 0.10–0.18 |

Finish form, $L = 4.5$ mm, $f = 0.1$ mm/rev, $n = 350$ rpm, $t_m = 8$ s, overlap

Fifth station:

Ream, $L = 81$ mm, $f = 0.18$ mm/rev, $n = 350$ rpm, $t_m = 77$ s
Separate reaming attachment required to allow additional reamer
Feed of 0.08 mm/rev relative to that of the central slide
Fine face, $L = 4.5$ mm, $f = 0.1$ mm/rev, $n = 350$ rpm, $t_m = 8$ s, overlap

Sixth station:

Cut-off, $L = 15$ mm, $f = 0.05$ mm/rev, $n = 350$ rpm, $t_m = 52$ s

The time of reaming is found to be the longest, and therefore, it determines the machine productivity. Assuming idle time $t_a = 3$ s, then the cycle time $T_{cyc}$ (floor-to-floor time) can be calculated as follows:

$$T_{cyc} = \left(t_m\right)_{max} + t_a$$
$$= 77 + 3 = 80\,s$$
$$= 1.33\,min$$

And the hour production rate = 60/1.33 = 45 pieces/h.

## 8.7 REVIEW QUESTIONS AND PROBLEMS

8.7.1	Mark true or false:
   [ ] Swiss-type automatics are best suited to turning long slender parts in mass production.
   [ ] In automatics, the main camshaft rotates one revolution per machining cycle.
   [ ] In a turret automatic screw machine, the auxiliary shaft rotates more slowly than the main camshaft.
   [ ] In Swiss-type automatics, turning occurs near to the guide bush.
   [ ] Threading and parting off can be performed simultaneously on Swiss-type automatics.
   [ ] Draw-in collet chucks produce the most accurate parts on automatics.
   [ ] The spindle rotational speed in automatics considerably affects the cycle time of a product.
   [ ] The productivity of six single-spindle automatics is exactly the same as that of a six-spindle automatic of the same size.

8.7.2	What are the necessary measures to reduce the cycle time in automatics?

8.7.3	List some important rules to be considered when operating a Swiss-type automatic.

8.7.4    What are the main operation features of a Swiss automatic?

8.7.5    What is the distinct advantage of a Swiss automatic over a single-spindle automatic screw machine?

8.7.6    "The material feeding in a Swiss-type automatic is not an auxiliary movement." Discuss this statement briefly.

8.7.7    What spring collets are available for bar automatics? What type do you recommend for:
- Single-spindle automatic
- Multispindle automatic?

8.7.8    What is a long part? On which machine can it be produced, and why?

8.7.9    A single-spindle bar automatic is set as shown in the following table to produce the same component at each setting.

| Setting | Spindle Speed (rpm) | Camshaft Speed (rpm) |
|---|---|---|
| First setting | 2000 | 2 |
| Second setting | 1500 | 3 |

8.7.10   Compare the two settings from the following points of view:
- Productivity
- Accuracy and surface finish

8.7.11   What are the two main types of multispindle automatic?

8.7.12   In an automatic screw machine, at what speed does the auxiliary control shaft revolve? List the functions it performs. How is the speed of the camshaft set up? List the functions of the main camshaft.

8.7.13   Mention one of the special attachments that can be provided on automatic screw machine. Why are these special attachments sometimes necessary?

## REFERENCES

Acherkan, N 1969, *Machine tool design*, vols. 1–4, Mir Publishers, Moscow.

Boguslavsky, BL 1970, *Automatic and semi-automatic lathes*, Mir Publishers, Moscow.

Browne, JW 1965, *The theory of machine tools*, vols. 1 and 2, 1st edn, Cassel and Co. Ltd, London, UK.

Chernov, N 1975, *Machine tools*, Mir Publishers, Moscow.

VEB-Drehmaschinenwerk/Leibzig, Pittlerstr, 1967, 26, Germany, Technical Information Prospectus Number 1556/e/67.

Index-Werke AG, Esslingen/Neckar, Germany.

Maslov, D, Danilevesky, V, & Sasov, V 1970, *Engineering manufacturing processes*, Mir Publishers, Moscow.

ASM International *Metals handbook* 1989, Machining, vol. 16, ASM International, Materials Park, OH.

Technical Data, Mehrspindel-Drehautomaten, 538–70083. GVD Pittler Maschinenfabrik AG, Langen bei Frankfurt/M, Germany.

# 9 Numerical Control and Computer Numerical Control

## 9.1 INTRODUCTION

In conventional or manually operated machine tools, the process starts from the part drawing, and the machinist is responsible for the entire job. The machinist determines the machining strategy, sets up the machine, selects proper tooling, chooses machining feeds and speeds, and manipulates machine controls to cut a part that will pass inspection. It is clear that using this method of machining involves a considerable number of decisions that influence the accuracy and surface finish of the machined part.

Numerical control (NC) is a system that uses prerecorded information prepared from numerical data to control a machine tool or the machining process. NC describes the control of machine movements and various other functions by instructions expressed as a series of numbers and initiated via an electronic control system. Figure 9.1 shows operator-controlled and numerically controlled machine tools.

Computer numerical control (CNC) is the term used when the control system includes a computer. Figure 9.2 shows the difference between NC and CNC of machine tools. Manufacturing areas of NC, CNC, and direct numerical control (DNC) include flame cutting, riveting, punching, piercing, tube bending, and inspection. NC and CNC are particularly suitable for the manufacture of a small number of components needing a wide range of work, such as those with complex profiles or a large number of holes. They are also suitable for batch work. In NC machining, the part programmer analyzes the drawing, decides the sequence of operations, and prepares the manuscript in a language that the NC system can understand. As shown in Figure 9.3, the NC system consists of data input devices, a machine control unit (MCU), a servo drive for each axis of motion, a machine-tool operative unit, and feedback devices. The program written and stored on the tape is read by the tape reader, which is a part of the MCU. The MCU translates the program and converts the instructions into the appropriate machine-tool movements. The movement of the operative unit is sensed and fed back to the MCU. The actual movement is compared with the input command, and the servo motor operates until the error signal is zero.

The history and development of NC dates back to 1952, when the first NC conventional milling machine was demonstrated at the Massachusetts Institute of Technology (MIT). In 1957, aircraft manufacturers installed a milling machine—the beginning of NC technology—that was used for machining complex profiles for the aircraft and aerospace industries. Drilling machines, jig borers, lathes, and other NC

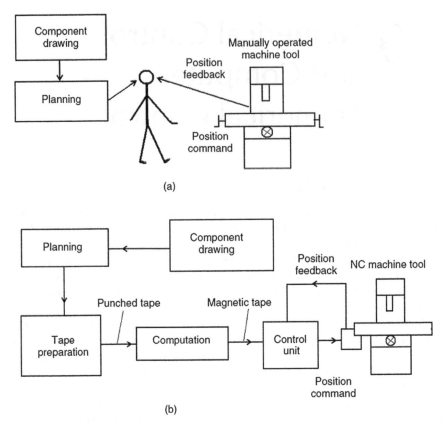

**FIGURE 9.1** Operator-controlled and numerically controlled machine tools: (a) manual machine tool and (b) NC machine tool.

machine tools were soon developed with less tooling and more operations performed in the same setup, and involvement of the operator in controlling the machine was avoided. NC machining centers and turning centers then appeared and gave machine designers and builders a chance to improve NC-machined products in terms of accuracy and surface quality.

The development in the electronics industry played a key role in the growth and acceptance of NC machine tools. Since the 1960s, smaller electronic components such as transistors, resistors, and diodes have increased the reliability and reduced the size and cost of machine tools. The development of integrated circuits in 1965 led to a further reduction of the size and cost of the control units and provided the basis for the use of minicomputers in CNC and DNC machining.

Earlier systems of NC machines consisted of a specially built control unit permanently connected to the machine tool. They are relatively inflexible, as they are special-purpose machine tools. Developments in the area of miniaturization and integration of circuits have led to the introduction of new, small, and powerful computers that are used to control the machine tools (CNC) instead of a conventional controller. The advantages of CNC are related to the control system, which allows a

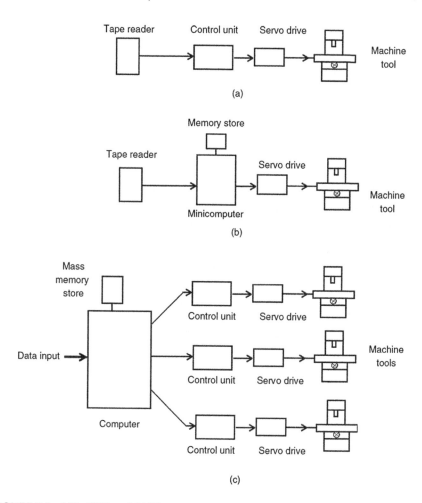

**FIGURE 9.2** NC, CNC, and DNC concepts: (a) conventional NC, (b) CNC, and (c) DNC.

great deal of flexibility, unobtainable with NC. DNC involves controlling more than one machine using the same computer and data transmission lines. The major advantage of CNC and DNC over NC is that punched tapes are not used directly to control the machine tool. Instead, all information flows from a computer that interfaces with each MCU (see Figure 9.2).

NC machines cost approximately 5 to 10 times as much as conventional machines of the same size, depending on the capacity of the control system and accessories. Figure 9.4 shows the total cost against the total quantity of parts being produced using different machining methods. At a volume of zero, the fixed cost of machining by NC includes tape preparation and setup in addition to the costs related to the design and fabrication of holding fixtures whenever required. When using conventional machine tools, this cost includes the design and fabrication of tooling, fixtures (when required), and setup. Manual preparation and machine adjustments require more time than tape preparation. For special-purpose and automatic machines,

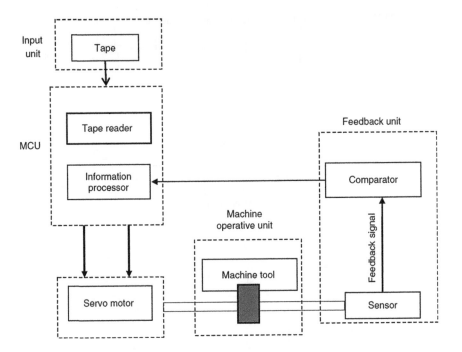

**FIGURE 9.3**  Main components of the NC system.

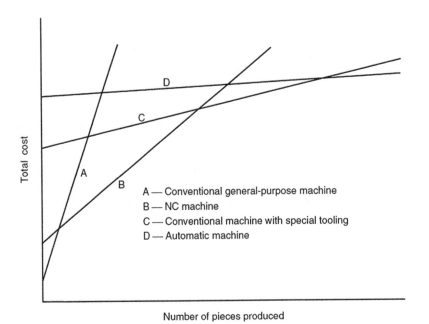

**FIGURE 9.4**  NC cost compared with other methods.

the design and fabrication of special tooling, manual setup, and adjustment of the machine are expensive.

With NC flexibility, the setup costs are often less than for conventional machines, smaller lot sizes are economical, and less floor space is needed for materials in process and storage. Referring to Figure 9.4, it is obvious that NC cannot compete with fixed-program special-purpose machines and tools when producing large quantities of pieces. In this regard, cam-operated automatic machines are simple, direct, and fast for turning operations, while transfer machines are specialized for machining certain products more effectively than NC machines for large quantities. NC cannot compete in terms of the machining cost with the special-purpose machines used for mass production. Their ultimate benefit is achieved when machining small and medium-size runs. Generally, NC can be used when:

- The tooling cost is high compared with the machining cost by conventional methods.
- The setup time is large in conventional machining.
- Frequent changes in tooling and machine setting are required.
- Parts are produced intermittently.
- Complex-shaped components are needed.
- Parts are expensive, and human errors are costly.
- Design changes are frequent.
- 100% inspection is required.

Advantages of NC include the following:

1. *Greater flexibility.* With NC, a wide variety of operations can be performed, changeovers from one run to another through tape or program changes can be made rapidly, and design changes to parts can be made rapidly through minor changes to the part program.
2. *Elimination of templates, models, jigs, and fixtures.* The NC control tape takes over the job of locating the cutting tools, which eliminates the design, manufacture, and use of templates, jigs, and fixtures.
3. *Easier setups.* By using simpler work-holding and locating devices, the operator does not have to set table limit stops or dogs, or depend on the feed screw dials when setting up for machining.
4. *Reduced machining time.* Machining with NC allows the use of a wider range of speeds and feeds than conventional machine tools. Optimum selection of feed rates and cutting speeds is ensured. The NC equipment can also move from one cutting operation to the next faster than the operator, which significantly reduces the total machining time.
5. *Greater accuracy and uniformity.* During NC machining, no human errors are possible, and machining of the same part is performed in the same way through the stored tape or program, which improves the uniformity and interchangeability of the machined parts. Therefore, inspection time is greatly reduced and is necessary for the first piece only, in addition

to random checks for critical dimensions. Hence, scrape and rework are greatly reduced or completely eliminated by using NC.

6. *Greater safety.* The operator is not as closely involved with the actual machining operations as with conventional machine tools. As the tape is checked out before actual production runs, there is less chance of machine damage that may cause human injuries.

7. *Conversion to the metric system.* An NC system can be converted to accept either inch or metric inputs.

Disadvantages of NC include the following:

1. NC follows programmed instructions that can lead to machine destruction if not properly prepared.

2. NC cannot add any extra machining capability to the machine tool, as no more power from the original drive motor and no more table travel than originally built into the machine tool can be added.

3. NC machines cost 5 to 10 times more than conventional machines of the same working capacity. The machine, therefore, cannot remain idle and needs special maintenance.

4. The skills required to operate an NC are usually high because of the sophisticated technology involved, which requires part programmers, tool setters, punch operators, and maintenance staff who are more educated and well-trained than conventional machine operators.

5. Special training for personnel in software and hardware is very important for successful adoption and growth of the NC technology.

6. NC requires high investments in terms of wages of highly skilled personnel and expensive spare parts.

## 9.2  COORDINATE SYSTEM

### 9.2.1  MACHINE-TOOL AXES FOR NC

In NC, the standard axis system is used to plan the sequence of positions and movements of the cutting tool. A drilling machine can be described using two or three of the axes $X$, $Y$, and $Z$. There are three rotational axes $a$, $b$, and $c$ around $X$, $Y$, and $Z$, respectively. The right-hand rule, shown in Figure 9.5, defines the relative positions of $X$, $Y$, and $Z$; Figure 9.6 shows the direction of positive rotation $a$, $b$, and $c$ around the $X$-, $Y$-, and $Z$-axes. The machine axis and motion nomenclature are published according to the Electronics Industry Association (EIA) standard. Figure 9.7 shows the designation of some typical machine tools. Accordingly, in NC turning and cylindrical grinding machines, $X$- and $Z$-axes are only required, where $X$ is in the radial direction and $Z$ is in the axial direction of the workpiece (WP).

### 9.2.2  QUADRANT NOTATION

As shown in Figure 9.8, a quadrant is a quarter of a circle in the Cartesian coordinate. Quadrants are numbered counterclockwise (ccw) from first to fourth. The

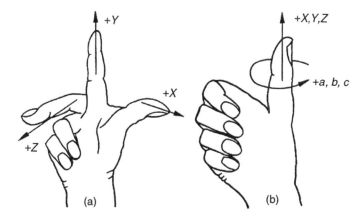

**FIGURE 9.5** Right-hand rule: (a) relative positions of X, Y, and Z axes; (b) rotation a, b and c around X, Y, and Z axes.

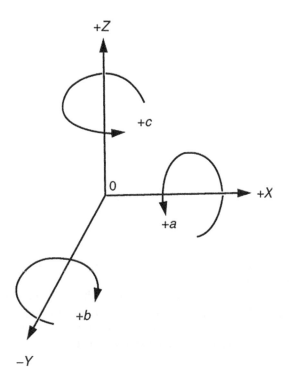

**FIGURE 9.6** Relative positions and direction of rotation.

positive and negative signs are taken from the zero point (0, 0, 0), where Z is positive in the direction perpendicular to the paper. In most NC machines, the work is carried out in the first quadrant. The programmer, therefore, must be familiar with the use of signs when programming in specific quadrants.

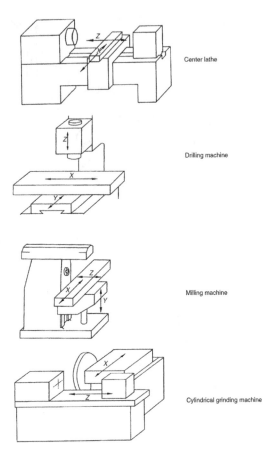

Center lathe

Drilling machine

Milling machine

Cylindrical grinding machine

**FIGURE 9.7**  Standard axes of some NC machines. Note: $Z$ is the direction of the machine spindle.

### 9.2.3 Point Location

This is used for locating points in the $X–Y$ plane. As shown in Figure 9.9, point A = 2,4 means that point A is located at $X = 2$ and $Y = 4$ from the zero point. The programmer should specify the correct dimension and the proper plus or minus sign for the hole or the point location in relation to the established zero point and the quadrant used. If all points are in the third quadrant, the minus sign can be avoided by the MCU. Figure 9.9 shows the locations of the following tabulated points:

| Point | X | Y |
| --- | --- | --- |
| A | 2 | 4 |
| B | −3 | 2 |
| C | −3 | −4 |
| D | 3 | −3 |

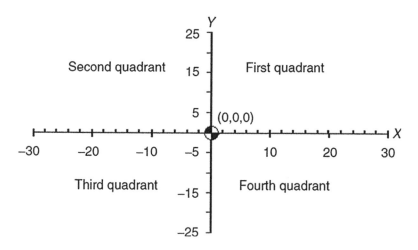

**FIGURE 9.8** Quadrant notation.

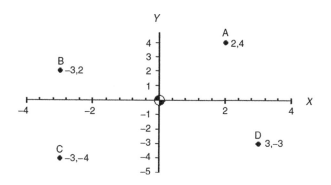

**FIGURE 9.9** Point location.

The centerline of the machine spindle is usually taken as the $Z$-axis, which is positive in the direction from the WP toward the tool. The plane formed by the $X$- and $Y$-axes, as in algebra, is perpendicular to the $Z$-axis. Although machine tools of two or three axes can be easily programmed, machine tools of four or five axes require computer assistance in writing the NC programs.

### 9.2.4 ZERO POINT LOCATION

The zero point location is where $X$, $Y$, and $Z$ intersect and the point from which all coordinate dimensions are measured. This point can be either fixed by the manufacturer (fixed zero) or determined by the programmer (floating zero). In the fixed zero location, the point of $X = 0$ and $Y = 0$ is located at a specific point on the machine table and cannot be changed. Accordingly, the coordinates of the center of the hole in Figure 9.10 are (20, 25). Floating zero is found in some NC machine tools where the programmer can select the location of the zero point at any convenient spot on

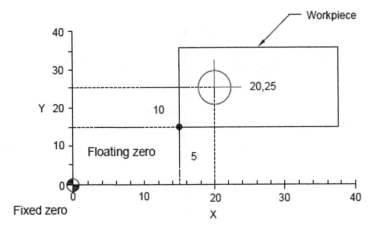

**FIGURE 9.10**   Fixed zero and floating zero.

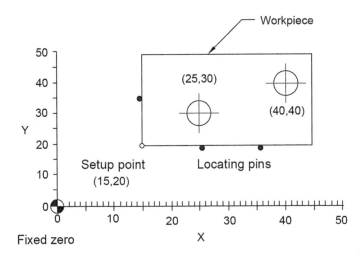

**FIGURE 9.11**   Setup point.

the machine table. Accordingly, the center of the hole location is (5, 10). Figure 9.10 shows fixed and floating zeros.

### 9.2.5   SETUP POINT

The setup point is actually on the WP or the fixture holding the WP, which tells the setup person where to place the part or the fixture (holding the part) on the machine table. Hence, holes and other machining operations are performed in the correct locations on the part when the tape or program is used. As can be seen in Figure 9.11, the setup point may be the intersection of two previously machined edges. In other situations, it may be the center of a previously machined hole, or a dowel or a hole on a given location on the fixture. The setup point must be accurately located in relation

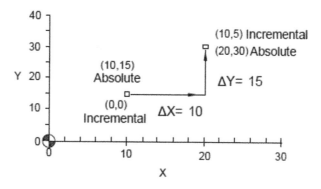

**FIGURE 9.12**   Absolute and incremental positioning.

to the zero point. In some cases, the zero and the setup points may coincide with each other in one point. The setup point should keep the part in a convenient place on the machine tool that ensures the ease of loading and unloading of the machined part.

### 9.2.6  ABSOLUTE AND INCREMENTAL POSITIONING

In absolute positioning, the tool locations are always defined in relation to the zero point. This setup is easy to check and correct, and programming mistakes affect only one line of the NC program.

In incremental positioning, the next tool position or location is defined with reference to the previous tool location, which is usually considered to be (0, 0). In such a system, any mistakes will affect all subsequent programmed positions. To check the incremental positioning, the tool must return to the original position of the program. Figure 9.12 shows an example of the absolute and incremental positioning methods.

## 9.3  MACHINE MOVEMENTS IN NUMERICAL CONTROL SYSTEMS

NC control systems are built to provide specific movements, such as simple movements used in drilling holes or complex ones used in the milling of dies and mold cavities. These movements include the following:

*Point-to-point (PTP) NC*: As shown in Figure 9.13, when drilling holes in a
   WP, the following steps are performed:
   1.   The spindle goes to the specific hole location on the WP (X, Y) position.
   2.   The tool then stops to perform drilling, reaming, boring, taping, coun-
        terboring, and countersinking at the programmed feed rate.
   3.   When the operation is finished, the tool goes to the next location, stops,
        and performs another operation.
   4.   The tool does not come in contact with the WP while moving from one
        point to another at a speed of 2.54 m/min in a path that is not important.
*Straight-cut NC*: This movement is used during the machining of successive
   shoulders in a WP or cutting rectangular shapes on the milling machine,

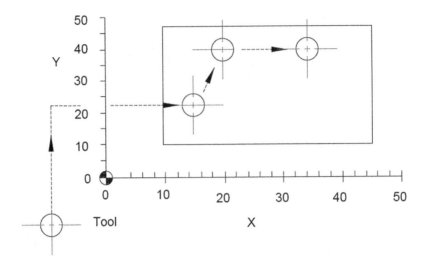

**FIGURE 9.13**   PTP control.

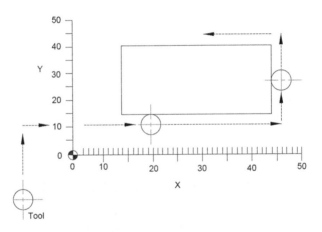

**FIGURE 9.14**   Straight-cut NC.

as shown in Figure 9.14. Such a system is equipped to control the feed rate as the tool travels from one point to another. Tool movements are restricted to lines parallel to the coordinate axes of the machine or at 45° to the axes. NC machine tools are often equipped with PTP systems in addition to the straight-cut movement that can be used for hole drilling and simple milling operations.

*Contouring (continuous-path) NC*: This movement is used for machining contours and other complex shapes. According to Figure 9.15, the tool moves at a controlled feed rate in any direction in the plane described by two axes. The cutting tool motion is limited only by the number and range of axes under control (three to five axes). The method by which a continuous-path

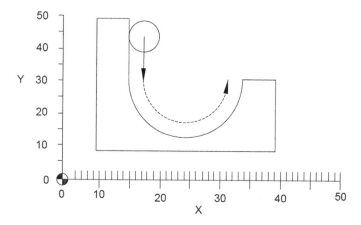

**FIGURE 9.15**   Contouring NC.

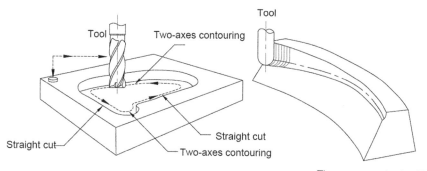

**FIGURE 9.16**   Examples of straight and contouring NC.

system moves the tool from one programmed point to the next is called *interpolation*; *resolution* is the minimum movement that can be commanded by the NC control unit. It is the table or the slide movement resulting from a single pulsed command. If a single pulse produces 0.025 mm of slide (table) movement, the resolution is then 0.025 mm. Machine tools are available at different resolutions depending upon the degree of precision required by the user of the machine. The smaller the resolution, the higher the possible accuracy. PTP systems are normally built with 0.026 mm, and contouring NC systems are built with 0.0025 mm. Figure 9.16 shows typical profiles cut by straight-cut and contouring NC.

*Combination systems*: Although a PTP system is effective for drilling, taping, and boring operations, straight-cut NC is effective for face milling. Combination systems may include PTP and straight-cut systems, which are common in machine tools used for milling, drilling, and boring. Additionally, PTP and continuous-path systems are used for machining profiles in addition to the work done by the PTP system.

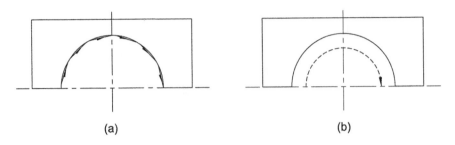

**FIGURE 9.17**   (a) Linear and (b) circular interpolations.

## 9.4   INTERPOLATION

Interpolation is the method of getting from one programmed point to the next so that the final WP shape is a satisfactory approximation of the programmed design. Today, continuous-path NC control systems can be supplied with four general types of interpolation: linear, circular, parabolic, and cubic interpolation.

> *Linear interpolation.* The cutting tool motion between two points is controlled in a straight line. Curves are then broken into a series (number) of straight-line tool movements that is sufficient to produce an approximation of the desired WP shape within the given tolerance. The smaller the tolerance, the more points are required to define the desired shape, as shown in Figure 9.17a.
> *Circular interpolation.* For machining an arc, the points needed are the coordinates of the center point and the start and finish points. A code is required to specify the direction of the cut in addition to the desired feed rate (Figure 9.17b).
> *Parabolic interpolation.* Produces parabolic tool paths with the minimum inputs and uses fewer blocks than circular interpolation.
> *Cubic interpolation.* Developed with the use of computers, whereby sophisticated cutter paths can be produced with few input data points.

## 9.5   CONTROL OF NUMERICAL CONTROL MACHINE TOOLS

The movement of NC machine tools is controlled using automatic control systems, which include the MCU, the drive motors, and other equipment. The main function of the control unit is to read and interpret instructions, store information until the time comes to use it, and send signals to the machine tool to get the appropriate movement for creating the finished WP. The two types of control systems are as follows:

> 1. *Open-loop control system.* As shown in Figure 9.18, this system is used in machine tools to perform specific movements without any check on whether the desired movements actually take place. Such a control system is simple and inexpensive. Because there is no provision for checking the actual movements, they must have extremely accurate, reliable, and responsive drive mechanisms, such as stepping motors. In this case, the frequency

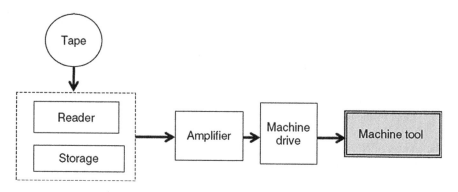

**FIGURE 9.18** Open-loop control system.

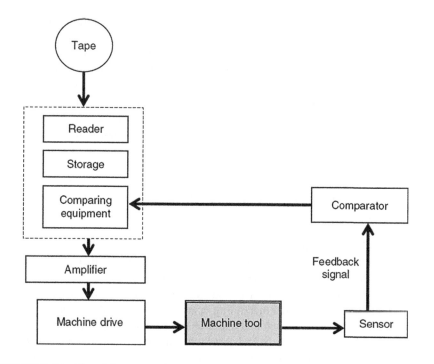

**FIGURE 9.19** Closed-loop control system.

and the total number of pulses determine the direction (+ or − charge), speed, and distance of travel resulting from the stepping motor.

2. *Closed-loop control system.* In such a system (Figure 9.19), the actual movement is checked by the feedback system. The measured slide or tool movement is then compared with the original input instruction, and the difference produces an error signal that is used to drive the system. In this system, the servo mechanisms are used to control the machine-tool movements to move the slide or the table to the desired position. In a closed-loop

control system, linear transducers, fitted to the machine table, provide the necessary feedback required for the servomotors to position the worktable accurately in accordance with the requirements of the program. Rotary transducers measure the angular displacement of a machine rotary element such as a lead screw or the spindle of the lathe machine. This measurement is essential for synchronizing the rotation of the WP and the axial movement of the tool during screw cutting on the lathe.

The control of NC machine tools includes spindle rotation, slide movements, tooling, work holding, and supplementary functions:

1. *Slide movements.* The accuracy of machined parts is affected by the operator's skill, especially when positioning the WP and the tool in the correct position relative to each other. Machine tools may have more than one slide, and therefore, the slide required to move must be identified. The plane of the slide movement may be horizontal, transverse, or vertical. These planes are referred to as axes and are designated by the letters $X$, $Y$, and $Z$. The $Z$-axis always relates to a sliding motion parallel to the spindle axis. The rate of travel in millimeters per revolution or in millimeters per minute is proportional to the revolutions per minute of the servomotor. Therefore, controlling that motor will in turn control the slide movement. The motor is controlled electronically via the MCU in a variety of ways such as paper tape, magnetic tape, and computer link.

2. *Control of spindles.* Rotary movements are controllable via the machining program. They are identified by the letters $a$, $b$, or $c$ (see Figure 9.6). Machine-tool spindles are driven directly or indirectly by electric motors. The degree of automatic control over these movements includes stopping and starting and the speed and direction of the rotation. The speed of the spindle is often infinitely variable and will automatically change as cutting takes place to maintain a programmed surface speed.

3. *Control of tooling.* NC machine tools may incorporate in their design turrets or magazines that hold a number of cutting tools. The machine controller can be programmed to cause indexing of the turret or the magazine to present a new tool to the machining operation or to facilitate tool removal and replacement when automatic tool-changing devices are involved.

4. *Control of work holding.* NC machine tool control can extend to loading the WP by the use of robots and securely clamping it by activating hydraulic or pneumatic clamping systems.

## 9.6  COMPONENTS OF NUMERICAL CONTROL MACHINE TOOLS

NC machine tools are made from specially designed parts, which ensures the production of accurate machined WPs. These parts include the following:

1. *Machine-tool structures.* Cast iron (CI) is widely used for building NC machine-tool structures, as it possesses adequate strength and rigidity in addition to a high tendency to absorb vibrations. Complex shapes can also be

produced by casting of one-piece box construction, which is heavily ribbed to promote rigidity and stabilized by an appropriate heat treatment. For large machines, fabricated steel structures are used that ensured a reduction in weight while ensuring adequate strength and rigidity. The general use of steel structures is, however, limited by the problems of making complex structures and the low tendency to vibration damping. Concrete is used as a machine tool-foundation; it has low cost and good damping characteristics.

2. *Machine spindles.* The machine-tool spindles are subjected to a radial load that may cause deflection. Additionally, spindle assembly is subjected to an axial load acting along its axis. Inadequate spindle support leads to dimensional inaccuracies, poor surface finish, and chatter. Spindle overhang of the turning and other horizontal machines must be kept to a minimum, as shown in Figure 9.20. Vertical machining spindles may slide up and down, which causes them to become extended, thus raising the risk of deflection. To overcome this problem, the spindle assembly is made to move up and down with minimum overhang (Figure 9.21a). To avoid the possible twist of the spindle housing that is located between two substantial slideways, a bifurcated or two-pillar structure is used (Figure 9.21b).

3. *Lead screws.* NC machines are fitted with recirculating ball screws that replace the normal sliding motion with a rolling motion, resulting in reduction of the frictional resistance. As shown in Figure 9.22, the balls make opposing point contact, which eliminates backlash. Ball screws, in comparison to Acme screws, offer longer life, less wear, less frictional resistance, less necessary drive power, higher traversing speeds, no stick/slip effect, and more precise positioning over the total life of the machine.

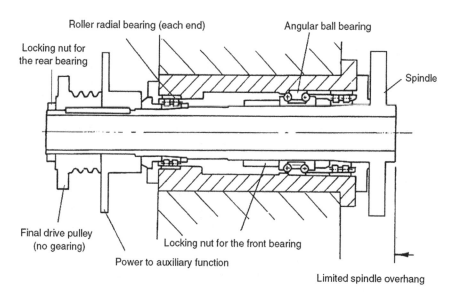

**FIGURE 9.20**   Spindle assembly for NC machine tools.

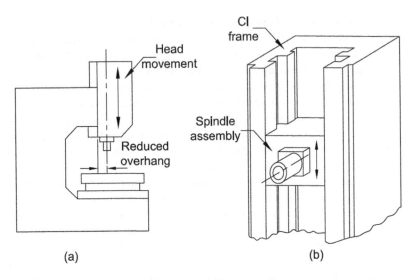

**FIGURE 9.21** Minimizing spindle deflection: (a) spindle assembly with reduced overhang; (b) bifurcated or two-pillar structure.

4. *Machine slides.* Machine-tool slides must be smooth and have minimum frictional resistance and low wear to ensure dimensional accuracy. Slides of flat bearing surfaces are widely used for NC machine tools. Such surfaces are usually hardened, ground, and coated with polytetrafluoroethylene (PTFE). The coated material has a low coefficient of friction plus a tendency to retain lubricant with superior load-carrying capacity. In some machines, the flat bearing is replaced by the rolling action of balls or rollers. Such an arrangement reduces the frictional resistance and thus also reduces the power required to achieve movement.

5. *Spindle drives (speed).* The majority of NC machine tools use electric rather than hydraulic motors. Electric motors provide sufficient power and speed for a wide range of applications. The main spindle speed may reach 5000 rpm when using diamond tools in turning. It may attain 20,000 rpm in some other applications. Such high speeds require special ceramic bearings. The maximum speed depends on the power of the drive motor, the type of bearings used, and the lubrication system. Although 5 kW is normally available, high power, in the range of 20–30 kW, is available for high machining rates. Alternating current (ac) motors have not been generally used for driving the NC spindles directly, because specialized and expensive electrical equipment is required to provide high power with accurate stepless variable speed. It is necessary to have a variable-speed unit to obtain the speed variation of the spindle required. Direct current (dc) motors can, however, supply a sufficient power with stepless variable speeds.

6. *Slide drives (feed).* The operative units that provide the feed movement are not as powerful as those used for driving the main spindles, and feed motors of 1 kW are adequate. Additionally, the feed rates are in the range 5–200

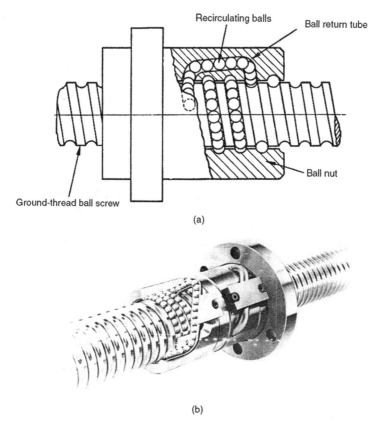

Recirculating balls

Ball return tube

Ball nut

Ground-thread ball screw

(a)

(b)

**FIGURE 9.22** External recirculating ball screw: (a) schematic diagram; (b) photographic image.

mm/min during machining and 5 m/min during rapid positioning. Such feed rates can be obtained using a screw and nut driving system where a screw of 5 mm pitch rotates at 1–40 rpm during machining and 1000 rpm during rapid positioning. It is essential that the movement provided by the feed motors is controlled very precisely and accurately. Generally, dc motors are used in closed-loop control systems for moving the tools or WP under precise control. Open-loop systems use stepper motors where the drive unit receives a direction input (clockwise [cw] or ccw) and pulse inputs. For each pulse received, the drive unit manipulates the motor voltage and current, causing the motor shaft to rotate by a fixed angle (one step). The lead screw converts the rotary motion of the motor shaft into linear motion of the WP or tool feed.

7. *Power units for ancillary services.* ac induction motors are generally used for coolant pumps, chip removal equipment, and driving hydraulic motors, where the only control required is on/off switching.

8. *Positional feedback.* Rotary-type synchronic systems transmit angular displacement to voltage signals. Such systems are composed mainly of the

rotor and the stator. The rotor rotates with the lead screw, while the stator is fixed around its periphery. The stator winding is fed with the electrical power at a rate that is determined by the MCU in response to the digital information related to the required slide movement, received via the part program. As the lead screw rotates, a voltage is induced in the rotor that will vary according to the angular position of the lead screw in relation to the stator windings. Information related to the induced voltage is fed back to the control unit, which counts the number of revolutions of the lead screw, thus confirming that the movement achieved corresponds to the original instructions.

9. *Optical gratings.* Transducers of this type transmit linear movement as a voltage signal in the form of a series of pulses. Two optical gratings are used; one is fixed to the main frame of the machine, and the other is attached to the moving slide. The number of pulses collected from the photo transmitter is fed back to the control unit as a confirmation that the correct movement has been made.

## 9.7 TOOLING FOR NUMERICAL CONTROL MACHINES

The most important points to be considered are:

*Tool materials.* Although high-speed steel (HSS) tools are used for small-diameter drills, taps, reamers, end mills, and slot drills, the bulk of tooling for NC machining involves the use of cemented carbides. Hardness and toughness are necessary requirements for a tool material. In this regard, HSS tools possess high toughness but are not hard and therefore cannot be used for high material removal rates. The hardness of cemented carbides is almost equal to that of diamond. However, lack of toughness presents a major problem, which can be improved by the addition of cobalt to the tungsten carbides (WCs). Titanium and tantalum carbides are also used. Coated and nanocoated tools provide high wear resistance and thus increase the tool life by up to five times.

*Solid carbide tools.* These are used when the WP material is difficult to machine using HSS tools. Solid carbide milling cutters of 1.5 mm diameter, small drills of 0.4 mm diameter, and reamers as small as 2.4 mm diameter are available. Such tools should be short, mounted with minimum overhang, and used on vibration-free NC machine tools.

*Indexable inserts.* These have the correct cutting geometry and precise dimensions and are located in special holders or cartridges. Such inserts do not require resharpening and ensure rapid replacement. The inserts are indexable; that is, as the cutting edge becomes blunt, the insert is moved to a new position to present a new edge to the machining process. A facility for the control of swarf is ensured by forming a groove in the insert that works as a chip breaker.

*Tool turrets.* Automatically indexable turrets, shown in Figure 9.23, are used to accommodate cutting tools. These turrets are programmed to rotate to a

**FIGURE 9.23**   Indexable tool turrets.

new position so that a different tool can be presented at work. Indexable turrets are used in the majority of turning centers as well as some NC milling and drilling machines. Turrets are now available that can accommodate 8 to 10 tools. Some machines have two turrets; one is in use, while the other is loaded with tools for a particular job and attached to the machine when required. Turrets fitted to NC drilling and milling machines have to rotate their tools at a predetermined speed, as they act as a spindle. Tool stations are numbered according to the tooling stations available. When writing the part programs, the programmer provides each tool with a corresponding number in the form of a letter T followed by the corresponding numerical identity in two digits: T01, T02, and so on.

*Tool magazines.* A tool magazine, shown in Figures 9.24 and 9.25, is indexable storage used on a machining center to store tools not in use. They are available as rotary drum and chain types. When the tool is called into use, the magazine indexes by the shortest route to bring the tool to a position where it is accessible to a mechanical handling device. At the end of use, the tool is returned to its slotted position in the magazine before calling the next tool. Rotary drums with 12–24 stations are available, and 24–180 stations are available for the chain type.

*Tool replacement.* Cutting tools should be replaced when affected by wear or breakage. Tool changes must be made rapidly. The replaced tool must be of identical dimensions to the original one, which is achieved by using a qualified or preset tooling. Temporary modifications could also be achieved by offsetting the tool from its original datum. The preset tooling concept, shown in Figure 9.26a, is used for both turret and spindle-type machines. For NC machines, the cutting tool is preset to a specific length and diameter while it is off the machine using special fixtures and gauges. The tool length

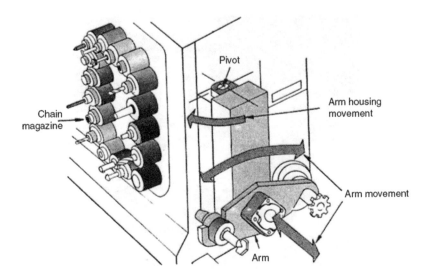

**FIGURE 9.24** Chain-type magazine with automatic tool changer.

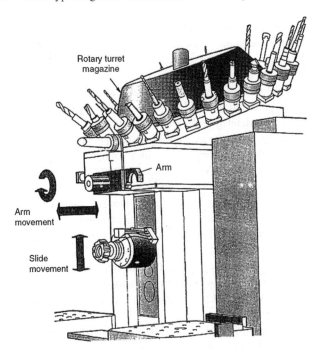

**FIGURE 9.25** Rotary-type magazine with automatic tool changer.

is used by the part programmer to develop the Z-axis coordinate. Preset tool holders and boring bars are available for many NC turning machines. Once these are preset to the appropriate dimensions, inserts can be changed, and WP tolerances are maintained by minor adjustment of the tool offset switches.

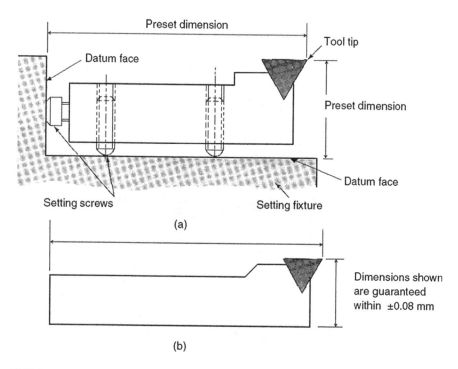

**FIGURE 9.26**   Qualified and preset NC tools: (a) preset tool and (b) qualified tool.

Qualified tool holders for NC lathes are ground to standardized dimensions at close tolerances, and no presetting is required (Figure 9.26b). The qualified tool holder is usually inserted into the turret tool block and tightened in position. The dimensions provided by the manufacturers of the qualified tool holder are used by the part programmer, and minor adjustments are easily made by tool offset switches.

## 9.8   TYPES OF NUMERICAL CONTROL MACHINE TOOLS

The most important types of NC machine tools are:

1. *NC drilling machines.* An NC drilling machine holds, rotates, and feeds the drilling tool into the WP. They are available in a wide range of types and sizes that are built with single spindle or multiple spindles. Some machines are equipped with turrets and others with tool-changing mechanisms. Either two or three axes are available, and some drilling machines are even capable of performing milling operations.
2. *NC milling machines.* These machines (Figure 9.27) are used to machine flat surfaces and produce contours and curved surfaces. The orientation of the spindle may be horizontal or vertical and provided with a single spindle or multiple spindles. Milling machines with two perpendicular spindles allow machining a hole and a vertical surface simultaneously. Such a facility is useful when machining large components, as shown in Figure 9.28.

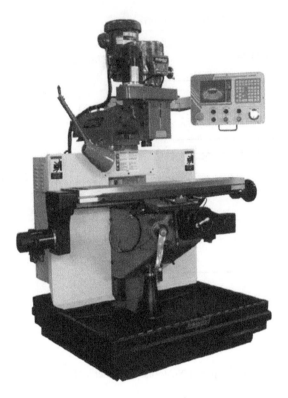

**FIGURE 9.27**    Typical CNC milling machine. (From Hardinge Inc.)

On the other hand, simultaneous machining of the vertical and horizontal planes could be achieved, in this setup, by replacing the boring tool in the vertical spindle by a face milling cutter. NC milling machines may have from two to five axes under tape control. NC milling machines can do some of the work normally performed on NC drilling machines, such as drilling, boring, and tapping.

3. *NC turning machines.* Lathes are primarily used for producing cylindrical shapes in addition to cutting tapers, boring, drilling, and thread cutting. NC lathes are equipped with either straight-cut or continuous-path control systems. Most NC lathes produced today are equipped with continuous-path control and circular interpolation. They are capable of tool offset so that the machine operator can make fine adjustments in the cutting tool location to achieve the required part size. Figure 9.29 shows a typical NC turning machine.

4. *NC machining centers.* Machining centers perform a wide range of operations that include milling, drilling, boring, tapping, countersinking, facing, spot facing, and profiling. Machining centers are able to change the cutting tools automatically, which allows most of the machine time to be devoted to the cutting operation. Most NC machining centers have three axes. In a four-axes system, the fourth axis is used to rotate the table, which

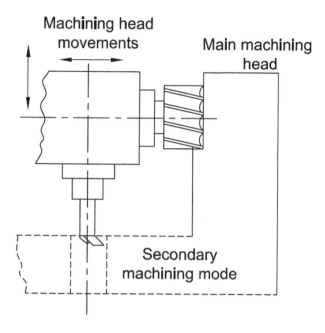

**FIGURE 9.28** Machining a hole and a vertical surface simultaneously. (From Gibs, D., *An Introduction to CNC Machining*, ELBS Cassell Publishers Ltd., London, 1988. With permission.)

**FIGURE 9.29** CNC lathe QUEST 10/56 of Hardinge Inc. (From Hardinge Inc., Berwyn, PA. With permission.)

enables the machining of four sides of a part. In many cases, it is possible to machine a part completely without removing it from the machine. Figure 9.30 shows a typical machining center.

5. *NC turning centers.* These machines combine the features of bar-type, chucking-type, and turret lathes. They are built with four axes of control

and are also equipped with continuous-path NC systems with circular interpolation. Their design may include a slanted or vertical bed rather than the
horizontal one normally used with conventional center lathes. The capabilities of turning centers can be extended by providing two turrets, such that
two tools can cut simultaneously. Power-driven tool holders (which rotate
when the WP is stationary) permit milling of flats, keyways, and slots in
addition to the drilling of holes offset from the machine axis. Figure 9.31
shows a typical turning center. The use of tooling magazines extends the
range of tooling that may be used, as shown in Figure 9.32.

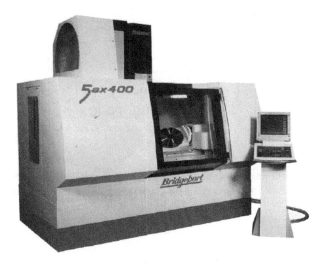

**FIGURE 9.30** Five-axes vertical machining center (5ax400) of Hardinge Incorporation.
(From Hardinge Inc.)

**FIGURE 9.31** CNC (Super Quadrex 250M) turning center. (From MAZAK Corporation.)

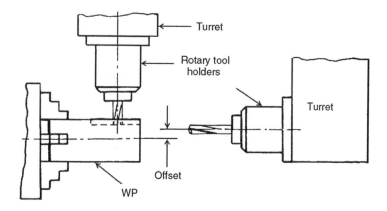

**FIGURE 9.32**   Additional tooling facilities. (From Gibs, D., *An Introduction to CNC Machining*, ELBS Cassell Publishers Ltd., London, 1988. With permission.)

## 9.9   INPUT UNITS

Data can be input into the MCU using one of the following methods:

*Manual data input (MDI).* This method is normally used for setting the machine and editing the program as well as writing complete simple programs. For NC machines with noncomputerized control units, data recording facilities are not often available. In the case of CNC machines, the computer retains the data so that it can be transferred to a recording medium such as magnetic tape or disk or transferred back to the machine when required.

*Conversational MDI.* This method involves the operator pressing appropriate keys on the control console in response to questions that appear in the visual display unit (VDU). This method is faster than methods that require the use of data codes.

*Punched tape.* Punched tapes are of standard 1 in. width. They use eight 0.072 in. holes across the width of the tape and one 0.046 in. sprocket/feed hole between tracks 3 and 4. Punched tapes can be read inexpensively, are less sensitive to handling, are inexpensive to purchase, and require less equipment for manufacturing and less costly space for data storage.

The binary coded decimal (BCD), shown in Figure 9.33, is used for coding digits on the tape. Accordingly, five of the eight tracks per channel are assigned the numerical values 0, 1, 2, 4, and 8 so that any numerical value from 0 to 9 can be represented in one row of the tape. The combination of punched holes per bit in the tape establishes the values associated with that row.

The EIA RS-244-A system and the American Standard Code for Information Interchange (ASCII) RS-358 systems are available for NC and are currently used for coding the numbers on the tape, as shown in Figure 9.33.

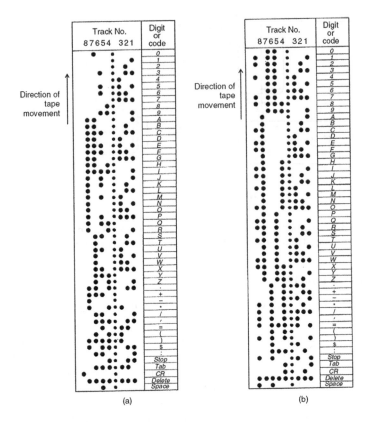

**FIGURE 9.33**    (a) EIA RS-244-A and (b) RS-358 (ASCII) coding systems.

The EIA RS-244-A system is commonly used, while ASCII-coded input is optional in many of today's NC systems. It should be mentioned here that the RS-244-A coding system involves the use of odd parity, in which track 5 makes certain that an odd number of holes (not including the sprocket hole) appears on every row of the tape, whereas the ASCII subset uses even parity, in which an extra hole is added to track 8 in the tape to ensure an even number of holes in each row. The BCD code format is the same in both code systems. All numerical and alphabetical codes along with some special characters and function codes are available in both systems.

*Magnetic tape.* Magnetic tapes, in the form of cassettes, are widely used for transmitting data. They require expensive equipment for program recording and reading. The programmer cannot see the recorded data, and therefore recording errors cannot be seen as they can on punched tapes. Magnetic tape requires special storage space and must be handled carefully.

*Portable electronic storage unit.* In this method, the data transferred into the storage unit away from the machine shop are carried to the machine and connected to the MCU, and data are then transferred. Data transfer is high, and the capacity of such units is high, so that a number of programs can be accommodated at a time.

*Magnetic disk input via computer.* In this method, it is possible to transfer data stored on a floppy disk into the computer and hence into the MCU. Similarly, data on the control unit can be extracted and recorded. The rate at which the data can be transferred or retrieved using a disk is faster than when using a tape, and the storage area is also much greater.

*Master computer.* The prepared program stored on the memory of a master computer is transferred to the microcomputer of the MCU when required. Such an arrangement, also described earlier, is what is known as DNC.

## 9.10   FORMS OF NUMERICAL CONTROL INSTRUCTIONS

The following are forms describing numbers:

1. *Decimal number system.* The number $(657)_{10} = 7 \times 10^0 + 5 \times 10 + 6 \times 10^1$
$$= 7 + 50 + 600 = 657$$

2. *Binary (base 2) number system.* This system is made up of two basic digits, 0 and 1:

| Decimal | Binary |
|---|---|
| $10^0 = 1$ | $2^0 = 1$ |
| $10^2 = 10$ | $2^1 = 2$ |
| $10^2 = 100$ | $2^2 = 4$ |
| $10^3 = 1000$ | $2^3 = 8$ |

### Illustrative Example 1

Convert 327 to binary:

| 2 | 327 | 1 | Least significant digit |
|---|---|---|---|
| 2 | 163 | 1 | |
| 2 | 81 | 1 | |
| 2 | 40 | 0 | |
| 2 | 20 | 0 | |
| 2 | 10 | 0 | |
| 2 | 5 | 1 | |
| 2 | 2 | 0 | |
| 2 | 1 | 1 | Most significant digit |
| | 0 | | |

hence, $(327)_{10} = (101000111)_2$.

**Illustrative Example 2**

Convert (101000111) to decimal:

$$\left(101000111\right)_2 = 1\times 2^8 + 0\times 2^7 + 1\times 2^6 + 0\times 2^5 + 0\times 2^4 + 0\times 2^3 + 1\times 2^2 + 1\times 2^1 + 1\times 2^0$$

$$= 256 + 0 + 64 + 0 + 0 + 0 + 4 + 2 + 1$$

$$= \left(327\right)_{10}$$

Computers cannot work with the complexity of the decimal system, because they are single electronic devices that can only sense the numbers 0 and 1, which can represent the presence (1) or absence (0) of voltage, light, transistor, or magnetic field, as follows:

| | | |
|---|---|---|
| Voltage | On (1) or | Off (0) |
| Light | On (1) or | Off (0) |
| Transistor | On (1) or | Off (0) |
| Magnetic field | On (1) or | Off (0) |

NC systems can understand the numbers 0 and 1, which in electrical terms correspond to on or off when sensing pressure, magnetism, light, or voltage. In NC systems, the command is given to the MCU in blocks of data, where the *blocks* are made up of a collection of words, arranged in a definite sequence, to form a complete NC instruction that can be understood by the machine. A *word* is a collection of characters used to form a part of an instruction. A *character* is a collection of bits that represent a letter, number, or symbol. A *bit* is a binary digit with a value of 0 or 1 depending on the presence or absence of a hole in a certain row and column on the tape.

## 9.11   PROGRAM FORMAT

Tape format is the general sequence and arrangement of the coded information on a punched tape. According to the EIA standard, it appears as words made of individual codes written in horizontal lines. The most common type of tape format in current use is the word address format. However, some earlier control systems still use the fixed block or the tab sequential format.

1. *Word address format.* Each element of information is prefixed by an alphabetical character. The alphabet acts as an address that tells the NC system what it must do with the numbers that follow the prefix. If the word remains unchanged, it need not be repeated in the next block:

    N001    X2.000    Y2.500    $F_1$2.50    S573    EOB

2. *Fixed block format.* Contains only numerical data, arranged in a sequence with all codes necessary to control the machine appearing in every block. The instructions are given in the same sequence, and all instructions are

given in every block, including those unchanged from the preceding blocks. It has no word address letter to identify individual words:

001    2.000    2.500    2.50    573

3. *Tab sequential format.* In this format, a block is given the same sequence as in the case of the fixed block format, but each word is separated by a tab character. If the word remains unchanged in the next block, the word need not be repeated, but a tab code is required to keep the sequence of words. Because the words are written in a set order, the address letters are not required:

001    TAB2.000    TAB2.500    TAB2.50    TAB573    EOB

The EIA standard RS-274-A defines the various standard word addresses and describe their use as shown in Table 9.1:

1. *Sequence number function.* This is the first word of a block, which is represented by the letter N followed by three digits.
2. *Preparatory functions.* The word addresses or G codes relate the various capabilities or functions of particular NC machine tools. These are used as prefixes in developing the NC words used in the programs to command specific machine functions, as shown in Table 9.2.
3. *Dimensional data function.* This is represented by a symbol followed by five to eight digits, as shown in Table 9.1.
4. *Feed rate function.* This is expressed by the letter $F_1$ plus three digits. The digits may represent the feed rate in millimeters per minute, millimeters per revolution, or the magic-three method (explained in the following section).
5. *Tool selection.* Information regarding the tool is given by a word prefixed by the letter T followed by a numerical code for the tool in use.
6. *Spindle speed function.* This is specified in rotational speed in revolutions per minute or the surface speed in meters per minute and is given by the letter S followed by the speed required.
7. *Miscellaneous functions.* In the word address format, miscellaneous functions are represented by the letter M followed by a numerical code for the function required. They are used to command miscellaneous or auxiliary functions of the machine, such as turning on the coolant and starting the spindle in conjunction with the first move of the machine. The standard miscellaneous functions are listed in Table 9.3.

## 9.12   FEED AND SPINDLE SPEED CODING

### 9.12.1   FEED RATE CODING

During milling operations, the feed rate is expressed in millimeters per minute or inches per minute (ipm). Additionally, the feed rate is expressed in millimeters per

**TABLE 9.1**
**EIA RS-274-A Standard Word Addresses**

| Code | Function |
|------|----------|
| a | Angular dimension around $X$-axis |
| b | Angular dimension around $Y$-axis |
| c | Angular dimension around $Z$-axis |
| d | Angular dimension around special axis, or third feed function[a] |
| e | Angular dimension around special axis, or second feed function[a] |
| f | Feed function |
| g | Preparatory function |
| h | Unassigned |
| i | Distance to arc center or thread feed parallel to $X$ |
| j | Distance to arc center or thread feed parallel to $Y$ |
| k | Distance to arc center or thread feed parallel to $Z$ |
| l | Do not use |
| m | Miscellaneous function |
| n | Sequence number |
| o | Rewind application stop |
| p | Third rapid traverse dimension or tertiary motion dimension parallel to $X$[a] |
| q | Third rapid traverse dimension or tertiary motion dimension parallel to $Y$[a] |
| r | Third rapid traverse dimension or tertiary motion dimension parallel to $Z$[a] |
| s | Spindle speed |
| t | Tool function |
| u | Secondary motion dimension parallel to $X$[a] |
| v | Secondary motion dimension parallel to $Y$[a] |
| w | Secondary motion dimension parallel to $Z$[a] |
| x | Primary $X$ motion dimension |
| y | Primary $Y$ motion dimension |
| z | Primary $Z$ motion dimension |

[a] When d, e, p, q, r, u, v, and w are not used as indicated, they may be used elsewhere.

revolution or inches per revolution (ipr) in the case of turning machines. Generally, feed rates can be expressed by one of the following methods:

1. *Four-digit field.* This coding process represents the number of digits the system can accept. Accordingly, 12.3 ipm or mm/min will be coded by $F_1 0123$, and 999.9 ipm will be coded by $F_1 9999$.
2. *Inverse time feed rate coding.* In this case, the feed rate number is expressed as the ratio of the feed rate to the distance traveled, according to the following equation:

$$\text{Feed rate number} = \frac{\text{Feed rate (ipm)}}{\text{Distance traveled (in.)}}$$

## TABLE 9.2
## Some Common Preparatory Codes and Functions

| Code | Function |
|---|---|
| G00 | PTP positioning |
| G01 | Linear interpolation |
| G02 | Circular interpolation arc cw |
| G03 | Circular interpolation arc ccw |
| G04 | Dwell |
| G05 | Hold |
| G08 | Acceleration |
| G09 | Deceleration |
| G17 | $X–Y$ plane selection |
| G18 | $Z–X$ plane selection |
| G19 | $Y–Z$ plane selection |
| G33 | Thread cutting, constant lead |
| G40 | Cutter compensation cancel |
| G41 | Cutter compensation left |
| G42 | Cutter compensation right |
| G80 | Fixed cycle cancel |
| G80–G89 | Fixed cycles as selected by manufacturers |

## TABLE 9.3
## Some Miscellaneous or Auxiliary Functions and Codes

| Code | Function |
|---|---|
| M00 | Program stop |
| M01 | Optional (planned stop) |
| M02 | End of program |
| M03 | Spindle start cw |
| M04 | Spindle start ccw |
| M05 | Spindle stop |
| M06 | Tool change |
| M07 | Coolant no. 2 on (mist) |
| M08 | Coolant no. 1 on (flood) |
| M09 | Coolant off |
| M10 | Clamp |
| M11 | Unclamp |
| M13 | Spindle cw and coolant on |
| M14 | Spindle ccw and coolant on |
| M15 | Motion + |
| M16 | Motion − |
| M30 | End of tape |
| M32–M35 | Constant cutting speed |

For a feed rate of 20 ipm and a distance of 2.6 in.,

$$\text{Feed rate number} = \frac{20\,\text{ipm}}{2.6\,\text{in}} \approx 7.7\,\text{min}$$

If the control system accepts inverse time feed rate coding and a four-digit feed rate field, then 7.7 ipm will be expressed by $F_1 0077$. Similarly, for a feed rate of 50 mm/min and a distance of travel 6.0 mm,

$$\text{Feed rate number} = \frac{50\,\text{mm}\,/\,\text{min}}{6\,\text{mm}} \approx 8.3\,\text{min}$$

The feed rate number becomes $F_1 0083$.

3. *Coded feed rate.* Such a code is used for low-cost PTP NC systems in which a fixed relation known from a chart supplied by the manufacturer is used. For example, 20 ipm or 50 mm/min will be coded by $F_1 10$, and so on.

4. *Magic-three method.* In this method, 3 is added to the number of digits on the left of the decimal of the numerical value of the feed or speed in metric or imperial units, the addition thus obtained providing the first digit of the feed value. The next two digits in the coded value are the first two digits of the numerical value of the feed. As an example, a feed rate of 35.5 ipm will be coded as follows. Add 3 to the number of digits (2). So, the first digit of the feed is 5. The next two digits of the feed code will be the first two digits of the numerical value of the feed, which is 35, so the feed rate code will be $F_1 535$. Similarly, 3.55 ipm will be coded as $F_1 435$, and 0.35 ipm will be coded as $F_1 335$. However, if the feed rate is less than 1 ipr, then the rule is modified by subtracting the number of zeros after the decimals from 3 to provide first digit of the coded value. The next two digits of the coded value will be the first two nonzero digits in the feed rate. For example, the feed rate of 0.087 ipm will be coded as $F_1 287$.

### 9.12.2 Spindle Speed Coding

During NC, spindle speeds can be coded using one of the following methods:

1. *Direct revolutions per minute.* In this case, the spindle speed code will have the same value preceded by the letter S. Hence, 1500 rpm will be coded as S1500.

2. *Coded format.* The spindle speed coding is performed through a chart that is supplied by the machine builder; for example, S10 represents a spindle speed of 146 rpm.

3. *Magic-three method.* The magic-three method is applied as described earlier for spindle speed coding. Hence, a rotating speed of 7 rpm will be coded as S470, 500 rpm as S650, and 1500 rpm as S715.

## 9.13   FEATURES OF NUMERICAL CONTROL SYSTEMS

NC and CNC controls can be equipped with a variety of features available as standard or optional equipment. These can add generally to the capabilities of the machine tool. The following are common features of NC and CNC control systems:

1. *Feed and spindle speed override.* This feature allows deviation from the programmed feed rate or spindle speed to increase production rate or to reduce the tool wear. Feed rates can be varied as a percentage between 0% and 125% of the programmed rate. As an example, if 250 mm/min is programmed on the tape, and the feed override ratio is 80%, the actual feed will be 200 mm/min. For spindle speeds, 80–100% of the programmed revolutions per minute is possible.
2. *Mirror imaging.* This function is also called axis inversion and can be controlled by a simple on/off switch. It is used to machine left- and right-hand parts from the same tape, as shown in Figure 9.34.
3. *Scaling.* As shown in Figure 9.35, a range of components can be machined, varying in size, from one set of programmed data.
4. *Rotation.* This technique is mainly required for milling and drilling operations. It enables the cutter path to be rotated by an angle and repeated if required at another angle, as shown in Figure 9.36. This makes the programming of complex shapes relatively easy.
5. *Jog.* The jog facility enables the machine operator to manually move the machine slides through the control console. Jog is used for establishing a datum at the initial setting of the machine, stopping an automatic sequence, moving the machine slides for WP measurements, and tool changing due to breakage. It is desirable to restart the program from the point at which it was interrupted. Therefore, most control systems have the facility to return from jog, which returns the machine slides to their original position via a button on the control console.

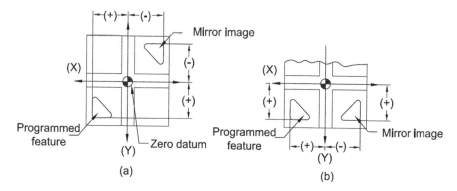

**FIGURE 9.34**   Mirror imaging: (a) mirror image in two axes and (b) mirror image in one axis.

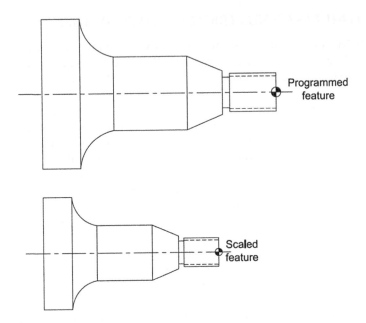

**FIGURE 9.35**   Scaling.

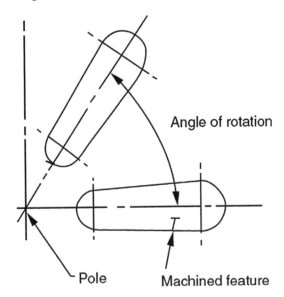

**FIGURE 9.36**   Rotation.

6. *Position displays.* These are used for setting the machine and inspecting parts on the machine tool. Readouts can be in the imperial (inch) or metric (millimeter) scales.

7. *Switchable input format.* Using a selector switch on the control console, the operator can choose either EIA tape format (RS-244-A) or ASCII tape

format (RS-358). A warning arises if the switch is in the wrong position. Automatic selection of the code can be made depending on the parity check, for which the selector switch is not necessary.

8. *Switchable imperial or metric input.* Using a switch on the control console, imperial or metric format can be selected. Different results will arise when this switch is set incorrectly.

9. *Tape readers.* Mechanical tape readers can read 10–150 characters (rows) per second and are commonly used in NC drilling machines. Photoelectric readers can read 150–600 rows/s in the case of NC turning and milling machines. As wear may arise in the case of mechanical readers, photoelectric readers are more reliable.

10. *Tool offsets.* These are used in lathe applications where minor changes are made to the program longitudinal and cross-feed motions. In such a case, the operator avoids over- or undersize resulting from the tool wear or from minor variations on the size and shape of the tools when they are changed.

11. *Tool length compensation.* This is used in some milling, drilling, and taping machines to compensate for different tool lengths of drills, boring bars, reamers, and milling cutters. As shown in Figure 9.37, the difference in length with respect to the presetting tool is recorded and is manually entered and stored with the associated tool number. Whenever these tools are called by the program instructions, their respective compensation values are activated and automatically taken into account in the tool motion.

12. *Cutter diameter compensation.* This allows the use of a cutter diameter that is different from the diameter used in developing the original part program.

13. *Operator control features.* These features include on/off, MDI, manual jog control for the machine axes, sequence number search, sequence number display, and the slide hold to inspect cut or tool conditions and then restart the cycle.

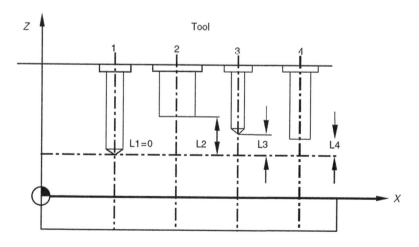

**FIGURE 9.37**   Tool length offset.

14. *Canned cycles.* These are control instructions found in many PTP control systems and in some continuous path systems. This feature allows common repetitive machining patterns, such as drilling, milling, threading, and turning, to be done automatically with a single command.

## 9.14 PART PROGRAMMING

The part program is a computer program containing a number of lines, instructions, or statements called NC blocks that describe the detailed plan of machining instructions proposed for the part. It is written using vocabulary understood by the MCU in terms of standard words, codes, and symbols. Part programs can also be written in higher languages, such as automatically programmed tools (APT), adaptation of APT(ADAPT), extended subset of APT(EXAPT), and so on. These programs can be converted into the machine-tool language with the help of processors. Programming in APT is mostly processed with the help of computers and is thus known as computer-assisted part programming. Part programs can also be directly developed using computer-aided design/manufacturing (CAD/CAM) systems such as Unigraphics and ProEngineer or CAM systems such as Master CAM, Surf CAM, and others.

The NC part programming procedure includes the following steps:

1. Process planning, which determines the sequence of operations to be performed during machining of the part.
2. Part programming, which is concerned with the documentation of the planned sequence of operations in a special format known as the manuscript.
3. Tape preparation on the basis of instructions written in the program manuscript prepared by the part programmer.
4. Tape verification and checking by machining foam or a plastic material or running the tape through a computer program that indicates the contents and errors in the tape.
5. The corrected tape is then used for actual production.

The data required for part programming include the following items:

1. Machine-tool specifications
2. Specifications of tools
3. WP geometry and material specifications
4. Speed and feed rate tables

Using the part drawing, the programmer determines the sequence of operations, speeds, and feeds for various operations and determines the magnitude of various motions decided. He prepares the planning sheet and writes the instructions in a coded format for the MCU. The part programs are written in blocks using the standard codes and following the sequence shown in Figure 9.38.

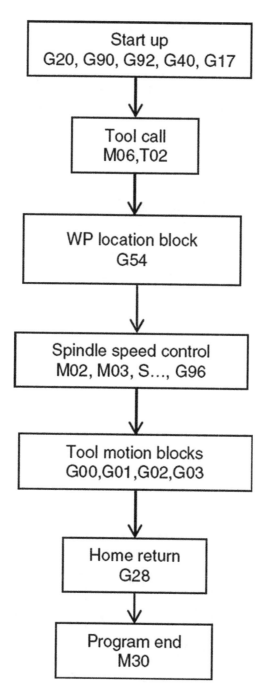

**FIGURE 9.38**  Part programming procedure.

*Program plane identifier.* As shown in Figure 9.39, G17 is used for programming in *X, Y;* G18 in *X, Z;* and G19 for programming in *Y, Z* planes.

*Tool diameter compensation.* The control unit offsets the path of the tool so that the part programmer can program the part as it appears. In this case, the same program can be used with different cutters. The amount of offset equals the cutter radius (Figure 9.40). The following G codes are used:

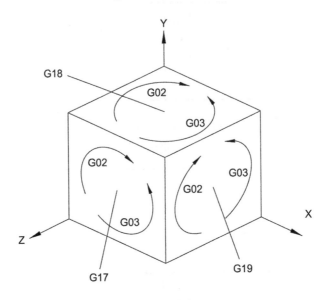

**FIGURE 9.39**   Plane identifiers G17, G18, and G19.

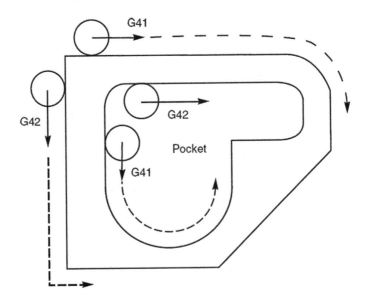

**FIGURE 9.40**   Tool diameter compensation: right G42, left G41, and G40 to cancel.

- G41, for cutter diameter compensation to the left
- G42, for cutter diameter compensation to the right
- G40, to cancel cutter diameter compensation

*WP coordinate setting G54.* This setting describes the distance from the tool tip at the home position to the Z zero position of the part, as shown in Figure 9.41. The use of G02 and G03 codes is shown in Figures 9.42 and 9.43. G02 and G03 are used for cutting arcs in cw and ccw directions, and G01 is used for cutting straight lines.

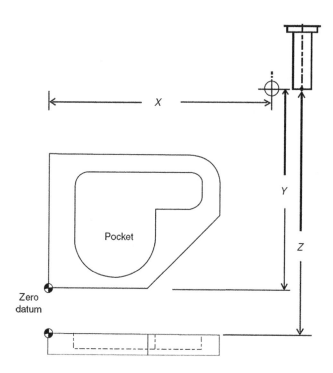

**FIGURE 9.41** WP coordinate setting (G54).

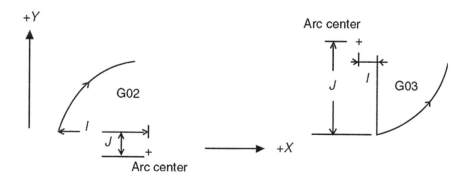

**FIGURE 9.42** Cutting arcs using G02 and G03 codes.

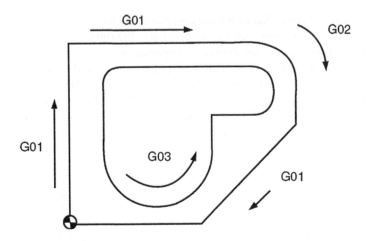

**FIGURE 9.43**   Linear interpolation (G01) and circular interpolation (right G02 and left G03).

## 9.15   PROGRAMMING MACHINING CENTERS

### 9.15.1   PLANNING THE PROGRAM

Planning for NC part programs should consider the following:

1. *Part drawing.* This drawing includes the shape, tolerances, material requirements, surface finish, and product quality. There are several methods of part dimensioning. In the centerline method (Figure 9.44), the part centerline is used as a reference point for dimensioning the component. The coordinate or base line method (Figure 9.45) enables easy

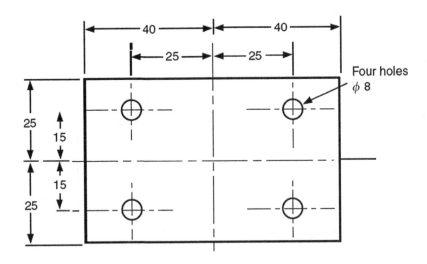

**FIGURE 9.44**   Centerline dimensioning.

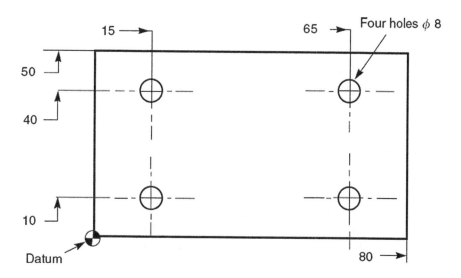

**FIGURE 9.45** Coordinate or baseline dimensioning.

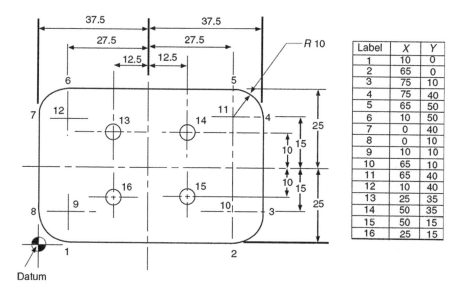

| Label | X | Y |
|-------|-----|-----|
| 1 | 10 | 0 |
| 2 | 65 | 0 |
| 3 | 75 | 10 |
| 4 | 75 | 40 |
| 5 | 65 | 50 |
| 6 | 10 | 50 |
| 7 | 0 | 40 |
| 8 | 0 | 10 |
| 9 | 10 | 10 |
| 10 | 65 | 10 |
| 11 | 65 | 40 |
| 12 | 10 | 40 |
| 13 | 25 | 35 |
| 14 | 50 | 35 |
| 15 | 50 | 15 |
| 16 | 25 | 15 |

**FIGURE 9.46** Labeling of positions. (From Thyer, G. E., *Computer Numerical Control of Machine Tools*, Industrial Press Inc., New York, 1988. With permission.)

transcription of the various features of the part. The dimensions should be specified in imperial (inches) or metric (millimeters) units. An alternative method of dimensioning that is convenient for ease of programming is to label with letters and numbers all the relevant features of the component, as shown in Figure 9.46.

2. *Machine tool used.* The part programmer should be aware of the machine size, power, accuracy, tooling, capacity, and number of axes. Specifically, he should be aware of the following points:
   a. Power of the drive motors, which decide the material removal rates and area of cut
   b. The possible movement for all the machine axes
   c. The possibility of machining the whole part in one setup
   d. The spindle speeds available
   e. Machine rigidity and accuracy capabilities
   f. The number of different tools needed, and whether the machine carousel will accept these tools
   g. Types of tools and the tool-holding facilities available
   h. The availability of the machine when required
3. *Work holding.* Whenever possible, the use of standard work devices is recommended, with preparing fixtures to hold the part if required.
4. *Part datum location.* The datum of the drawing should be taken as the program datum. Generally, the choice of part datum depends on the type of component, required machining operations, WP holding, and the direction of cutting forces.
5. *Selecting the program tooling.* Decide the tooling ahead of time. In this regard, especially ground tools should be ready. Standard tools made of carbide or specially coated tools are recommended. Cost is a major factor, which includes cycle time, tolerance, surface finish, and quality of parts needed.
6. *Program plan.* This plan is an outline of the machining steps to be done on the part. In small job shops, the programmer can do the process plan. In larger shops, the process plan comes from the engineering area and includes each step in manufacturing the part.

The part configuration will specify the sequence of operations. Generally, the recommended procedure for machining is as follows:

   a. Face-mill the top surface
   b. Rough-machine the profile of the part
   c. Rough bore
   d. Drill and tap
   e. Finish profile surfaces
   f. Finish bore
   g. Finish reaming

The operator determines the most economical way to produce the WP using proper sequence, tools, cutter path, work-holding devices, and machining conditions. The G and M codes for machining centers are listed in Tables 9.4 and 9.5.

## TABLE 9.4
## Commonly Used G Codes for Machining Centers

| Code | Function | Condition |
|------|----------|-----------|
| G00 | PTP positioning/rapid traverse | Modal |
| G01 | Linear interpolation at a feed rate | Modal |
| G02 | Circular interpolation cw | Modal |
| G03 | Circular interpolation ccw | Modal |
| G28 | Zero or home position | Nonmodal |
| G40 | Cutter compensation cancel | Modal |
| G41 | Cutter compensation left | Modal |
| G42 | Cutter compensation right | Modal |
| G43 | Tool height offset | Modal |
| G49 | Tool height offset cancel | Modal |
| G54 | WP coordinate preset | |
| G80 | Fixed cycle cancel | Modal |
| G81 | Canned drilling cycle | Modal |
| G83 | Canned peck drilling cycle | Modal |
| G84 | Canned tapping cycle | Modal |
| G85 | Canned boring cycle | Modal |
| G90 | Absolute coordinate positioning | Modal |
| G91 | Incremental positioning | Modal |
| G92 | WP coordinate preset | |
| G98 | Canned cycle initial point return | Modal |
| G99 | Canned cycle R point return | Modal |

Modal—active unless unchanged.

## TABLE 9.5
## M Codes for Machining Centers

| Code | Function | Condition |
|------|----------|-----------|
| M00 | Program stop | Nonmodal |
| M01 | Optional (planned stop) | Nonmodal |
| M02 | End of program | Nonmodal |
| M03 | Spindle start cw | Modal |
| M04 | Spindle start ccw | Modal |
| M05 | Spindle stop | Modal |
| M06 | Tool change | Nonmodal |
| M07 | Coolant no. 2 on (mist) | Modal |
| M08 | Coolant no. 1 on (flood) | Modal |
| M09 | Coolant off | Modal |
| M30 | End of program and reset to the top | Nonmodal |
| M40 | Spindle low range | Modal |
| M41 | Spindle high range | Modal |
| M98 | Subprogram call | Modal |
| M99 | End subprogram and return to main program | Modal |

Modal—active unless unchanged.

### 9.15.2 CANNED CYCLES

Machining centers provide many canned cycles, including:

A. *Drilling cycle G81*. For drilling a hole, the operation sequence is as follows:

- Rapid positioning of the X- and Y-axis using G00 code
- Positioning of the Z-axis to a clearance plane using G00 code
- Feeding the tool down to the required depth using G01 code
- Feeding back to the initial Z position

Using canned drilling cycle G81 (Figure 9.47), a set of programmed instructions eliminates the need for many lines in programming. Canned drilling cycles are cancelled using G80 code. To drill a hole located at X, Y, and Z starting from reference point $R_0$, and at a feed rate $F_1$, use the following block:

$$N\ldots G81\,X\ldots Y\ldots Z\ldots R\ldots F_1\ldots$$

where

    $R$ = initial level position
    $F_1$ = NC feed rate

B. *Peck drilling cycle G83*. Used for drilling a hole of depth three to four times the diameter at $Q$ steps. As shown in Figure 9.48, the following block is used:

$$N\ldots G83\,X\ldots Y\ldots Z\ldots R\ldots Q_s\ldots F_1\ldots$$

C. *Canned tapping cycle G84*. The following block is used:

$$N\ldots G84\ X\ldots Y\ldots Z\ldots R\ldots F_1\ldots$$

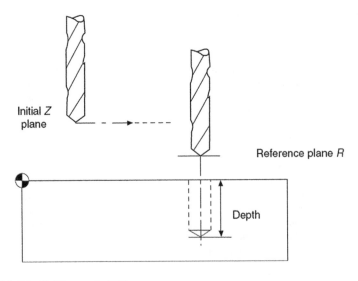

**FIGURE 9.47** Drilling cycle G81.

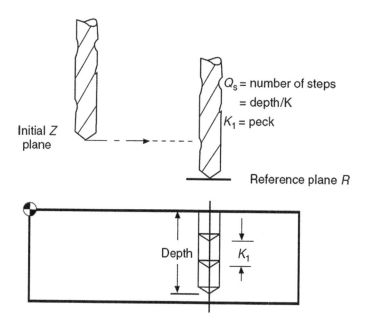

**FIGURE 9.48** Peck drilling cycle G83.

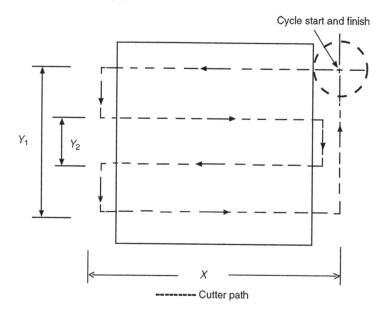

**FIGURE 9.49** Face milling cycle G77.

D. *Face milling cycle G77.* This cycle considerably reduces the number of blocks required to face mill a surface. The path followed by the center of the cutter is shown in Figure 9.49:

$$N\ldots G77\,X\ldots Y_1\ldots Y_2\ldots F_1\ldots$$

where
  $X$ = incremental distance to be milled along $X$-axis, cutter center to cutter center
  $Y_1$ = incremental distance to be milled along $Y$-axis, cutter center to cutter center
  $Y_2$ = $Y$-axis "stepover" value. Maximum stepover is the diameter of the cutter; for efficient cutting, a more practical value is 70–80% of the cutter diameter. Last stepover is automatically adjusted by control to satisfy $Y_1$....
  $F_1$ = feed rate

E. *Slot milling cycle.* Figure 9.50 shows a slot milling cycle that is similar to the face milling cycle.

F. *Pocket milling cycle G78.* This cycle is used to mill a rectangular pocket; the path followed by the cutter centerline is shown in Figure 9.51. The cutter has to be positioned in the center of the pocket and at the desired depth before the cycle is activated. The cycle requires the following format:

$$N \ldots X_1 \ldots X_2 \ldots X_3 \ldots Y_1 \ldots Y_2 \ldots F_1 \ldots F_2 \ldots$$

where
  $X_1$ = distance from the center of the pocket to wall along $X$-axis ($L/2$) minus the cutter radius ($D/2$):
    $= L/2 - D/2$
  $X_2$ = stepover value on $X$-axis. Stepover is the amount the cutter moves for each cut. Maximum stepover is the diameter of the cutter
  $X_3$ = step over for final boundary cut. If it is not programmed, default stepover value for the final cut will be 0.5 mm
  $Y_1$ = distance from the center of the pocket to wall along $Y$-axis ($W/2$) minus cutter radius $D/2$:
    $= W/2 - D/2$
  $Y_2$ = stepover on $Y$-axis. If not programmed, stepover in $Y$ will be same as $X$ stepover
  $F_1$ = NC feed rate for clearing pocket

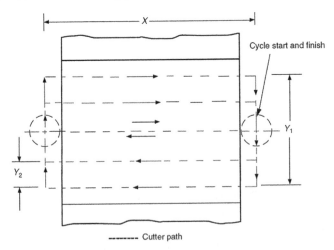

FIGURE 9.50   Slot milling cycle G77.

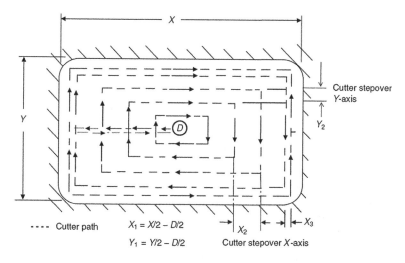

**FIGURE 9.51**   Pocket milling cycles G78.

$F_2$ = NC feed rate for final cut. If $F_2$ not programmed, default feed rate will be 1.5 times $F_1$

G. *Hole milling cycle G79.* This cycle is used for milling large circular holes, as shown in Figure 9.52, and is written as follows:

$$N \ldots G79 J \ldots F_1 \ldots$$

where $J$ is the radius of hole to be milled minus cutter radius and $F_1$ is the NC feed rate.

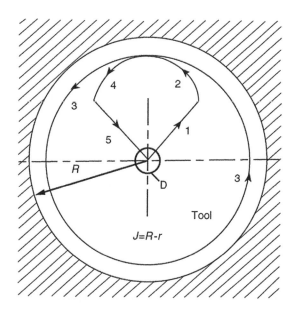

**FIGURE 9.52**   Hole milling cycles G79.

## Illustrative Example 3

Write the part program for the part shown in Figure 9.53 for milling, drilling, and tapping.

- Use canned cycles where appropriate.
- Preset the absolute tool reference at $X = 5$, $Y = 5$, and $Z = 1$ in.
- Use letters A, B, C, D, and so on to describe the tool paths.

### SOLUTION

| | | |
|---|---|---|
| 0010 | G90 G20 | Absolute or inch programming |
| 0020 | G40 | Cutter compensation cancel |
| 0030 | M06 T01 | Change to tool 1 |
| 0040 | G54 X5.0 Y5.0 Z 1.0 | WP preset position |
| 0050 | S300 M03 | Rotate spindle cw at 300 rpm |
| 0060 | G00 X-1.0 Y-1.0 | Rapid positioning |
| 0070 | Z0.1 | |
| 0080 | Z-0.65 | |
| 0090 | G41 X0.0 Y0.0 | |
| 0100 | X2.5 Y5.0 | |
| 0110 | G01 Y 5.0 $F_1$5 | Linear interpolation to point B |
| 0120 | X2.5 Y5 | To point C |
| 0130 | Y4.25 | To point D |
| 0140 | G03 X3.25 Y3.5 I .75 J0 | Circular interpolation ccw to point E |
| 0150 | G01 X3.875 | Linear interpolation to point F |
| 0160 | G02 X4.375 Y3.0 I0J-0.5 | Circular interpolation cw to point G |
| 0170 | X 3.875 Y2.5 I-0.5 J0 | Circular interpolation cw to point H |
| 0180 | G01 X3.5 | Linear interpolation to point I |
| 0190 | G03 X3.25 Y 0.5 I0 J-2 | Circular interpolation ccw to point J |
| 0200 | G01 X1.25 Y0.0 | Linear interpolation to point K |
| 0210 | X0.0 Y0.0 | Linear interpolation to point A |
| 0220 | G40 Y-1.0 | Ramp off |
| 0230 | G00 Z0.1 | Rapid above |
| 0240 | G00 X0.5 Y2.5 | Rapid positioning |
| 0250 | G81 Z-0.65 $F_1$3 | Drill hole 1 at point L |
| 0260 | Y3.0 | Drill hole 2 at point M |
| 0270 | Y3.5 | Hole 3 at point N |
| 0280 | X3.75 Y3.0 | Hole 4 at point O |
| 0290 | G80 | Cancel |
| 0300 | G28 | Home |
| 0310 | M06 T03 | Change tool |
| 0320 | M03 S100 | Rotate spindle |
| 0330 | G00 X3.75 Y3.0 | Rapid positioning |
| 0340 | Z0.1 | Rapid positioning |
| 0350 | G84 Z-0.65 $F_1$4 | Tap at point O |
| 0360 | G80 | Tap cancels |
| 0370 | G28 | Home |
| 0380 | M05 | Spindle stop |
| 0390 | M30 | Rewind the program |

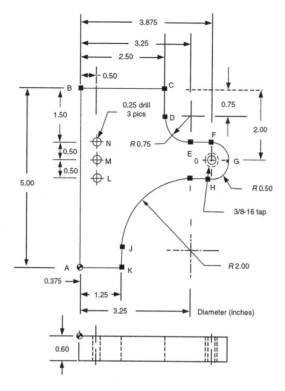

**FIGURE 9.53**   Part drawing. (From Senerstone, J. and Kuran, K., *Numerical Control Operation and Programming*, Prentice-Hall, New York, 1997. With permission.)

## 9.16   PROGRAMMING TURNING CENTERS

### 9.16.1   PLANNING THE PROGRAM

For successive programming, the following points should be considered:

a. *Tool consideration.* Use 80° diamond insert for roughing operations, 35° diamond insert for finishing and grooving processes, and 60° diamond insert for threading.

b. *Sequence of operations.* Adopt the following recommended sequence of operations:
   * Facing
   * Rough turning of the profile of the part
   * Finish turning of the profile of the part
   * Drilling
   * Rough boring
   * Finish boring
   * Grooving
   * Threading

c. *Commonly used codes.* Tables 9.6 and 9.7 show the commonly used G and M codes.

d. *Return home (safe position) G28.* This enables turret indexing (Figure 9.54) away from the WP.

e. *WP coordinate setting G92.* This tells the machine where the part zero location is with respect to the machine, as shown in Figure 9.55. For this purpose, the following code is used:

$$N\ldots G92\,X\ldots Z\ldots$$

f. *Circular interpolation cw G02 and ccw G03.* As shown in Figure 9.56, G02 and G03 use the same concept used in programming machining centers, taking into consideration that the machine movements are in the $X$ and $Z$ directions, and therefore, the letters I and K will be used instead of I and J.

g. *Tool nose radius compensation G41 and G42.* This facility permits the use of the same program for a variety of tool types. The exact radius of the tool is entered into the offset file. It may be right G42 or left G41, as shown in Figure 9.57.

---

**TABLE 9.6**

**Commonly Used G Codes for Turning Centers**

| Code | Function | Condition |
|------|----------|-----------|
| G00 | Rapid positioning | Modal |
| G01 | Linear positioning at a feed rate | Modal |
| G02 | Circular interpolation cw | Modal |
| G03 | Circular interpolation ccw | Nonmodal |
| G28 | Zero or home position | Modal |
| G40 | Tool nose radius compensation cancel | Modal |
| G41 | Tool nose radius compensation left | Modal |
| G42 | Tool nose radius compensation right | Modal |
| G50 | WP coordinate setting/maximum spindle revolutions per minute setting | Modal |
| G20 | Inch programming | Modal |
| G75 | Grooving cycle | |
| G76 | Threading cycle | |
| G90 | Absolute coordinate positioning | Modal |
| G91 | Incremental positioning | Modal |
| G92 | WP coordinate preset | Modal |
| G96 | Constant surface speed footage | |
| G77 | Revolutions per minute input | Modal |
| G98 | Feed rate per minute | Modal |
| G99 | Feed rate per revolution | Modal |

Modal—active unless unchanged.

**TABLE 9.7**

**Common M Codes for Turning Centers**

| Code | Function |
|------|----------|
| M00 | Program stop |
| M01 | Optional (planned stop) |
| M03 | Spindle start cw |
| M04 | Spindle start ccw |
| M05 | Spindle stop |
| M08 | Coolant no. 1 on (flood) |
| M09 | Coolant off |
| M30 | End of program and reset to the top |
| M41 | Spindle low range |
| M42 | Spindle high range |
| M43 | Subprogram call |

Tool code calls the tool and the offset, which accommodates tool wear or exact sizing of the tool.

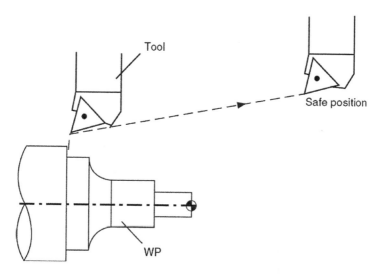

**FIGURE 9.54** Return to home position G28.

### 9.16.2 Canned Turning Cycles

a. *Rough turning cycle G71.* For rough turning, use the following block details:

$$N...G71\,P...Q...U...W...D...F_1...$$

b. *Finish turning cycle G70.* For finish turning, use the following block details:

$$N...G70\,P...Q...F_1...$$

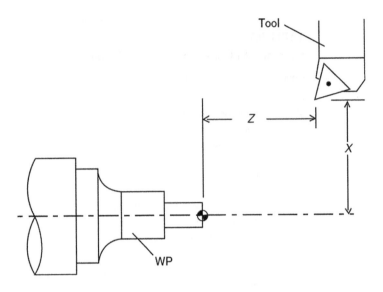

**FIGURE 9.55**   WP coordinate setting G92.

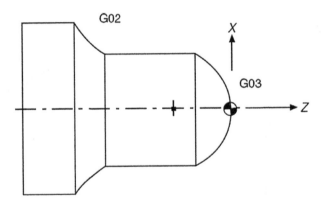

**FIGURE 9.56**   Turning arcs using G02 and G03 codes.

where

    $P$ = line number for the start of profile
    $Q$ = line number for the end of profile
    $U$ = allowance left for finishing in the $X$-axis
    $W$ = allowance left for finishing in the $Z$-axis
    $D$ = depth removed per pass
    $F_1$ = NC feed rate

c. *Peck drilling cycle G74.* This cycle, shown in Figure 9.58, can be called by the following block:

$$N...G74\,X0.0\,Z...F_1...K$$

where $K$ is the depth of the peck.

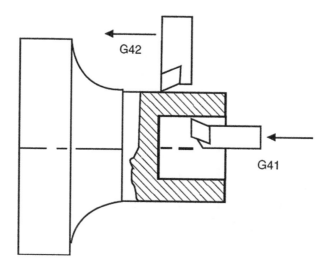

**FIGURE 9.57**   Tool nose radius compensation: right G42 and left G41.

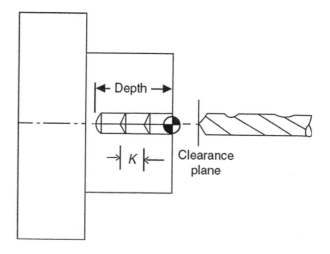

**FIGURE 9.58**   Peck drilling cycle G74.

d. *Grooving cycle G75.* As shown in Figure 9.59, the following block applies:

$$N...G75\,X...Z...F_x...I_x...K_z$$

where
$X$ = diameter at the bottom of the groove
$Z$ = end position of the groove
$I_x$ = depth of cut on the $X$-axis
$K_z$ = depth of cut on the $Z$-axis
$F_x$ = the retract of the grooving tool

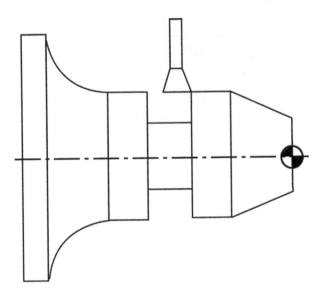

**FIGURE 9.59**   Grooving canned cycle G75.

e. *Thread cutting cycle (G76).* The following block can be used:

$$N...G76...X...Z...K...D...F_1...A$$

where

$X$ = diameter at the bottom of the thread
$Z$ = end position of the thread
$K$ = thread height
$D$ = depth of cut in the first pass
$F_1$ = thread lead (pitch for a single start)
$A$ = included thread angle

### Illustrative Example 4

Program the part shown in Figure 9.60. You are requested to

1. Use canned cycles whenever possible
2. Preset the absolute tool reference at $X = 6.0$ and $Z = 10.0$ in.
3. Choose the proper tool from the following list:

| Tool Number | Tool Description | Operation Performed |
|---|---|---|
| 1 | 80° diamond | Rough turning |
| 2 | 35° diamond | Finish turning |
| 3 | 60° diamond | Threading |
| 4 | 0.5 mm diameter | Drilling |

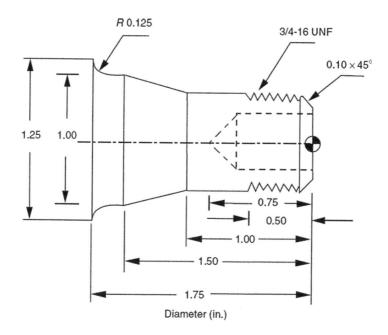

**FIGURE 9.60** Part drawing. (From Senerstone, J. and Kuran, K., *Numerical Control Operation and Programming*, Prentice-Hall, New York, 1997. With permission.)

## SOLUTION

| | | |
|---|---|---|
| 0010 | G90 G20 | |
| 0020 | G40 | |
| 0030 | T0101 | |
| 0040 | G92 X6.0 Z 10.0 | |
| 0050 | G96 S400 M03 | |
| 0060 | G00 G42 X1.3 Z0.1 | |
| 0070 | G71 P0080 Q0120 U0.03 W 0.001 D 0.04 $F_1$0.01 | Rough turning cycle |
| 0080 | G00 X .55 | |
| 0090 | G01 X.75 Z-.1 | |
| 0100 | G01 Z-1.0 | |
| 0110 | G01 X1.0 Z-1.5 | |
| 0120 | G01 Z-1.625 | |
| 0130 | G02 X1.25 Z-1.75 I.125 K0 | |
| 0140 | G28 | |
| 0150 | T0202 | |
| 0160 | G00 G42 X1.3 Z.1 | |
| 0170 | G70 P080 Q 0120 $F_1$0.008 | Finish turning cycle |
| 0180 | G40 G00 X2.0 Z2.0 | |
| 0190 | G28 | |

| 0200 | T0303 | |
|------|-------|---|
| 0210 | G92 X 6.0 Z10.0 | |
| 0220 | G96 S300 M0S | |
| 0230 | G00 X.75 Z.2 | |
| 0240 | G76 X.75 Z-.5 K0.053 D0.02 $F_1$0.065 A60 | Threading cycle |
| 0250 | G28 | |
| 0260 | T0404 | |
| 0270 | G92 X6.0 Z 10.0 | |
| 0280 | G96 S800 M03 | |
| 0290 | G00 X0 Z.2 | |
| 0300 | G74 X0 Z-.75 $F_1$0.01 $K_1$.125 | Drilling cycle |
| 0310 | G28 | |
| 0320 | M30 | |

## 9.17 COMPUTER-ASSISTED PART PROGRAMMING

### 9.17.1 AUTOMATICALLY PROGRAMMED TOOLS LANGUAGE

In complicated PTP jobs and contouring applications, manual part programming becomes tedious and is subject to possible errors. Many part programming language systems have been developed to automatically perform most of the calculations that the programmer is usually tasked with. This, in turn, saves time and results in a more accurate and more efficient part program. The use of computer-aided part programming is justified when the part is of a complex shape or when the part is simple but the program required is too long. Additionally, computer-aided part programming is justified when the NC machine is complex, such as automatic tool-changing machining centers and four-axes NC lathes. The most widely used automatic programming system is known as APT. APT is one of the alternative methods for part programming, the common tasks of which include the following steps:

1. Definition of WP geometry, which can be performed by defining the elements forming the part and identifying each element in terms of dimensions and location
2. Specification of tool path or operation sequence

Computer-assisted part programming generates the cutter positions based on APT statements, and the tool path is directed to various point locations and along the surfaces of the WP to carry out machining. An APT program includes language statements that fall in the following four categories:

1. *Geometry statements*. Any part is composed of basic geometric elements that can be described by points, lines, planes, circles, cylinders, and other mathematically defined surfaces. Each geometric element must be identified by the part programmer,

as well as the dimensions and location of the element. In APT programming language, the following statements may be used:

| | | |
|---|---|---|
| Point | P1 = POINT/5.0,4.0,0.0 | X, Y, Z coordinate values |
| | P2 = POINT/INTOF, L1, L2 | Intersection of two lines |
| Line | L3 = LINE/P3,P4 | Between two points |
| | L4 = LINE/P5, PARALEL, L3 | From point 5 and parallel to line 3 |
| Plane | PL1 = PLANE/P1, P4, P5 | Defined by three points |
| | PL2 = PLANE/P2, PARALEL, PL1 | From point P2 and parallel to plane PL1 |
| Circle | CIRCLE/CENTER, P1, RADIUS, 5.0 | Centers at P1 and the radius is 5 |

2. *Motion statements.* After defining the WP geometry, the programmer constructs the path that the cutter will follow to machine the part. The tool path specification involves a detailed step-by-step sequence of cutter moves. Such cutter moves are made along the geometric elements, which have been defined earlier. In APT language, the following statements could be used for describing the tool movements:

| | |
|---|---|
| FROM/TARG | From target |
| FROM/-2.0,-0.20,0.0 | From X, Y, Z |
| GOTO/P1 | Go to point P1 |

For PTP motions, the following statements are used:

| | |
|---|---|
| GOTO/P2 | Move to point P2 |
| GOTO/2.0,7.0,00 | Absolute |
| GODELTA/2.0,7.0,0.0 | Incremental |

For continuous motions as shown in Figure 9.61, the tool motion is guided by the following three surfaces:

- *Drive surface.* Guides the sides of the cutter.
- *Part surface.* The bottom of the cutter rides.
- *Check surface.* Stops the movements of the tool (TO, ON, PAST, TANTO), as shown in Figure 9.62.

Figure 9.63 shows the motion commands, which include GOLFT, GORGT, GOFWD, GOBACK, GOUP, and GODOWN.

3. *Postprocessing statements.* These statements contain the machine instructions that are passed unchanged into the cutter location data (CLDATA) file to be dealt with by the postprocessor. It operates the spindle speed, feed rate, and other features of the machine tool, such as adding coolant and stopping the program. Some

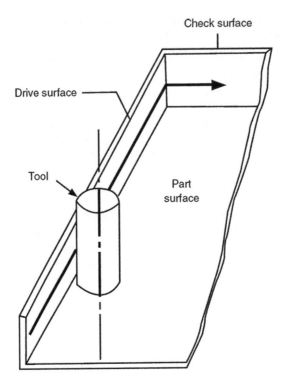

**FIGURE 9.61**   APT definitions of drive, part, and check surfaces.

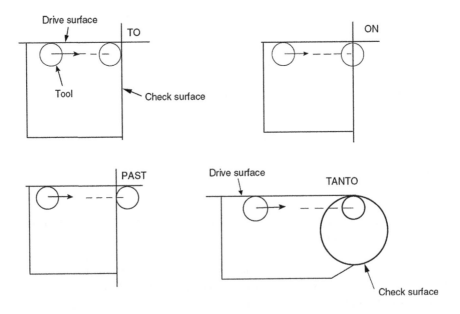

**FIGURE 9.62**   Use of APT modifier words TO, ON, PAST, and TANTO.

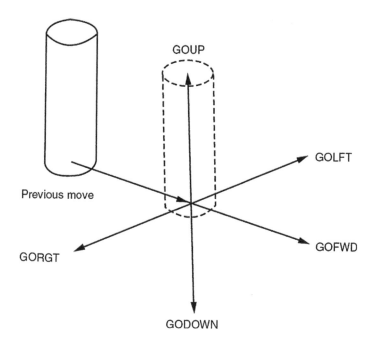

**FIGURE 9.63** APT motion commands.

common postprocessor statements are as follows, where the slash indicates that some descriptive data are needed:

| | |
|---|---|
| COOLANT/ | For coolant control (ON or OFF) |
| RAPID | To select rapid cutter motion |
| SPINDL/ | To select spindle on/off, speed and direction of rotation |
| FEDRAT/ | To select feed rate |
| TURRET/ | To select cutter number |

4. *Auxiliary statements.* These statements provide additional information to the APT processor, giving part name, tolerances to be applied, and so on.

### 9.17.2 PROGRAMMING STAGES

Figure 9.64 shows the main steps of computer-assisted part programming, which can be summarized as follows:

1. Identify the part geometry, general cutting motions, feeds, speeds, and cutter parameters.
2. Code the geometry, cutter motions, and general machine instructions into the part programming language (APT).
3. Process the source to produce a machine-independent list of movements and ancillary machine control information, and CLDATA file.

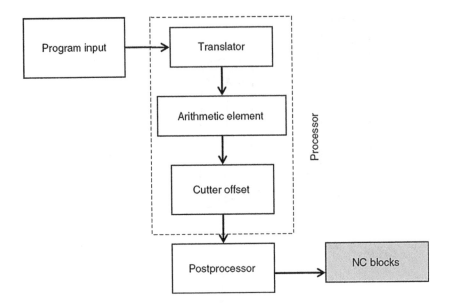

**FIGURE 9.64**   Steps of computer-assisted part programming.

4. Postprocess CLDATA to produce the machine control data (MCD) for a particular machine.
5. Transmit the MCD to the machine, and test.

The computer's job in computer-assisted part programming consists of the following steps:

1. *Input translation.* Converts the coded instructions contained in the program into computer usable form, ready for further processing.
2. *Arithmetic calculations.* Degenerates the part surface using subroutines that are called by the various part programming language statements. The arithmetic unit frees the programmers from the time-consuming calculations.
3. *Cutter offset computation.* Offsets the tool path from the desired part by the radius of the cutter. This means that the part programmer defines the exact part outline in the geometry statements.
4. *Postprocessing.* This is a separate computer program that prepares the punched tape or the program for a specific machine tool. The input to the processor is the output from the previous three steps, and the output is the NC tape or program written in the correct format for the machine tool to be used.

### Illustrative Example 5

Program the part shown in Figure 9.65 using APT.

## SOLUTION

---

Initial Auxiliary and Postprocessor Statements

| PARTNO | H9253 |
|---|---|
| | MACHIN/MILL.1 |
| | INTOL/.01 |
| | OUTOL/.01 |
| | CUTTER/15.0 |

*Geometry Statements*

| START | = POINT/−80.0, −80.0, 40.0 | |
|---|---|---|
| P0 | = POINT/0.0, 0.0, 0.0 | |
| P1 | = POINT/160.0, 0.0, 0.0 | |
| P2 | = POINT/160.0, 160.0, 0.0 | |
| P3 | = POINT/0.0, 160.0, 0.0 | |
| L1 | = LINE/P0, P1 | |
| L2 | = LINE/P1, P2 | |
| L3 | = LINE/P2, PARLEL, L1 | |
| L4 | = LINE/P3, P0 | |
| PL1 | = PLANE/P0, P1, P3 | |

Start spindle and coolant

| | SPINDL/800 |
|---|---|
| | COOLNT/ON |
| | FEDRAT/40.0 |

*Motion Statements*

| | FROM/START | |
|---|---|---|
| | GO/TO, L1 TO, PL1, TO, L4 | 1 |
| | GORGT/L1, PAST, L2 | 2 |
| | GOLFT/L2, PAST, L3 | 3 |
| | GOLFT/L3, PAST, L4 | 4 |
| | GOLFT/L3, PAST, L4 | 4 |
| | GOLFT/L4, PAST, L1 | 5 |

Select rapid feed and return to start

| | RAPID |
|---|---|
| | GOTO/START |

Turn off coolant and spindle, end section, and program

| | SPINDL/OSS |
|---|---|
| | END |
| | FINI |

---

# 9.18 CAD/CAM APPROACH TO PART PROGRAMMING

## 9.18.1 COMPUTER-AIDED DESIGN

CAD is a technology that involves a computer in the design process. It enables the engineer to develop, change, and interact with the graphical model of a part.

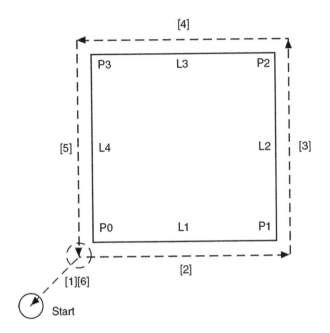

**FIGURE 9.65**   Part drawing.

Computers are strong in the areas of graphics, calculations, analysis, modeling, and testing. During CAD stages, the engineer draws the part on the screen. This information can be used to create a program to machine the part. The designer must work closely with the manufacturing experts to establish some standards for the design.

CAD allows different layers to be created. This technology allows other software to take the part geometry from each layer and assign different tools to machine it. The computer allows the design to be viewed and tested before manufacturing. CAD systems stress-test parts to meet the strength requirements of the application. Graphics capabilities allow three-dimensional (3-D) viewing of parts from any angle. CAD systems can also export the CAD part file in Drawing Exchange File (DXF) format. Using CAD systems, the following advantages can be secured:

1. Increase the productivity of the designers
2. Create better designs
3. Reduce redundant effort
4. Allow easy and rapid modification of prints
5. Enable integration of engineering and manufacturing

### 9.18.2   COMPUTER-AIDED MANUFACTURING

CAM utilizes computers in the manufacturing stage. Such a modern technique has the following features:

- It allows the programmer to develop a model that represents the part and the machining operations.

- The programmer can interact with the model graphically to make the necessary adjustments and modifications before the CNC code is generated.
- CAM software reads the DXF, which contains the part geometry and the levels or layers that the geometry exists on.
- CAM software utilizes a job plan to assign the correct tool path to each layer.
- The job plan knows the work material that will be used, so that it can calculate speeds and feeds for each tool.

The integrated CAD/CAM approach prepares the part program directly from the CAD part geometry, either by using NC programming commands called in the CAD/CAM system or by passing the CAD geometry into a dedicated CAM program. Using CAD/CAM systems, the CAD drawings can therefore be changed to CNC programs. The CAD/CAM approach has the following advantages:

1. No need to encode the part geometry and the tool motion
2. Allows the use of interactive graphics for program editing and verification
3. Displays the programmed motions of the cutter with respect to the WP, which allows visual verification of the program
4. Allows interactive editing of the tool path with the addition of the tool moves and standard cycles
5. Incorporates the most sophisticated algorithms for part programming generation

The programming steps for CAD/CAM approach are as follows:

1. The aspects of the part geometry that are important for machining purposes are identified (and perhaps isolated on a separate level or layer); geometry may be edited or additional geometry added to define boundaries for the tool motion.
2. Tool geometry is defined; for instance, by selecting tools from a library.
3. The desired sequence of machining operations is identified, and tool paths are defined interactively for the main machining operations.
4. The tool motion is displayed and may be edited to refine the tool motion, or other details may be added for particular machining cycles or operations.
5. A CLDATA file is produced from the edited tool paths.
6. The CLDATA file is postprocessed to MCD, which is then transmitted to the machine.

### 9.18.2.1  Postprocessor
The postprocessor takes the part geometry and job plan and writes the code that the specific machine will understand.

### 9.18.2.2  Simulation
The program can be written and tested before it is actually run, and jigs, fixtures, and clamping can be shown during simulation to determine whether there are any

potential problems. The simulation shows the machining time. Operations can then be adjusted to optimize production offline, which keeps the machine operating.

### 9.18.2.3 Download the CNC Programs

The program is sent to the machine manually, using a tape or disk, or electronically, using a communication cable.

## 9.19 REVIEW QUESTIONS

9.19.1 Mark true or false:

[ ] Circular interpolation G02 and G03 are limited to 90°.

[ ] Contouring NC can perform PTP and straight cuts.

[ ] DNC allows a computer to control one machine only.

[ ] Errors in programming using incremental positioning are more dangerous than in absolute programming.

[ ] A fixed zero is easier for the programmer than a floating zero.

[ ] G00 requires a feed number during part programming.

[ ] G01 performs machining of straight lines at any number of axes.

[ ] G02 performs circular interpolation in ccw direction.

[ ] GOBACK is a motion statement in computer-assisted part programming.

[ ] The magic-three method is used for programming $X$-, $Y$-, and $Z$-coordinates.

[ ] NC machines can understand decimal and binary numbers.

[ ] NC adds many machining capabilities that are not built into the machine tool.

[ ] NC is a form of programmable automation by cams and mechanisms.

[ ] NC is an advanced machining method.

[ ] NC machine tools can produce shapes and cam profiles as easily as circular ones.

[ ] NC machines can be used to produce large numbers only.

[ ] The NC user does not necessarily speak the language of the machine control system.

[ ] Parity check in tape code is helpful in winding the tape.

[ ] Polar coordinates provide the ability to generate a 360° arc in a smaller number of blocks than circular interpolation.

[ ] Rapid positioning requires a feed number during part programming.

[ ] Row 5 is used as the parity check in ISO tape code set.

[ ] Straight-cut NC can produce contours.

[ ] The character is a collection of words that forms a block.

[ ] The setup point can be chosen as the zero point on the machine.

[ ] Tape readers used in NC applications are photoelectric and electromechanical.

[ ] Turning on NC machines can be programmed at a constant surface speed or constant revolutions per minute.

9.19.2 Explain the following using neat sketches:
- Absolute and incremental programming
- Types of NC system

9.19.3 Convert 101000111 to decimal and 1997 to binary.

9.19.4 Express in the magic-three method: 1525 rpm–0.035 mm/min.

9.19.5 Show, using sketches, each of the following:
- DNC
- Advantages of CAD/CAM systems
- Steps of computer-assisted part programming
- 3-D NC contouring

9.19.6 Choose the right answer to finish each sentence:

   a. Machining instructions that are coded and provided to an NC machine to produce a finished WP are called:
   A. Integrated circuits
   B. Input media
   C. Output data
   D. Part programs

   b. The decimal number 92 converts to the binary number:
   A. 1101110
   B. 1001110
   C. 1010101
   D. 1011100

   c. The magic-three method code for a desired feed rate of 42 ipm is:
   A. $F_1542$
   B. $F_1545$
   C. $F_1742$
   D. $F_1745$

   d. Which control system would probably be used to control the path of a cutting tool in five axes?
   A. Incremental NC system
   B. Continuous-path NC system
   C. Straight-cut NC system
   D. PTP NC system

   e. A prominent feature of machining centers is:
   A. Circular interpolation
   B. An automatic tool changer
   C. The BCD format
   D. Tool length compensation

9.19.7 Write the necessary part program for the part shown in Figure 9.66.

9.19.8 Program the part shown in Figure 9.67:
- Use canned cycles where appropriate.
- Preset the absolute tool reference at $X = 100$ and $Z = 50$ mm from the zero point.
- Use letters a, b, c, d, and so on to describe the tool paths.
- Choose the proper tool from the following table:

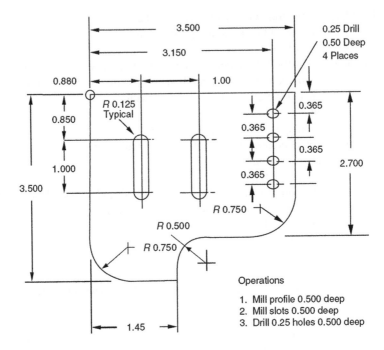

**FIGURE 9.66** Part drawing. (From Senerstone, J. and Kuran, K., *Numerical Control Operation and Programming*, Prentice-Hall, New York, 1977. With permission.)

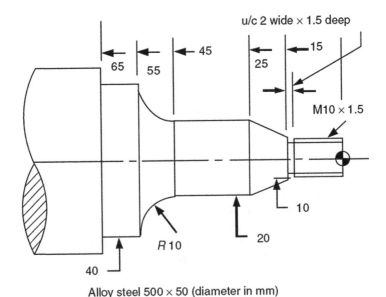

**FIGURE 9.67** Part drawing. (From Thyer, G. E., *Computer Numerical Control of Machine Tools*, Industrial Press Inc., New York, 1988. With permission.)

| Tool Number | Tool Description | Operation |
|---|---|---|
| 1 | 80° diamond | Rough turn |
| 2 | 35° diamond | Finish turn |
| 3 | 60° diamond | Threading |

## REFERENCES

Gibs, D 1988, *An introduction to CNC machining*, 2nd edn, ELBS Cassell Publishers Ltd., London. Harding Inc.

Senerstone, J & Kuran, K 1997, *Numerical control operation and programming*, Prentice-Hall, New York.

Thyer, GE 1988, *Computer numerical control of machine tools*, 1st edn, Industrial Press Inc., New York.

# 10 Automated Manufacturing Systems

## 10.1 INTRODUCTION

In the early days of the eighteenth century, the first industrial revolution started when attempts were made to substitute muscle power by mechanical energy. Machine tools such as boring machines, lathes, drill presses, copying lathes, turret lathes, and milling machines were introduced into the production of goods. Geared and automatic lathes were introduced in the 1900s. Mass production techniques and mechanized transfer machines were developed between 1920 and 1940. These systems had fixed mechanisms and were designed to produce specific products. These developments were best represented in the automobile industry and were characterized by high production rates at low cost (Figure 10.1). Since that time, productivity has become a major concern; productivity is often defined as the use of all resources, such as materials, energy, capital, labor, and technology, or it may be defined as the output/labor hour. It is basically a measure of operating efficiency.

Mechanization was related to the first industrial revolution and reached its peak by the 1940s, when most manufacturing operations were carried out on traditional machinery, such as lathes, milling machines, and automatic lathes, which required skilled operators and lacked flexibility. Each time a different product was manufactured, the machine had to be retooled. Furthermore, new products with complex shapes required tedious work from the operator to set the proper processing parameters. Mechanization refers to the use of various mechanical, hydraulic, pneumatic, or electrical devices to run the manufacturing process. In a mechanized system, the operator still directly controls the particular process and checks each step of machine operation.

The next step after mechanization was automation, derived from the Greek word *automatos* (self-acting); this word was first used in 1945 by the U.S. automobile industry to indicate automatic handling and processing of parts in production machines.

The world is now passing through the second industrial revolution, with fantastic advances occurring continuously in the fields of electronics and computer technology. The computer is substituting for the human brain in controlling machines and industrial processes. A major breakthrough in automation was the invention of the first digital electronic computer (1943), followed by the first prototype of a numerically controlled machine tool (1952). Since this historic development, rapid progress has been made in automating most aspects of manufacturing, including the introduction of computers to enhance automation using computer numerical control (CNC), adaptive control (AC), industrial robots, and computer-integrated manufacturing

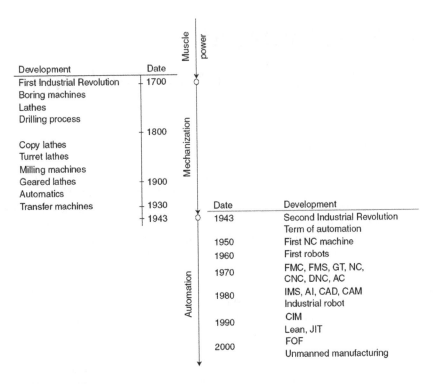

**FIGURE 10.1**   Mechanization and automation of MS.

systems (CIMSs), including computer-aided design (CAD), computer-aided engineering (CAE), and computer-aided manufacturing (CAM) (Figure 10.1).

The manufacturing situation today has made the mass production of any component economically possible; however, industry in many cases demands variety in products in small lots. Economical production methods suitable for smaller lots should be followed. Further, higher accuracies are required at lower cost. To meet these requirements, there is a rapidly growing need for improved communication and feedback between manufacturing and design processes, integrating them into a single system capable of being optimized as a whole. The use of computerized integrated manufacturing (CIM) is the answer to meeting these requirements and objectives. Computers therefore have an important role to play, especially in job shop and batch production manufacturing plants, which constitute an important domain of overall manufacturing activity. Traditional job shop and batch manufacturing suffers from drawbacks such as:

- Low equipment utilization
- Long lead times
- Inflexibility to market needs
- High inventory
- Dependence upon highly skilled operators
- Poor quality control
- Increased indirect cost

It is estimated that in traditional batch production, only 5–10% of the time is used on machines, and the rest is spent in moving and waiting. Out of the total time on the machine, only 30% is machining time, the rest being for positioning, gauging, and idling (Jain and Gupta, 1993). These shortcomings can be overcome through the use of the following:

- Material handling equipment
- Feedback systems and continuous flow process
- Computers for process control and data collection, planning, and decision making to support manufacturing activities

It must be understood that a computer cannot change the basic metal working processes. It can only influence their control and their sequences so that down-time is kept to a minimum. It provides quick reflexes, flexibility, and speed to meet the desired results. The computer enables detailed analysis and accessibility to accurate data necessary for the integration of the manufacturing system (MS). For a plant to produce diversified products of best quality at enhanced productivity and lower prices, it would be essential for all elements of manufacturing (design, machining, assembly, quality assurance, management, and material handling) to be computer-integrated, both individually and collectively (Jain and Gupta, 1993).

This chapter emphasizes the importance of flexibility in machine, equipment, tooling, and production operations in order to be able to respond to the market and to ensure on-time delivery of high-quality products to attain customer satisfaction. Important developments during the past three decades have had a major impact on modern manufacturing, among which are group technology (GT), cellular manufacturing, flexible manufacturing systems (FMSs), and just-in-time (JIT) production.

## 10.2   MANUFACTURING SYSTEMS

The MS takes inputs and produces products for the customer (Figure 10.2). It is a complete set of elements that includes machines, people, materials, information, handling, equipment, and tooling. The system output may be consumer goods or services to user or inputs to some other processes. The materials are processed within the system and gain value as they are passed from machine to machine. MSs are very interactive and dynamic. An MS should be designed and integrated for low cost, superior quality, and on-time delivery.

It should be an integrated whole that is composed of integrated subsystems, each of which interacts with the entire system. System operation requires information gathering and communication within the decision-making processes that are integrated into the MS. Each company will have many differences resulting from discrepancies in subsystem combinations, people, product design, and materials (Degarmo et al., 1997).

MSs differ in structure or design. They may be classified into the following types:

1. *Job shop.* In this type, a variety of products are manufactured, which result in small lot sizes; often, these products are one of a kind. This is commonly done by specific customer order. Because the plant must perform a wide

variety of manufacturing processes, general-purpose production equipment is required. In a job shop, workers must have relatively high skill levels to perform a range of work assignments. Job shop products include space vehicles, aircraft, machine tools, special tools, and equipment. Figure 10.3 shows the functional or process layout of a job shop. Forklifts and handcarts are used for material handling.

2. *Flow shop.* This type has a product-oriented layout composed mainly of a flow line. When the volume gets very large, as in assembly lines, it is called mass production. Specialized equipment, dedicated to the manufacture of a particular product, is used. This system may be designed to produce a particular product of a family of products, using special-purpose high-investment machines rather than general-purpose ones. Figure 10.4 shows an automated production flow line consisting of a number of machines or stations arranged according to a certain configuration and linked to each other by conveyors to direct the part from one machine to other. The manual skill level tends to be lower than in a job shop. The time item spent at

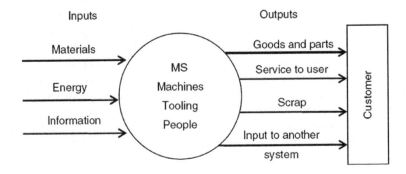

**FIGURE 10.2** Manufacturing system.

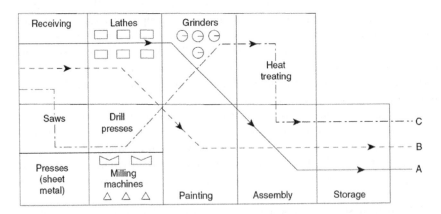

**FIGURE 10.3** Job shop MS—functional or process layout. (From Degarmo, E. P., Black, J. T., and Kohser, R. A., *Materials and Processes in Manufacturing,* Prentice Hall, New York, 1997.)

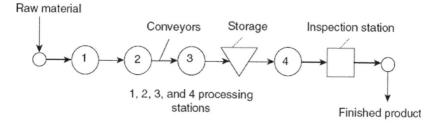

**FIGURE 10.4** Schematic of an automated production line (flood shop MS).

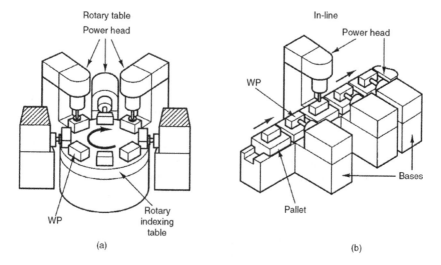

**FIGURE 10.5** Rotary (a) and line (b) transfer machines. (Courtesy of Heald Machine Company.)

each station is fixed and equal (balanced). Automated production flow lines are adapted for production at reduced labor cost, increased production rate, and reduced inventory cost. In this respect, two types of automated production lines (transfer units) are differentiated: straight and circular. The choice of either type depends upon the number of machining stations and the available area in the manufacturing plant. More stations are consumed in the straight type, whereas the second type does not require large areas (Figure 10.5).

3. *Project shop (fixed-position layout).* In this type of shop, the product must remain in a fixed position or location during manufacturing because of its size or weight. Materials, machines, and labor involved in the fabrication are brought to the site. Constructional jobs (buildings, bridges, and dams), locomotive manufacturing, aircraft assembly, and shipbuilding use a fixed-position layout. When the job is completed, the equipment is removed from the site.

4. *Continuous process.* This process finds application in oil refineries, chemical processing plants, and food processing. This system is sometimes called

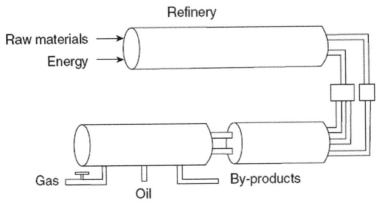

Continuous process layout

**FIGURE 10.6** Continuous process layout of a refinery. (From Degarmo, E.P., Black J.T., and Kohser, R.A., *Materials and Processes in Manufacturing,* Prentice Hall, New York, 1997.)

flow production if a complex single product is to be fabricated (television production, canning operations, and so on). Continuous processes are usually easy to control and efficient, and they are the simplest systems; however, they are also the least flexible systems (Figure 10.6).

5. *Cellular manufacturing.* This process is intended for producing parts one at a time in a flexible design. The cell capacity (cycle time) can be altered quickly to respond to rapid changes in market demand, thus allowing more product variety in smaller quantities, which is highly desirable. Figure 10.7 illustrates an example of a cellular unit comprising two machine tools, automated part inspection, and a serving robot.

Flexible cells are typically manned, but unmanned cells are beginning to emerge with a robot replacing the worker (Figure 10.8). Workpieces (WPs) are placed on the rotary feeder and are loaded and unloaded one by one to the CNC machine by the robot. Machining is performed according to numerical control (NC) command data (cutting information) stored in advance in the CNC machine-tool memory.

For best results, it is recommended not only to automate the local environment of the machine tool but also to automate the global environment, including the following activities:

- Management and provision of resources
- Preparation and transportation of WPs
- Supply and evolution of production data
- Inspection of WPs and machine tools

The flexibility is acquired when a computer is applied to control the mentioned global environment of production.

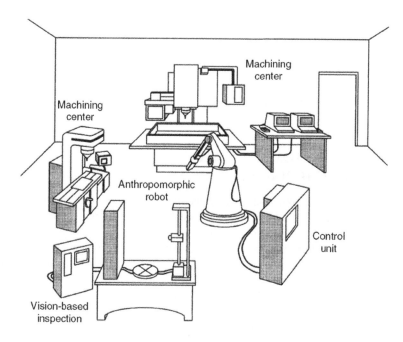

**FIGURE 10.7** Example of FMC composed of two machines, automated part inspection, and a serving robot. (From Kalpakjian, S. and Schmidt, S. R., *Manufacturing Processes for Engineering Material,* Pearson Education, Inc., NJ, 2003.)

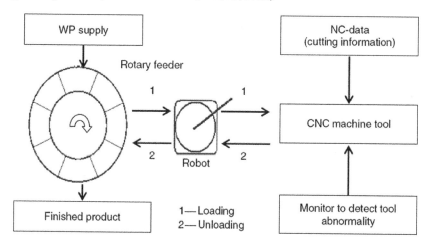

**FIGURE 10.8** Unmanned manufacturing cell.

The flexible manufacturing cell (FMC) process offers the following advantages:

- Flexibility for varied and small-quantity production
- Manned and unmanned operation
- Automatic operation of numerous processes

- Simple setup
- Easy operation
- Integration into FMS
- High operation economy
- Immediate stop if necessary
- Operation reliability depending on an adequate supply of NC command data

## 10.3  FLEXIBLE AUTOMATION–FLEXIBLE MANUFACTURING SYSTEMS

In earlier times, the automation of manufacturing processes was limited to mass production (fixed automation), which was feasible only for a large number of parts.

Another field of automation is the FMS, which is a highly automated MS comprising a collection of production devices, logically organized under a host computer and physically connected by a central transporting system. It has been developed to provide some of the economics of mass production to small-batch manufacturing.

The main advantage of an FMS is its high flexibility in terms of small effort and short time required to manufacture a new product. It is an alternative that fits in between the manual job shop and hard automated transfer lines (Figure 10.9). It is

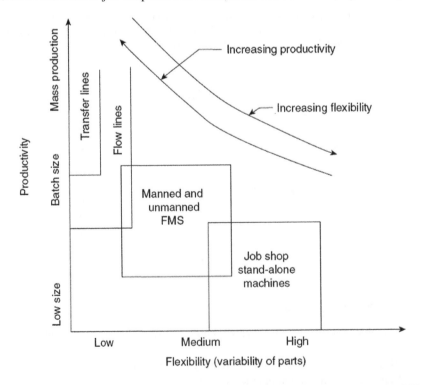

**FIGURE 10.9**  Flexibility against production rate of different MSs (intermediate flexibility/ productivity of FMSs).

best suited to applications that involve an intermediate level of flexibility and productivity (El-Midany, 1994).

FMS can be regarded as a system that combines the benefits of two other systems:
(a) the highly productive but inflexible transfer lines and (b) job shop production,
which fabricates a large variety of products on standalone machines but is inef-
ficient. In FMS, the time required for changeover to a different part is very short (1
min), thus making the quick response to product and market-demand variations a
major benefit of FMS.

## 10.3.1  Elements of Flexible Manufacturing System

The basic elements of an FMS are as follows:

- Workstations
- Automated handling and transport of materials and parts
- Control systems

*Workstations* are arranged to achieve the greatest efficiency in production, with an
orderly flow of materials and products through the system. The type of worksta-
tion depends on the type of production. For machining operations, these usually
consist of a variety of three- to five-axes machining centers and CNC machines
(milling, drilling, and grinding). They also include other equipment, such as that for
automated inspection (including coordinate measuring machines [CMMs]), assembly, and cleaning. For sheet-metal forming, punching, and shearing and forging, the
workstations of an FMS incorporate furnaces, trimming presses, heat-treatment
facilities, and cleaning equipment.

Machine tools may be equipped with either turret or tool changers for supplying
the desired tools for machining. A turret has a shorter cycle time and is preferred
when turning small components; indexing tools with a turret is the fastest method.
For larger components with longer cycle times, an automatic tool changer is required
where a bigger magazine can be used (Jain and Gupta, 1993).

*Handling and transport* are very important for the system's flexibility. Material
handling is controlled by a central computer and is performed by automated vehicles, conveyors, and various transfer mechanisms.

An industrial robot is best used for serving several machines in a production cell. A
robot should not be located in front of the machine tool, as it prevents manual control
and supervision. However, the automatic changing of WPs is best satisfied by using
handling equipment that is built integral with the machine tool. Such equipment is
called a computerized part changer (CPC). It is installed physically separate from the
machine tool to eliminate vibration to the machine when the parts are being handled by
the CPC. It consists of a portal, shuttle carriage, vertical slide, and gripping unit. The
jaws of the gripping unit can clamp external and internal parts. The gripping unit axis
can be positioned in four different angular positions: 0°, 90°, 180°, and 270°. Both the
rotary motion and the gripping action are hydraulically actuated (Jain and Gupta, 1993).

*Computer control* of FMS is the main brain, which includes various software and
hardware. This subsystem controls the machinery and equipment in workstations

and transportation of materials in various stages. It also stores data and provides communication terminals that display the data (Kalpakjian and Schmidt, 2003).

Because FMS involves a major capital investment, efficient machine utilization is essential; machines must not stand idle. Consequently, proper scheduling and process planning are crucial. Scheduling for FMS is dynamic, unlike that in job shops, where a relatively rigid schedule is followed. Dynamic scheduling is capable of responding to quick changes in product type and is thus responsive to real-time decisions (Kalpakjian and Schmidt, 2003).

In brief, FMS is an integrated system of computer-controlled machine tools and other workstations with an automated flow of information, WPs, tools, and so on. Control of FMS is achieved by computer-implemented algorithms that make all the operational decisions. This system is arranged so that the automated production of a group of complex WPs in any lot size, particularly small and medium batches, is possible (Figure 10.9). The FMS is usually planned by simulated techniques in which a model is drawn. This model is an idealized representation of the components, internal relations, and characteristics of a real-life system. Analysis of model behavior shows ways for improving the system by carrying out necessary changes (Jain and Gupta, 1993).

FMS offers the advantages of part cost reduction, throughput time reduction, increasing flexibility toward changes of the product mix, reduced inventory and lead times, and increased productivity. This system usually incorporates features such as adaptive controls, tool breakage detectors, and a tool life monitoring system. A major drawback of FMS is the unknown availability and reliability of the planned system (Jain and Gupta, 1993).

### 10.3.2 Limitations of Flexible Manufacturing System

The use of FMS is hindered due to the following reasons:

- High programming cost
- Lower degree of sophistication of fabrication and assembly processes
- Unavailability of reliable feedback devices for tool wear and breakage

### 10.3.3 Features and Characteristics

FMS features and characteristics are summarized as follows:

- FMS offers emerging cost and quality benefits for most engineering sectors requiring batch production.
- Batch production using conventional machine tools necessitates a minimum number of similar components to be produced economically. There is no batch size limitation in case of FMS; consequently, there is no need to lock up the money in extensive stocks of finished parts. The work in progress is reduced considerably, and the inventory cost is therefore eliminated.
- It is possible to produce at random all varieties of products planned by a firm. FMS is capable of quickly responding to any design changes or market demands in the product.

- FMS are usually equipped with robots or handling equipment. Software is developed to integrate CNC control and the handling systems. All necessary tools can be stored in a magazine.
- All part programs of different models are stored in the system memory. The system identifies the model program to be produced.
- FMS can be conceived in multiples of 15–20 minute operations. If a certain operation takes a longer time, multiples of similar machines can be installed in the line.
- Extensive use of touch triggers is made to minimize the operator's intervention in the line.
- Industrial robots are used for material handling (loading and unloading), inspection activities, and assembly operations.

### 10.3.4  NEW DEVELOPMENTS IN FLEXIBLE MANUFACTURING SYSTEM TECHNOLOGY

These developments dramatically boost the capabilities of FMS:

- Computerized tool-setting station
- Establishing tool information such as tool lengths, tool offsets, and so on by linear variable displacement transducer (LVDT)
- Automatic tool changer
- Monitoring of tool life
- Providing tool breakage detectors
- Providing tool compensation system and AC
- Increased use of robots and handling system
- Application of laser and fiber optic technology to check bore diameters and part surface location
- Spindle probes to check WP features like bore diameter and hole pattern location
- Improved software
- Fault analysis (vision system for online quality control)
- Swarf and coolant control
- Computerized simulation to establish efficiencies and programming facilities

## 10.4  COMPUTER-INTEGRATED MANUFACTURING

CIM is a recent technology that has been in development and trial since the 1990s. It comprises a combination of software and hardware for product design, production planning and control, production management, and so on in an integrated manner. It is a methodology and a goal rather than merely an assemblage of equipment and computers. Its effectiveness greatly depends on the use of large-scale integrated communications systems involving computers, machines, and their controls.

As with traditional manufacturing approaches, the purpose of CIM is to transform product designs and materials into sellable goods at a minimum cost in the shortest possible time. The CIM begins with the design of a product (CAD) and ends with the manufacture of that product (CAM). With CIM, the usual split between CAD and CAM is supposed to be eliminated (Degarmo et al., 1997).

CIM differs from the traditional job shop in the role the computer plays in the manufacturing process. CIMSs are basically a network of computer systems tied together by a single integrated database (DB). Using the information in the DB, a CIMS can direct manufacturing activities, record results, and maintain accurate data. CIM is the computerization of design, manufacturing, distribution, and financial functions into one coherent system (Degarmo et al., 1997).

CIM is an attempt to integrate the many diverse elements of discrete parts manufacturing into one continuous process-like stream. It can result in increased manufacturing productivity and quality and reduced production cost. It employs FMS, which saves the manufacturer from replacing equipment each time a new part has to be fabricated. The current equipment can be adopted to produce a new part, as long as it is in the same product family, with programmable software and some retooling. Thus, this system has the ability to switch from component to component with no down-time for changeover. This system requires NC lathes, machining centers, punch presses, and so on, which have the ability to be readily incorporated into a multimachine cell or fully integrated manufacturing system (IMS).

Multispindle CNC machines with greater horsepower, stiffness, and wider speed ranges are important for CIM. Automatic tool changers (to change the tool in the spindle and to renew dulled tools) are a must. A robot for handling WPs is another important machine-tool peripheral essential for the CIMS.

CIM is a very powerful concept and has the potential for great benefits; however, it is not easy to implement. Like any powerful and complex tool, it can be dangerous and costly if not implemented properly.

The main tasks involved in CIM can be separated into four areas (Figure 10.10):

1. *Product design,* for which an interactive CAD system allows drawing, analysis, and design to be performed. The computer graphics are useful to get the data from the designer's mind to be ready for interaction (Figure 10.11).
2. *Manufacturing planning,* where the computer-aided process planning (CAPP) helps to establish optimum manufacturing routines and processing steps, sequences, and schedules.

**FIGURE 10.10**  Main tasks of CIM.

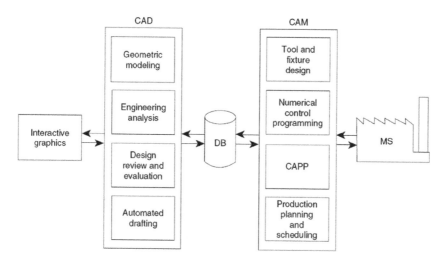

**FIGURE 10.11** Database to CAD/CAM.

3. *Manufacturing execution,* in which CAM identifies manufacturing problems and opportunities. Intelligence in the form of microprocessors is used to control machines and material handling and to collect data controlling the current shop floor (Figure 10.11).
4. *Computer-aided inspection (CAI) and computer-aided reporting (CAR), so as to provide a feedback control loop (Figure 10.10).*

Computer integration of these four tasks provides the most current and accurate information about manufacturing, thereby permitting better and tighter control, and enhances the overall quality and efficiency of the entire system. Improved communication among these activities results in enhanced productivity and accuracy if the designer considers limitations and manufacturing problems, and vice versa. The availability of current data permits instantaneous updating of production-control data, which in turn permits better planning and more effective scheduling. All machines are fully utilized, handling time is reduced, and parts move more efficiently through production. Workers become more productive and do not have to waste time in coordination and searching of previous data.

Figure 10.12 illustrates a block diagram of activities for a typical CIMS. With the introduction of computers, changes have occurred in the organization and execution of production planning and control through the implementation of material requirement planning (MRP), capacity planning, inventory management, shop floor control, and cost planning and control. Engineering and manufacturing DBs contain all the information needed to fabricate the components and assemble the products (Figure 10.12). The design engineering and process planning functions provide the inputs for the engineering and manufacturing DB, which includes all the data on the product generated during design, such as geometrical configurations, part lists, and material specifications (Figure 10.12). The bill of material is a key part of the DB. Figure 10.11 shows how a CAD/CAM DB is related to the design and manufacturing.

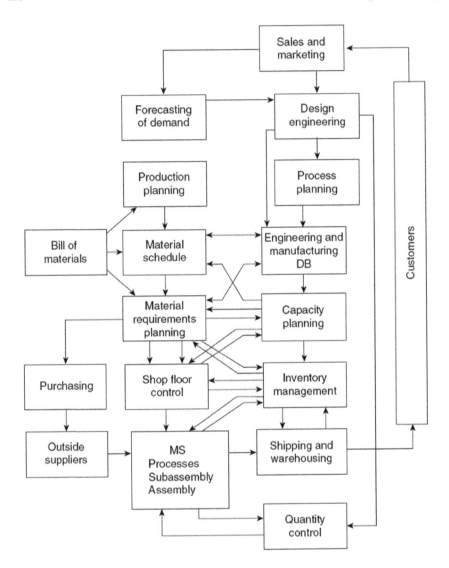

**FIGURE 10.12** Cycles of activities in CIMS. (From Degarmo, E. P., Black J. T., and Kohser, R. A., *Materials and Processes in Manufacturing,* Prentice Hall, New York, 1997.)

Included in the CAM is a CAPP module that acts as an interface between CAD and CAM (Degarmo et al., 1997).

Capacity planning is performed in terms of labor or machine hours available. The master schedule is transformed into material and component requirements using MRP. These requirements are then compared with available plant capacity. If the schedule is incompatible with capacity, adjustments must be made either in the master schedule or in plant capacity. The possibility of adjustments in the master schedule is indicated by the arrow in Figure 10.12 leading from capacity planning to the master schedule. The term "shop floor control" (Figure 10.12) refers to a system for monitoring the status of manufacturing activities and the plant floor and reporting the status to management so that effective control can be exercised. The cost planning and control system consists of a DB to determine the expected costs to manufacture a product. It also consists of cost collection and analysis software to determine what the actual costs of manufacturing are and how they compare with expected costs (Degarmo et al., 1997).

In order that a computer successfully controls a given production line, a DB has to be built up that provides the computer with the necessary information based on a deep understanding of the manufacturing process and the entire system. Normally, the FMS is managed by a computer through a distributed logic architecture operating at several levels.

A machine may have a dedicated program for its own functioning. It must simultaneously obtain commands from many computers that provide necessary information desired for functioning of the complete system. All such activities have to be properly coordinated. An intermediate level of control could be provided to take care of material handling. Parts in such a system will be loaded according to instructions from a supervisory computer. The computer also dispatches tooling according to the expected tool life for each tool and takes desired action at the appropriate time. It also generates a variety of related reports such as tool life data and production data.

CIM has the advantage of possessing intelligence to maximize the process performance, provided that all the parameters can be measured in real time. It has been established that in metal cutting, every parameter related to machining can be determined if the shear angle is known. It is possible to feed into the computer all the data to compute shear strain, strain rate, flow stress, coefficient of friction at the chip–tool interface, unit horsepower and the total power being consumed, cutting temperature, and so on (Jain and Gupta, 1993).

The intelligent machine tools included in the CIMS result in the following advantages:

- Increased accuracy
- Reduced scrap
- Reduction in manned operation
- Increased predictability of machine tool operation
- Less skill for setting up and operation on the machine
- Reduced machine down-time
- Reduced production cost
- Increased machine throughput
- Reduced setup time
- Reduced tooling due to better operation planning
- Increased range of materials and part geometries
- Increased quantity and quality of information exchange between the machine control and the part designer and between human and machine-tool control

CIM technology offers the following advantages:

- High rates of production with high precision
- Remarkable flexibility for producing diverse components in the same setup
- Easy and quick manipulation of software
- Uninterrupted production with less supervision or work handling
- Economical production, even in the case of moderate batch sizes
- Drastic reduction of lead times
- Drastic changes in product design
- Integrating and fine-tuning of all factory functions, such as raw and semi-finished materials flow, tooling, metal cutting, and inspection

### 10.4.1 COMPUTER-AIDED DESIGN

A major element of a CIMS is the CAD system, which involves any type of design activity that makes use of a computer to create, develop, analyze, optimize, and modify an engineering design.

The design-related tasks performed by a CAD system (Figure 10.11) are:

- Geometric modeling
- Engineering analysis
- Design review and evaluation
- Automated drafting

*Geometric modeling* is the most important phase of the design process, through which the designer constructs the graphical image of the object on the cathode ray tube (CRT).

*Engineering analysis* may involve stress calculations, finite element analysis for heat transfer computations, or the use of differential equations to describe the dynamic behavior of the system. Usually, general-purpose programs are used to perform these analyses.

*Design review and evolution* techniques check the accuracy of the CAD design. Other features of these techniques are checking and animation, which enhance the designer's visualization of the mechanism and help to avoid interference.

*Automated drafting* involves the creation of hard-copy engineering drawings from the CAD DB. It is also able to perform dimensioning automatically, generate cross-hatching, scale drawings, develop sectional views, enlarge a view for details, perform rotation of parts, and perform transformations such as oblique, isometric, and perspective views (Degarmo et al., 1997).

### 10.4.2 COMPUTER-AIDED PROCESS PLANNING

CAPP uses a computer to determine how a part is to be made. If GT is used, parts are grouped into part families. For each part family, a standard process plan is established and stored in computer files and then retrieved for new parts that belong to that family.

For a manufacturing operation to be efficient, all its diverse activities must be planned and coordinated; this task has traditionally been done by process planners. Process planning involves selecting methods of production, tooling, fixtures, machinery, sequence of operations, standard processing time for each operation, and methods of assembly. These choices are all demonstrated on a routing sheet (Table 10.1). When performed manually, this task is labor-intensive and time-consuming, and it also relies heavily on the process planner's experience.

These route sheets may include additional information regarding materials, estimated tooling time for each operation, processing parameters, and other details. It travels with the part from one operation to another.

CAPP is an essential adjunct to CAD/CAM, although it requires extensive software and good coordination with CAD/CAM as well as other aspects of IMSs. CAPP is a powerful tool for effective planning and scheduling operations. It is particularly effective in small-volume, high-variety parts production requiring machining, forming, and assembly operations.

Process planning activities are a subsystem of CAM (Figure 10.11). Several functions can be performed using these activities, such as capacity planning for plants to meet production schedules, inventory control, and purchasing (Kalpakjian and Schmidt, 2003).

CAPP offers many advantages, which can be summarized as follows:

- The standardization of process plans improves productivity, reduces lead times and costs of planning, and improves the consistency of product quality and reliability.

## TABLE 10.1
## Sample Routing Sheet in CAPP

Routing sheet
Customer's name: Midwest Valve Co.  
Quantity: 15

Part name: Valve body  
Part no.: 302

| Operation No. | Operation Description | Machine |
|---|---|---|
| 10 | Inspect forging, check hardness | RC tester |
| 20 | Rough machine flanges | Lathe no. 5 |
| 30 | Finish machine flanges | Lathe no. 5 |
| 40 | Bore and counter bore | Boring mill no. 1 |
| 50 | Turn internal grooves | Boring mill no. 1 |
| 60 | Drill and tap holes | Drill press no. 2 |
| 70 | Grind flange endfaces | Grinder no. 2 |
| 80 | Grind bore | Internal grinder no. 1 |
| 90 | Clean | Vapor degreaser |
| 100 | Inspect | Ultrasonic tester |

From Kalpakjian, S. and Schmidt, S. R., *Manufacturing Processes for Engineering Material*, Pearson Education, Inc., NJ, 2003.

- Process plans make use of GT to retrieve plans to produce new parts.
- Process plans can be modified to suit specific needs.
- Neat and legible routing sheets can be prepared more quickly.
- Many other functions, such as cost estimation and work standards, can be incorporated into CAPP.

### 10.4.3 COMPUTER-AIDED MANUFACTURING

CAM involves the use of computers to assist in all phases of manufacturing a product, including process and production planning, scheduling, manufacturing, quality control, and management. CAM is another major element of CIM. Because of the increased benefits, CAD and CAM are integrated into CAD/CAM systems. This integration allows the transfer of information from the design stage to the planning stage for manufacturing of the product without the need to manually reenter data on part geometry. The DB developed during CAD is stored and then processed further by CAM into necessary data and instructions for operating and controlling production machinery and material-handling equipment and for performing automated testing and inspection (Kalpakjian and Schmidt, 2003).

An important feature of CAD/CAM integration in machining is the capability to describe the cutting tool path for various operations such as NC turning, milling, and drilling. The programs are computer-generated and can be modified by the programmer to optimize the tool path and to visually check for possible tool collisions with clamps or fixtures or for other interferences. The tool path can be modified at any time to accommodate other shapes to be machined.

The tasks performed by a CAM system (Figure 10.11) are as follows:

- NC or CNC scheduling
- Production planning and scheduling
- Tool and fixture design
- CAPP

NC can use special computer languages. Today, APT and COMPACT II are the two most common language-based computer-assisted programming systems used in the industry. These systems take the CAD data and adapt them to the particular machine control unit to make the part (Degarmo et al., 1997).

## 10.5   LEAN PRODUCTION—JUST-IN-TIME MANUFACTURING SYSTEMS

Regardless of all that has been written about CIM, this technology is not widespread. What is called "lean production" appears to be more important to the future than CIM. It is evident that unless a company first adopts the approach of lean production, the conversion to CIM is likely to fail. In this approach, the functions of the production system, such as production control, inventory control, quality control, and machine-tool maintenance, are first to be integrated. Lean production has been developed and practiced by Toyota (the Toyota system) instead of CIM (Degarmo et al., 1997).

Integration of the production system functions into the MS requires commitment from top-level management and communication with everyone, particularly manufacturing. Total employee and union participation is absolutely necessary, but it is not usually the union leadership or the production workers who raise barriers to an integrated manufacturing production system (IMPS). It is those in middle management who have the most to lose in this systems-level change. In this respect, here are the preliminary steps:

1. All levels in the plant, from the production workers to the president, must be educated in the IMPS philosophy and concepts.
2. Top management must be totally committed to this venture, and everyone involved must be motivated.
3. Everyone in the plant must understand that cost, not price, determines the profit. The customer wants low cost, superior quality, and on-time delivery.
4. Everyone must be committed to the elimination of waste to reduce cost; this is fundamental for getting lean production.

### 10.5.1 STEPS FOR IMPLEMENTING THE IMPS LEAN PRODUCTION

Many companies have implemented IMPS (lean production) by converting a factory from a job shop–flow MS to a true IMS. The following steps are to be followed:

1. *Build foundation by forming U-shaped cells*—This is done to replace the production job shop. Restructure and reorganize the FMS, composed of cells that fabricate families of parts. Creating cells is the first step in designing an MS in which inventory control and quality control are integrated.
2. *Rapid exchange of tooling and dies (RETAD)*—Everyone on the plant floor must be told to reduce setup time by using single-minute exchange of die (SMED). A setup reduction team assists workers and foremen and demonstrates a project on the plant's worst setup problem. Reducing setup time is critical to reducing lot size.
3. *Integrating quality control*—A multifunctional worker can do more than operate machines. The worker is also an inspector who understands process capability, quality control, and process improvement. In lean production, every worker has the responsibility and the authority to make the product right the first time and every time and the authority to stop the process when something is wrong. The integration of quality control into the MS considerably reduces defects while eliminating the inspector. Cells provide for integration of quality control.
4. *Integrating preventive maintenance and ensuring machine reliability*—Installing an integrated preventive maintenance program makes machines operate reliably; moreover, it gives the workers the chance to maintain equipment properly.
5. *Leveling and balancing final assembly*—This is done by producing a mix of final assembly products in small lots. Each process, cell, and subassembly has essentially the same cycle time as the final assembly.

6. *Linking cells*—Integration of production control is realized by linking the cells, subassemblies, and final assembly elements, utilizing Kanban. All linked cells, processes, subassemblies, and final assemblies start and stop together in a synchronized manner. Thus, the integration of production control into the MS is realized.

7. *Integrating the inventory control*—The inventory levels are directly controlled by the people on the floor through the control of Kanban. This is the integration of the inventory control to reduce work in progress (WIP). The minimum level of WIP is determined by the percentage of defectives, the reliability of the equipment, and the setup time.

8. *Integrating the suppliers*—This involves educating and encouraging suppliers (vendors) to develop their own lean production system for superior quality, low cost, and rapid on-time delivery. They should deliver material to the customer when needed, where needed, without inspection.

9. *Automating and robotizing*—Solve problems by converting manned to unmanned cells, initiated by the need to solve problems in quality and reliability and to eliminate bottlenecks.

10. *Computerizing the whole production system*—Once the MS has been restructured into JIT MS and the critical functions well integrated, the company (MS) will find it expedient to restructure the rest of the company. It is basically restructuring the production system (PS) to be as waste-free and efficient as the MS.

### 10.5.2  JUST-IN-TIME AND JUST-IN-CASE PRODUCTION

As previously mentioned under lean production, the JIT production concept was first implemented by Toyota in Japan under the name Kanban (visible record) to eliminate wastage of materials, machines, capital, manpower, and inventory throughout the MS. The JIT philosophy is summarized as follows (Degarmo et al., 1997):

> Produce and deliver finished goods just-in-time to be sold, subassemblies just-in-time to be assembled into finished goods, fabricated parts just-in-time to go into finished goods, fabricated parts just-in-time to go into subassemblies, and purchased materials just-in-time to be transformed into fabricated parts.

To be more specific, JIT seeks to achieve the following goals (McMahon and Browne, 1998):

- Zero defects
- Zero setup time
- Zero inventories
- Zero handling
- Zero breakdowns
- Zero lead time
- Lot size of one

To achieve this goal, all elements of excess should be eliminated. Large safety stocks, long lead times, long setting times, large queues at machines, high scrap and rework levels, machine breakdowns, and so on should also be eliminated. JIT is not, therefore, a simple off-the-shelf solution to all manufacturing problems. If JIT is realized in the firm, the unnecessary inventories will be completely eliminated, making stores or warehouses unnecessary, inventory cost will be diminished, and the ratio of capital turnover will be increased. Consequently, JIT is sometimes called zero inventories, material as need, stockless production, or demand scheduling (El-Midany, 1994).

In traditional manufacturing, the parts are made in batches placed in inventory and used whenever necessary. This approach is known as the just-in-case (JIC) or push system, meaning that the parts are made according to a schedule and are kept in inventory to be used if and when they are needed. In contrast, JIT manufacturing is a pull system, meaning (as previously mentioned) that parts are produced to order and the production is matched with the demand for the final assembly of products.

In JIT, parts are inspected by the worker and used within a short period of time. Accordingly, the worker maintains a continuous production control, identifies defective parts immediately, and produces quality products. Implementation of the JIT requires that all manufacturing aspects be continuously reviewed and monitored so that all operations and resources that do not add value are eliminated.

Because the basic promise of JIT is to produce the kind of units needed in the quantities needed at the time needed, the system should depend on smoothing (leveling) of the MS, so it is necessary to eliminate fluctuation in the final assembly. This is called leveling or balancing the final assembly. The object of JIT is to make the same amount of product part every day. Balancing is making the output from the cells equal to necessary demand for the parts downstream. In summary, small lot sizes, made possible by setup reduction within the FMCs, single-unit conveyance within the cells, and standard cycle times are the keys to accomplishing a smoothed MS (Degarmo et al., 1997).

### Advantages of Just-in-Time

- Low inventory cost
- Fast detection of defects in production and hence, low scrap loss
- Reduced need for inspection and reworking of parts
- Production of high-quality parts at low cost

Implementation of JIT, as compared with FMS (Kalpakjian and Schmidt, 2003), realizes:

- Reduction of 20–40% in production cost
- Reduction of 60–80% in inventory
- Reduction of up to 90% in rejections
- Reduction of up to 90% in lead times
- Reduction of up to 50% in scrap, rework, and warranty cost
- Increase of 30–50% in direct labor productivity
- Increase of 60% in indirect labor productivity

## 10.6 ADAPTIVE CONTROL

AC systems for machine tools are a logical extension of CNC systems. The part programmer sets the processing parameters based on the existing knowledge of the WP material and various data on the particular manufacturing process. In CNC machines, these parameters are held constant during a particular process cycle. In AC, on the other hand, the system is capable of automatic adjustments during processing through closed-loop feedback control. It is therefore readily appreciated that this approach is basically a feedback system. A schematic of a typical AC configuration for a machine tool is shown in Figure 10.13. Accordingly, AC represents a process control that operates in addition to the CNC position or servo control system.

In manufacturing, several AC systems or strategies are distinguished (*Metals Handbook*, 1989):

1. *Adaptive control with optimization (ACO)*, in which an economic index of performance is used to optimize the process using online measurements. This strategy may involve maximizing material removal rate or improving surface quality.
2. *Adaptive control with constraints (ACC)*, in which the process is controlled using online measurements to maintain a particular process constraint (force, power, temperature, and so on). Referring to Figure 10.14, if the cutting force and hence the torque increases excessively (Figure 10.14a, b), the AC system changes the speed or the feed (cutter travel) to lower the cutting force to an acceptable level (Figure 10.14c). Without AC or without direct intervention of the operator (in the case of conventional machining), high cutting forces may cause the tools to chip or break, or the WP to deflect

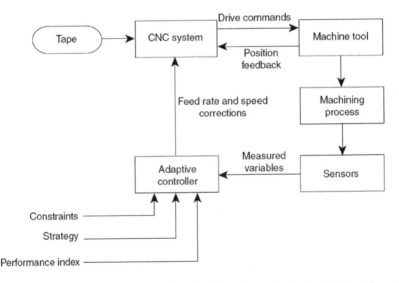

**FIGURE 10.13** Typical adaptive control configuration for a machine tool. (From Koren, J., *Computer Control of Manufacturing Systems*, McGraw-Hill, Tokyo, 1983. With permission.)

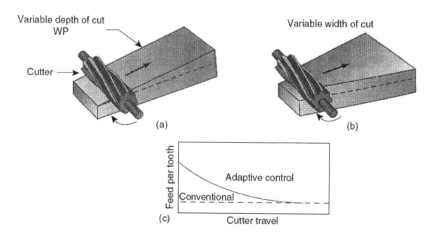

**FIGURE 10.14** Adaptive control in milling: (a), (b) increase of cutting force with cutter travel, and (c) AC is used to decrease the feed with cutter travel. (From Koren, J., *Computer Control of Manufacturing Systems,* McGraw-Hill, Tokyo, 1983. With permission.)

or distort excessively. As a result, the accuracy and surface finish would deteriorate.

3. *Geometric adaptive control (GAC),* in which the process is controlled using online measurements to maintain the desired dimensional accuracy or surface finish (Figure 10.15).

Response time must be short for AC to be effective. Currently, all AC systems are based on either ACC or GAC because the development and proper implementation of ACO is complex. ACC is well-suited to rough cutting, whereas GAC is used for finishing operations.

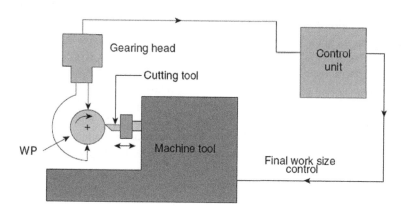

**FIGURE 10.15** In-process inspection of WP diameter in turning. (From Kalpakjian, S. and Schmidt, S. R., *Manufacturing Processes for Engineering Material,* Pearson Education, Inc., NJ, 2003.)

### 10.6.1 INTEGRATION OF AC INTO CAD/CAM/CIM SYSTEMS

This integration is an important issue in the future development of AC systems as well as their role in the CIM hierarchy. Such issues are extremely important for unmanned manufacturing and will require additional research to extend the current understanding of AC systems. Important issues include the interface between CAD and CAM and the application of expert systems (ESs) and other methods from artificial intelligence (AI) to AC systems, as well as process monitoring and diagnostics (*Metals Handbook*, 1989).

## 10.7 SMART MANUFACTURING AND ARTIFICIAL INTELLIGENCE

AI is the basic tool for smart manufacturing (SM). AI is an area of computer science concerning systems that exhibit some characteristics that are usually associated with intelligence in human behavior, such as learning, reasoning, problem solving, and understanding of language. Its goal is to simulate such human endeavors on the computer, and it represents a technique for solving problems in a better way than is available with conventional computer programs (CCPs). A CCP typically relies on an algorithmic solution in which a finite number of explicit steps produce the solution of a specific problem. These algorithms work fine for scientific or engineering calculations that are numeric in nature to produce satisfactory answers. In contrast, AI uses a heuristic—a rule-of-thumb search. It should be understood that an exhaustive search can only be used for relatively simple, well-defined problems. For complex and uncertain problems, exhaustive search routines become impractical.

A CCP is difficult to modify, but AI is usually easy to modify, update, and enlarge. In a CCP, information and control are integrated, but in AI, the control structure is usually separate from domain knowledge. CCP is often primarily numerical, but AI is concerned primarily with processing symbolic information in which some meaning other than a numerical value is attached to a symbol. The symbols in AI may represent a concept about a process or a condition related to it; the AI programs manipulate the relationships among such symbols and arrive at logical conclusions from these relationships (Jain and Gupta, 1993).

AI is having a major impact on all steps of manufacturing cycles, including design, automation, production planning, scheduling, and the overall economics of manufacturing operations. AI programs also find applications in diagnosis, monitoring, analysis, interpretation learning, consultation, instruction, conceptualization, prediction, debugging, and repair. AI packages costing approximately a few thousand dollars have been developed, many of which can now be used on personal computers for application in both office and shop floors. AI application in manufacturing generally encompasses ESs, natural language, machine vision, artificial neural networks (ANNs), and fuzzy logic.

### 10.7.1 EXPERT SYSTEMS

An ES, also called a knowledge-based system (KBS), is generally defined as an intelligent computer program that has the capability to solve real-life problems using knowledge base (KB) and inference procedures. The goal of ES is to develop the capability to conduct an intellectually demanding task in the way that a human expert would. ESs use a KB containing facts, data, definitions, and assumptions.

They also have the capability to follow a heuristic approach; that is, to make good judgments on the bases of discovery and revelation and to make high-probability guesses, just as a human expert would. The KB is expressed in computer codes, usually in the form of if-then rules, and can generate a series of questions; the mechanism for using these rules to solve problems is called an inference engine. ESs can also communicate with other computer software packages (Kalpakjian and Schmidt, 2003).

To construct ESs for solving the complex design and manufacturing problems encountered, the necessary elements include:

- A great deal of knowledge
- A mechanism for manipulating the knowledge to create solutions

Because of the difficulty involved in modeling the many years of experience of a team of experts and the complex inductive reasoning and decision-making capabilities of humans, including the capability to learn from mistakes, the development of KBSs requires much time and effort. ESs operate on a real-time basis, and their short reaction times provide rapid responses to problem. The programming languages most commonly used are C++, list processing (LISP), and programming logic (PROLOG). A significant development is ES-software shells or environments (framework). These shells are essentially ES outlines that allow a person to write specific applications to suit special needs. Writing these programs requires considerable experience and time (Kalpakjian and Schmidt, 2003).

Several ESs have been developed to be used in:

- Problem diagnosis in machines and equipment
- Modeling and simulation of production facilities
- CAD, process planning, and production scheduling
- Management of a company's manufacturing strategy

### 10.7.2 Machine Vision

In systems that incorporate machine vision, computers and software-implementing AI are combined with cameras and other optical sensors. These machines then perform operations such as inspecting, identifying, and sorting parts and guiding intelligent robots (Figure 10.16); in other words, operations that would otherwise require human involvement and intervention.

### 10.7.3 Artificial Neural Networks

ANNs are used in applications such as noise reduction in telephones, speech recognition, and process control in manufacturing. For example, they can be used to predict the surface finish of machined WPs on the basis of input parameters such as

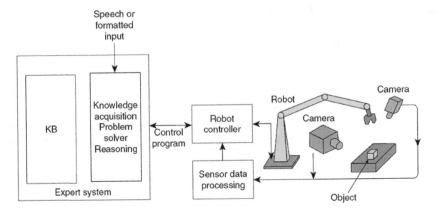

**FIGURE 10.16** ES as applied to an industrial robot guided by machine vision. (From Kalpakjian, S. and Schmidt, S. R., *Manufacturing Processes for Engineering Material,* 2003, Pearson Education, Inc., NJ, 2003.)

cutting force, torque, acoustic emission, and spindle acceleration. However, this field is still under development.

### 10.7.4 Natural-Language Systems

These systems allow a user to obtain information by entering English-language commands in the form of simple typed questions. Natural-language software shells are used in the scheduling of material flow in manufacturing and analyzing information in DBs. Major progress is being made to develop computer software with speech-synthesis and voice-recognition capabilities, thus eliminating the need to type commands on the keyboard.

### 10.7.5 Fuzzy Logic (Fuzzy Models)

Fuzzy logic is an element that has important applications in control systems and pattern recognition. It is based on the observation that people can make good decisions on the basis of nonnumerical information. Fuzzy models are mathematical means of representing vagueness and imprecise information (hence the term "fuzzy"). These models have the capability to recognize, manipulate, interpret, and use data and information that are vague or lacking precision, with fuzzy-logic methods including reasoning and decision making at a level higher than ANNs. Typical concepts used in fuzzy logic are: *few, more or less, small, medium, extreme,* and *almost all.*

Fuzzy-logic technologies and devices have been developed and successfully applied in areas such as intelligent robotics, motion control, image processing and machine vision, machine learning, and design of intelligent systems. Some applications of fuzzy logic include automatic transmissions of Lexus cars, automatic washing machines, and helicopters that obey vocal commands (Kalpakjian and Schmidt, 2003).

## 10.8   FACTORY OF THE FUTURE

The trend toward the automated factory seems unavoidable in modern industries. The integration of many new techniques adopted in MSs, such as CNC, machining centers, FMS, robots, material-handling equipment, and automatic warehouses, together with CAD, CAM, CAPP, and GT software, have brought the factory of the future (FOF) closer to reality. All manufacturing, material handling, assembly, and inspection will be done by computer-controlled machinery and equipment. Similarly, activities such as the processing of incoming orders, production planning and scheduling, cost accounting, and various decision-making processes performed by management will also be done automatically by computers. The role of human beings will be confined to activities such as overall supervision; preventive maintenance; upgrading of machines and equipment; receiving supplies of materials and shipping of finished products; provision of security for the plant facilities; programming, upgrading, and monitoring of computer programs; and maintenance and upgrading of computer hardware.

The reliability of machines, equipment, control systems, power supplies, and communications networks is crucial to full-factory automation. Without human intervention, a local or general breakdown in even one of these components can cripple production. The computer-integrated FOF must be capable of automatically rerouting materials and production flows to other machines and controlling other computers in the case of such emergencies.

An important consideration in fully automating a factory is the nature and extent of its impact on employment. Although forecasts indicate that there will be a decline in the number of machine- tool operators and tool-and-die workers, there will be a major increase in the number of people working in service occupations, such as computer technicians and maintenance electricians. Thus, the generally high-skilled, manual-effort labor force traditionally required in manufacturing may be trained or retrained to work in such activities as computer programming, information processing, implementation of CAD/CAM, and similar tasks. The development of more user-friendly computer software is making these tasks much easier.

In this respect, it should be recognized that the designation of "world-class," like "quality," is not a fixed target for the manufacturing country or the company to reach but rather, a moving target, rising to higher and higher levels as time passes. Manufacturing organizations must be aware of this moving target and plan and execute their programs accordingly (Kalpakjian and Schmidt, 2003).

## 10.9   CONCLUDING REMARKS RELATED TO AUTOMATED MANUFACTURING

1. Installations of FMSs are very capital-intensive; consequently, a thorough cost/benefit analysis must be conducted before a final decision is taken. This analysis should include:
   - Capital cost, energy, materials, and labor.
   - Market analysis for which products are to be produced.
   - Anticipated fluctuations in market demand and product type.

- Time and effort required for installing and debugging the system. An FMS can take 2–5 years to install and at least 6 months to debug.
2. Although FMS requires few, if any, machine operators, the personnel in charge of the total operation must be trained and highly skilled. These include manufacturing engineers, computer programmers, and maintenance engineers.
3. The most effective FMS applications have been in medium-volume, high-variety batch production (50,000 units/year). In contrast, high-volume, low-variety part production is best done by transfer machines.
4. CIMSs have become the most important means of improving productivity, responding to changing market demands, and better controlling manufacturing operations and management functions. Regardless of all that is written about CIM, this technology is not as widespread as that of lean production.
5. CAM is often integrated with CAD to transfer information from the design stage to the planning stage and to production; that is, CAD/CAM bridges the gap from design to production.
6. Advances in manufacturing operations, such as CAPP, computer simulation of manufacturing processes, GT, cellular manufacturing, FMSs, and JIT manufacturing, contribute significantly to the improvement in productivity.
7. Significant advances have been made in material handling, particularly with the implementation of industrial robots and automated conveyors.
8. The FOF appears to be theoretically possible. However, there are important issues to be considered regarding its impact on employment.
9. In the FOF, many of the functions of production system are integrated into the MS. This requires that the job shop MS is replaced with a linked cell or MS. The functions of production control, inventory control, quality control, and machine-tool maintenance are the first to integrate.

## 10.10  REVIEW QUESTIONS

10.10.1  Describe the difference between mechanization and automation. Give some typical examples of each.
10.10.2  Explain the difference between hard and soft automation. Why they are named as such?
10.10.3  Describe the principles and purpose of adaptive control. Give some applications in manufacturing in which you think it can be implemented.
10.10.4  Differentiate between ACO, ACC, and GAC.
10.10.5  What are the benefits and limitations of FMS?
10.10.6  Draw a sketch to show the idea of:
- An MS
- A flow-line manufacturing cell
- Unmanned FMC
- GAC for turning operation
10.10.7  What are the components of an MS?

10.10.8      List the benefits of CIMS.

10.10.9      Describe the principles of FMS. Why does it require a major capital investment?

10.10.10     What are the benefits of JIT production? Why it is called a pull system? What is a push system?

10.10.11     Differentiate between JIT and JIC production.

10.10.12     What is meant by the term "FOF"? Explain why humans will still be needed in the FOF.

10.10.13     What is Kanban? Why was it developed?

10.10.14     Describe the elements of AI. Why is machine vision a part of it?

10.10.15     Explain the principles of CAM, CAPP, and CIM to an older worker in a manufacturing facility who is not familiar with computers.

10.10.16     What is lean production? Enumerate and explain the main steps toward lean production.

## REFERENCES

Degarmo, EP, Black, JT & Kohser, RA 1997, *Materials and processes in manufacturing*, 8th edn, Prentice Hall, New York.

El-Midany, TT 1994, *Computer automated manufacturing and flexible technologies*, 1st edn, El-Mansoura University, Egypt.

Jain, RK & Gupta, SC 1993, *Production technology*, 13th edn, Khanna Publishers, New Delhi.

Kalpakjian, S & Schmidt, SR 2003, *Manufacturing processes for engineering material*, 4th edn, Prentice Hall Publishing Co., New York.

Koren, J 1983, *Computer control of manufacturing systems*, 1st edn, McGraw-Hill, Tokyo.

McMahon, C & Browne, J 1998, *CADCAM—principles, practice and manufacturing management*, 2nd edn, Addison-Wesley, Reading, MA.

ASM International, *Metals handbook* 1989, Machining, vol. 16, 9th edn, ASM International, Materials Park, OH.

Midwest Valve Company, Detroit, MI.

# 11 Machine-Tool Dynamometers

## 11.1 INTRODUCTION

Machining is still one of the most important techniques for shaping metallic and nonmetallic components. During machining, the cutting tool exerts a force on the workpiece (WP) as it removes the machining allowance in the form of chips. Empirical values for estimating the cutting forces are no longer sufficient to reliably establish the optimum machining conditions. Depth of cut, feed rate, cutting speed, WP materials, tool material and geometry, and cutting fluid are just a few of the machining parameters governing the amplitude and direction of the cutting force.

The optimization of a machining process necessitates accurate measurement of the cutting force by a special device called a machine-tool dynamometer, which is capable of measuring the components of the cutting force in a given coordinate system. It is a useful and powerful tool employed in a variety of applications in engineering research and manufacturing. A few examples of these applications are:

- Investigating the machinability of materials
- Comparing similar materials from different sources
- Comparing and selecting cutting tools
- Determining optimum machining conditions
- Analyzing causes of tool failure
- Investigating the most suitable cutting fluids
- Determining the conditions that yield the best surface quality
- Establishing the effect of fluctuating cutting forces on tool wear and tool life

The machine-tool dynamometer is not standard equipment or a device that can be used on every machine. Rather, it is equipment especially designed to fulfill some desired requirements that adapt a specific machine type operating at a specific range of machining conditions.

## 11.2 DESIGN FEATURES AND REQUIREMENTS

As in most design problems, a satisfactory dynamometer design involves a compromise in which the dynamometer structure allows the highest possible sensitivity at sufficient stiffness and rigidity that the geometry of the cutting process is maintained. At the same time, the dynamometer structure should maintain a high natural frequency to minimize chattering.

Cutting forces cannot be measured directly. Whenever a force acts on a material, it undergoes a certain deformation, which can be measured, and hence, the acting forces can be accordingly derived. Therefore, the principle on which all dynamometers are designed is to measure the deflections or strains induced in the dynamometer structure by the resultant cutting force. Dynamometer designs differ depending on whether the deflections of the structure are directly measured with displacement transducers or whether the induced strains in the structure are measured by strain gauges and their associated high-sensitivity equipment, which allow a dynamometer structure of sufficient stiffness to be used.

## 11.2.1 RAPIER PARAMETERS FOR DYNAMOMETER DESIGN

Rapier (1959) suggested two useful parameters that are used in comparing the efficiency of various dynamometer designs (Figure 11.1). These are:

1. Displacement ratio, $r_d$

$$r_d = \frac{y}{x} = \frac{\text{Displacement measured by gauge}}{\text{Tool displacement by the point of application of force}}$$

2. Tool displacement, $x$

    In a well-designed dynamometer of high stiffness, the displacement $x$ should be as small as possible for the reasons mentioned before. In addition, to obtain the maximum output, the displacement $y$ measured by the

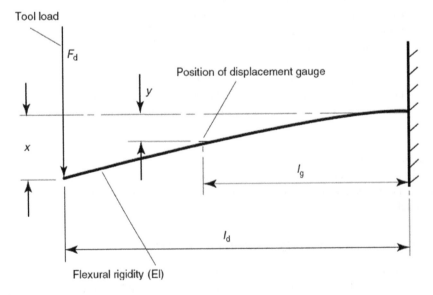

**FIGURE 11.1** Displacement ratio of a cantilever dynamometer. (From Rapier, A. C., Cutting Force Dynamometers, H. M. Stationary Office, NEL Plasticity Report, 158, 1959. With permission.)

gauge should be as large as possible. Thus, the ratio $r_d$ gives a guide to the dynamometer efficiency.

In most dynamometer designs, $r_d$ does not exceed unity, and thus, a value of $r_d$ approaching unity corresponds to an efficient design. For some designs, such as the slotted cantilever dynamometer, $r_d$ may exceed unity.

In Figure 11.1, for simplicity, the dynamometer is represented by a cantilever of uniform cross section (constant flexural rigidity = EI). The deflection/unit force is given by

$$\frac{x}{F_c} = \frac{l_d^3}{3EI} \tag{11.1}$$

where

$F_c$ = force applied (main cutting force component)
$l_d$ = cantilever length
EI = flexural rigidity of the cantilever (assumed constant)

The deflection per unit force at the displacement gauge arranged at a distance $l_g$ from cantilever support is given by

$$\frac{y}{F_c} = \frac{l_d \cdot l_g}{3EI} \left( 3 - \frac{l_g^2}{l_d} \right) \tag{11.2}$$

From Equations 11.1 and 11.2, $r_d$ can be calculated as follows:

$$r_d = \frac{1}{2} \left( \frac{l_g}{l_d} \right)^2 \left[ 3 - \frac{l_g}{l_d} \right] \tag{11.3}$$

Equation 11.3 is used to estimate the efficiency of a dynamometer design. In most cases, the ratio $l_g/l_d \approx 0.9$, and consequently, $r_d \approx 0.85$.

Rigidity, sensitivity, and accuracy are the most important requirements in dynamometer design. A good dynamometer should be sensitive and accurate to within ±1%. All machine tools operate with some level of vibration. In the case of milling and shaping machines, such vibrations may have large amplitudes, and therefore, the dynamometer should be rigid enough to withstand such vibrations. To avoid the effect of vibrations on the measured forces by the dynamometer, its natural frequency should be at least four times as large as the frequency of the exciting vibration ($f_e$). The dominating stiffness criterion is the natural frequency of the dynamometer, and it is given by

$$f_n = \frac{1}{2\pi} \sqrt{\frac{K}{m}} \, Hz \tag{11.4}$$

where

$K$ = spring constant (MN/m)
$m$ = mass of the dynamometer supported by spring (kg)

For a machine running at maximum speed of 4200 (rpm), therefore,

$$f_e = \frac{N}{60} = 70\text{Hz} \qquad (11.5)$$

Accordingly, the natural frequency of the dynamometer $(f_n)$ should be at least 280 Hz.

## 11.2.2  MAIN REQUIREMENTS OF A GOOD DYNAMOMETER

The primary requirements of a quality dynamometer are as follows:

- It should possess high stiffness and rigidity.
- It should possess high sensitivity, accuracy, and reliability.
- For any cutting process, it is desirable to measure the three force components (multichannel) in a set of rectangular coordinates.
- In multichannel dynamometers, the dynamometer should be so designed so that the force in any direction should give no reading in other direction; that is, there is no cross-sensitivity (cross talk) between the channels (3% cross-sensitivity is acceptable).
- If possible, dynamometers should always be manufactured from a single block of material, as the use of clamped or bolted joints, or pivots of any kind, gives rise to hysteresis caused by friction. Furthermore, sliding surfaces should be avoided because they introduce always unknown friction forces.
- It should be designed to provide the possibility of force recording using necessary bridges and multichannel recorders.
- The presence of cutting fluids makes waterproofing of the dynamometer essential.
- The dynamometer should be stable (giving consistently accurate readings with respect to time, temperature, and humidity). Once calibrated, it should only have to be checked occasionally.
- It is convenient to use a dynamometer having a linear calibration. If the calibration curve is not linear, it is then necessary to determine the zero point accurately.
- In addition, there are some special requirements that should be met, such as size, ruggedness, and adaptability to several jobs.

## 11.3  DYNAMOMETERS BASED ON DISPLACEMENT MEASUREMENTS

### 11.3.1  TWO-CHANNEL CANTILEVER (CHISHOLM) DYNAMOMETER

This is the simplest type of two-component turning dynamometer, an advanced design proposed by Chisholm (1955), shown in Figure 11.2. The cutting tool is supported at the end of the cantilever. The deflections in the vertical (main cutting

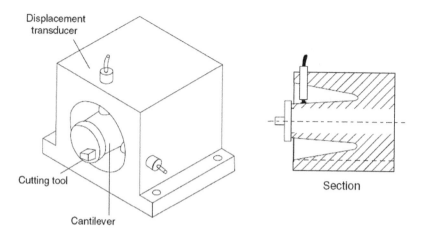

**FIGURE 11.2**   One-piece displacement turning dynamometer. (From Chisholm, A. W. J., Progress Report on the Wear of Cutting Tools, H. M. Stationary Office, MERL, Plasticity Report, 106, 1955. With permission.)

force) and horizontal (feed force) are measured by displacement gauges (dial gauge or inductive transducer). The hollow, tapered cantilever provides maximum stiffness with high natural frequency. Moreover, the dynamometer is totally machined from a solid piece of metal, as previously recommended. A Chisholm dynamometer is completely free from cross-sensitivity, and its displacement ratio $r_d$ is calculated according to Equation 11.3.

### 11.3.2   Two-Channel Slotted Cantilever Dynamometer

This two-channel turning dynamometer bends the structure about the weakest points, A and B in Figure 11.3. Two displacement transducers are arranged to measure the vertical force component (main cutting force $F_c$) and the horizontal force component (feed force $F_f$), respectively. In this particular design, the displacement ratio $r_d$ is different for each force component. For the vertical component $F_c$, $r_d$ is given by

$$r_d = \frac{q}{p} \tag{11.6}$$

Referring to Figure 11.3, it is accordingly clear that $r_d$ can be arranged to exceed unity, which means higher design efficiency. However, a disadvantage of this type of dynamometer is that considerable cross-sensitivity, reaching about 15%, may result.

## 11.4   DYNAMOMETERS BASED ON STRAIN MEASUREMENT

Many successful strain gauge dynamometers were developed by Shaw (1986), Hottinger-Baldwin Mep-technik (1989), Pahlitzsch and Spur (1959), and Youssef (1971).

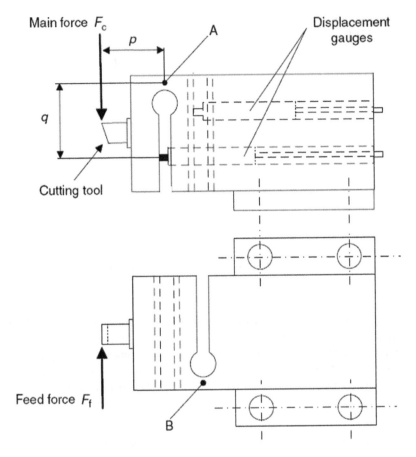

**FIGURE 11.3** Two-channel cantilever-type displacement turning dynamometer. (Modified from Boothroyde, G., *Fundamentals of Metal Machining and Machine Tools,* McGraw-Hill, New York, 1981.)

### 11.4.1 STRAIN GAUGES AND WHEATSTONE BRIDGES

Bonded-type strain gauges are sensitive sensors used to measure strains initiated from both tension and compressive static or dynamic loading. They are made of gauge wire in a flat form and of thickness ranging from 18 to 25 μm (Figure 11.4). The strain gauge is cemented to the sensitive member of the dynamometer after it has been properly cleaned and suitable cement has been applied to both the surface of the member and that of the gauge. There should be no air bubbles in the cement. After cementing, the gauge should be baked at about 90 °C to remove moisture and then coated with wax or resin to provide some mechanical protection and prevent moisture absorption. For a proper application of the strain gauge, the resistance between the strained member and the gauge should be at least 50 MΩ. Strain gauges used for such a purpose are usually very small in order to limit the dynamometer size.

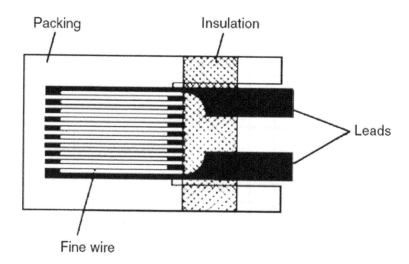

Packing    Insulation

Leads

Fine wire

**FIGURE 11.4**  Hottinger-Baldwin strain gauge.

The two important parameters supplied by the manufacturer of a strain gauge are:

1. The electric resistance, $R$, which is usually 120 $\Omega$
2. The gauge factor, $k_g$, which varies from about 1.75 to 3.5; frequently, $k_g = 2$

The gauge factor is defined according to

$$k_g = \frac{\Delta R / R}{\Delta l / l} = \frac{\Delta R}{R} \cdot \frac{1}{\varepsilon_s} \qquad (11.7)$$

where

$\varepsilon_s$ = elastic strain = $\dfrac{\Delta l}{l}$

$\Delta R$ = change of original resistance

The fine wires of the strain gauges are made of Nichrome V, Pt-alloy 1200, or Konstantan, which is an alloy of 55% Cu and 45% Ni. As an advantage, its gauge factor ($k_g$) is not affected by the temperature within a range of $\pm 100\ °C$, as shown in Figure 11.5. The change in wire resistance caused by the change of its length and cross section can be measured by the Wheatstone bridge with four active arms (Figure 11.6). No current will flow through the galvanometer (G) if the bridge is in balance when Equation 11.8 is satisfied:

$$\frac{R_1}{R_4} = \frac{R_2}{R_3} \qquad (11.8)$$

It is apparent from Equation 11.8 that although a single gauge only can be used, the sensitivity can be increased fourfold if two gauges, $R_1$ and $R_3$, are used in tension, while the others, $R_2$ and $R_4$, are used in compression. In practice, the four gauges are selected to be of the same resistance.

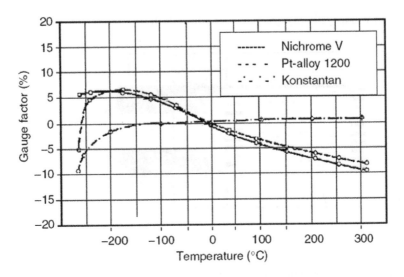

**FIGURE 11.5** Effect of temperature on gauge factors. (From Hottinger-Baldwin Meß-technik, Der Weg zum Meßgrößenaufnehmer Technical Data, Darmstadt, 1989. With permission.)

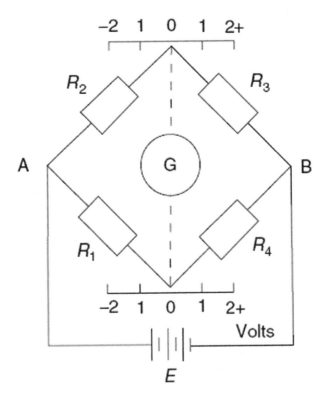

**FIGURE 11.6** Balanced Wheatstone bridge.

## 11.4.2 Cantilever Strain Gauge Dynamometers

Cantilever strain gauge dynamometers are mainly used on lathes and drilling machines.

1. *Turning dynamometer.* Figure 11.7 shows a particular design of a two-channel strain gauge lathe dynamometer. It is convenient to locate the strain gauges on a prepared gauging section far away from the tool cutting

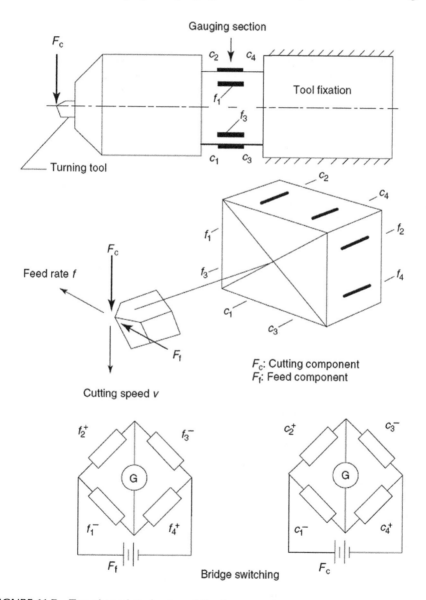

**FIGURE 11.7**   Two-channel strain gauge lathe dynamometer.

edge. The figure also shows how the strain gauges are switched to measure the vertical main force component $F_c$ and the horizontal feed force component $F_f$.

2. *Drilling dynamometer.* In any cutting process, as the force components acting on the cutting tool are equal in magnitude and opposite in direction to those acting on the WP, it is possible to measure these components from the tool side or from the WP side. Both possibilities are widely used, and each method has its advantages and disadvantages. In the first case, the dynamometer is attached to the tool, and the WP may have any geometry and dimension. However, the dynamometer has a complicated construction in the case of the drilling process because the drill rotates. A good design of this type, suggested by Pahlitzsch and Spur (1959), was produced by Hottinger, GmbH, in 1959. In the second possibility, the dynamometer is placed under the stationary WP to be drilled; the WP should be limited in size and weight.

Another drilling dynamometer has been suggested by Youssef (1971). It is composed of a star-shaped transducer (1), which is secured to the worktable (2) and simply supported on three frictionless supports (3) (Figure 11.8). A U-shaped element (4) fixed to the dynamometer base (5) prevents the transducer from rotation when subjected to drilling torque. It thus enables the drilling torque $M_Z$ to be transmitted to the transducer active arms contained in the U-elements. A set of four strain gauges is applied to the sides of the active arms to measure the drilling torque $M_Z$. Another set of gauges is applied on the top and the bottom of the transducer active arms to measure the axial thrust (feed force $F_y$). A locator (6) enables the table (2), transducer (1), and base (5) to be aligned with each other. Figure 11.8 also illustrates the switching of a Wheatstone bridge for measuring both components $M_Z$ and $T$. To achieve the best results, the strain gauges should be symmetrically placed on the transducer active arms. In drilling, it is necessary to measure only the axial thrust and the torque. Therefore, a two-component dynamometer is sufficient for this purpose.

### 11.4.3 OCTAGONAL RING DYNAMOMETERS

#### 11.4.3.1 Strain Rings and Octagonal Ring Transducers

A theoretical prediction of points of zero strain in circular rings under radial and tangential loads has been established and verified by many investigators. It gave the best results regarding the separation of the effects of two mutually perpendicular forces. Such characteristics are very important in dynamometer design.

Consider the stress analysis of a circular ring under the action of vertical and horizontal loads $F_y$ and $F_x$, respectively. The problem is solved as a statically undetermined structure, and the bending moment distributions $M_y$ and $M_x$, due to the application of the vertical and horizontal force components, is tabulated as follows (Youssef, 1971). $R_1$ is the ring radius and $\theta$ is the location angle (Figure 11.9).

$$M_y = -\frac{F_y}{2} R_1 \sin\theta + \frac{2R_1}{\pi} \frac{F_y}{2}$$

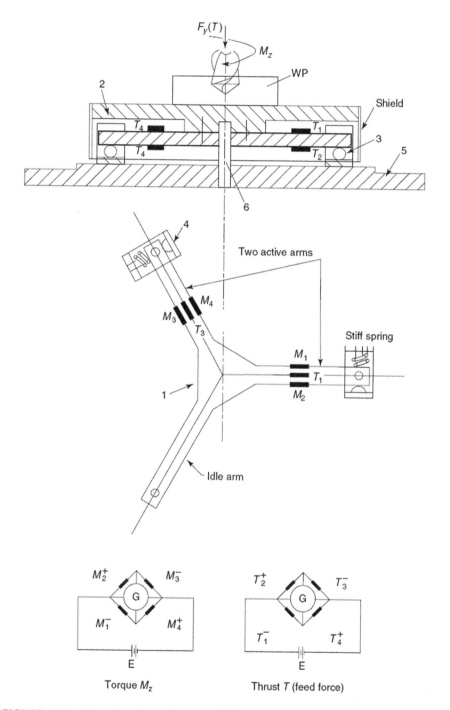

**FIGURE 11.8**  Two-channel strain gauge drilling dynamometer. (From Youssef, H. A., Design of machine tool dynamometers, *Bulletin of the Faculty of Engineering,* Alexandria University, Vol. X, pp. 279–303, 1971. With permission.)

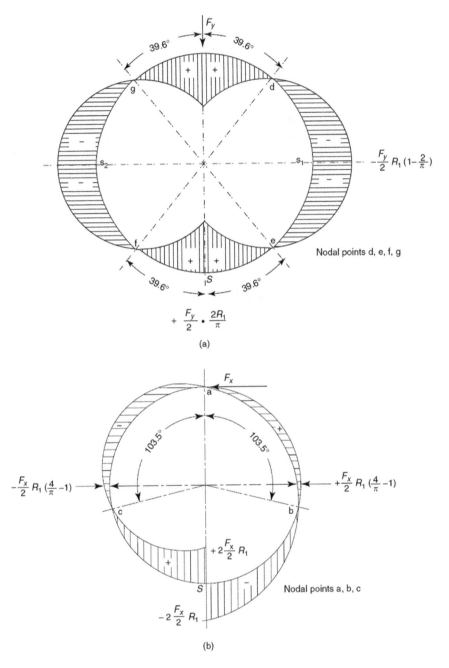

**FIGURE 11.9** Distribution of moments $M_y$ and $M_x$ on the strain ring: (a) stress distribution due to $F_y$ and (b) stress distribution due to $F_x$. (From Youssef, H. A., Design of machine tool dynamometers, *Bulletin of the Faculty of Engineering*, Alexandria University, Vol. X, pp. 279–303, 1971. With permission.)

| $\theta$ | 0 | $\pi/2$ | 39.6° | 140.4° | $\pi$ |
|---|---|---|---|---|---|
| $M_y$ | $\dfrac{2R_1}{\pi}\cdot\dfrac{F_y}{2}$ | $-\dfrac{F_y}{2}R_1\left(1-\dfrac{2}{\pi}\right)$ | 0 | 0 | $\dfrac{2R_1}{\pi}\cdot\dfrac{F_y}{2}$ |

$$M_x = -\frac{F_x}{2}R_1\left(1-\cos\theta\right)+\frac{4}{\pi}\frac{F_x}{2}R_1\sin\theta \qquad \text{for } \theta = 0 - \pi$$

$$M_x = -\frac{F_x}{2}R_1\left(1-\cos\theta\right)+\frac{4}{\pi}\frac{F_x}{2}R_1\sin\theta \qquad \text{for } \theta = \pi - 2\pi$$

| $\theta$ | 0 | $\pi/2$ | 103.5° | $\pi$ | $\pi$ | 256.5 | $3\pi/2$ | $2\pi$ |
|---|---|---|---|---|---|---|---|---|
| $M_x$ | 0 | $-\dfrac{F_x}{2}R_1\left(\dfrac{4}{\pi}-1\right)$ | 0 | $-\dfrac{F_x}{2}R_1$ | $2\dfrac{F_x}{2}R_1$ | 0 | $-\dfrac{F_x}{2}R_1\left(\dfrac{4}{\pi}-1\right)$ | 0 |

Again referring to Figure 11.9, if the vertical force $F_y$ is applied, maximum strain occurs along the horizontal centerline at points $s_1$ and $s_2$, whereas points of zero strain are located at positions = 39.6° from the vertical (points d, e, f, and g, respectively, in Figure 11.9a). If a horizontal force $F_x$ is applied, the strain nodes related to this force are located at positions $\theta = 0$, 103.5°, and 256.5° (points a, b, and c, respectively) (Figure 11.9b).

Such a stress condition is very important in the design of machine-tool dynamometers. The strain gauges are mounted on the ring transducer at positions corresponding to the nodes described earlier; specifically, in the following manner:

1. Gauges for measuring vertical force $F_y$ are mounted at nodal points b and c at positions $\theta = 103.5°$ and 256.5° (Figure 11.10).
2. Gauges for measuring horizontal force $F_x$ are mounted at nodal points d, e, f, and g at position $\theta = 39.6°$ from the vertical axis of the ring (Figure 11.10). Accordingly, the cross talk between the components $F_x$ and $F_y$ is totally eliminated. The corresponding bridge switching is shown in the same figure.

Because the nodal points are difficult to lay out accurately on a circular ring transducer, an octagonal transducer with a circular hole is more practical (Figure 11.11). It is stiffer than a ring transducer of the same minimum thickness. For these reasons, most machine-tool dynamometers use the octagonal transducer's section. Moreover, it is easier to secure. Of course, an amount of cross-sensitivity should be expected, as the angles of application of the strain gauge are changed to 45° instead of 39.6° for measuring the horizontal component $F_x$ and changed to 90° instead of 103.5° for measuring the vertical component $F_y$.

### Illustrative Example

Design a ring transducer to measure horizontal and vertical force components, each up to 250 kg.

### SOLUTION

According to Figure 11.9a and b, it is clear that the maximum stress occurs at the lowest ring cross section S.

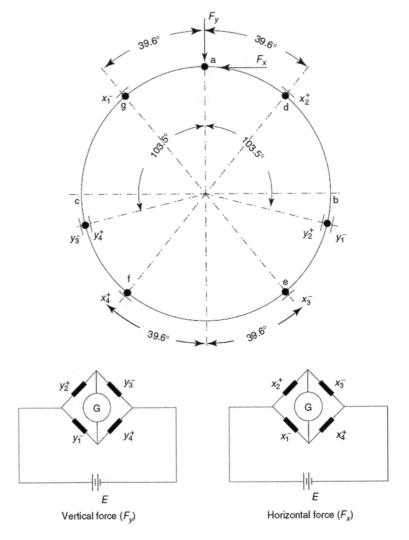

**FIGURE 11.10** Application of strain gauges at nodal points of strain ring to totally eliminate the cross talk between horizontal and vertical components.

$$M_s = \frac{2R_1}{\pi} \times \frac{F_y}{2} + 2\frac{F_x}{2}R_1 = R_1\left[\frac{F_y}{\pi} + F_x\right]$$

Assume a mean radius of transducer ring $R_1$ = 40 mm and a width $l$ = 80 mm. Therefore,

$$M_s = \frac{40}{1000}\left(\frac{250}{\pi} + 250\right) = 13.2 \text{ kg m}$$

$$= 13.2 \times 10^3 \text{ kg mm}$$

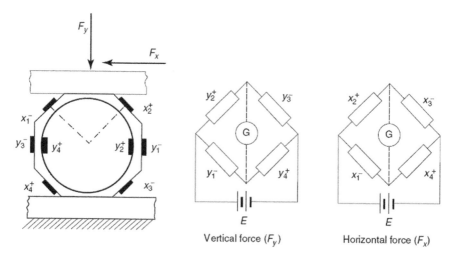

**FIGURE 11.11**   Octagonal transducer and related strain bridges.

Assume that the transducer is made of steel 40 Mn 4 of allowable bending strength $f_b$ = 35 kg/mm². Therefore, the minimum transducer thickness $(t_r)$ is determined from the equation

$$f_b = \frac{M_s}{lt_r^2 / 6}$$

$$35 = \frac{6 \times 13.2}{80 \times t_r^2} \times 10^3$$

from which $t_r$ = 5.3 mm.

### 11.4.3.2   Turning Dynamometer

A two-component strain gauge dynamometer with a stretched octagonal transducer is illustrated in Figure 11.12. It is used in orthogonal cutting operations. The flat sides of the stretched octagonal transducer facilitate mounting of the strain gauges. In the same figure, the strain bridges for measuring the main and feed force components are visualized.

### 11.4.3.3   Surface Plunge-Cut Grinding Dynamometer

This is also a two-component dynamometer, based on the foregoing principles, and is applied to measure the main cutting force $F_x$ and the thrust force component $F_y$ in a plunge-cut surface-grinding operation (Figure 11.13).

The stretched octagonal transducer is made of a single piece of aluminum to provide a sensitive element of low mass and high natural frequency. By using two gauges from each half ring in each of the bridge circuits, the dynamometer becomes independent of the point of application of the load between the half rings. This is a very important characteristic, especially in a dynamometer on which the WP is clamped.

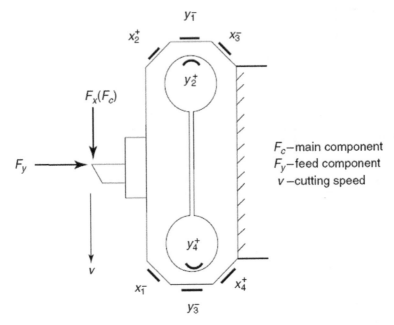

Two-component turning dynamometer

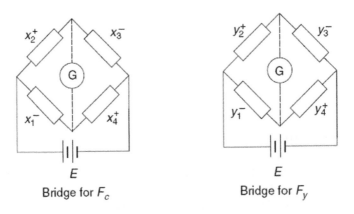

Bridge for $F_c$          Bridge for $F_y$

**FIGURE 11.12** Two-component turning dynamometer with a stretched octagonal transducer used for orthogonal cutting.

### 11.4.3.4 Milling Dynamometers

The two-component milling dynamometer is very similar to the surface plunge-cut grinding dynamometer. It is used for two-dimensional milling, as illustrated in Figure 11.14. In three-dimensional milling, a helical milling cutter is used. In this case, a third force component is generated in the axial direction, ($F_a$) and a three-component milling dynamometer, as illustrated in Figure 11.15, is used to measure the three components.

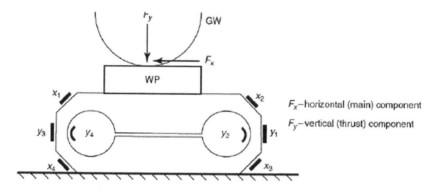

**FIGURE 11.13** Two-component plunge-cut surface-grinding dynamometer.

$F_x$—horizontal (main) component

$F_y$—vertical (thrust) component

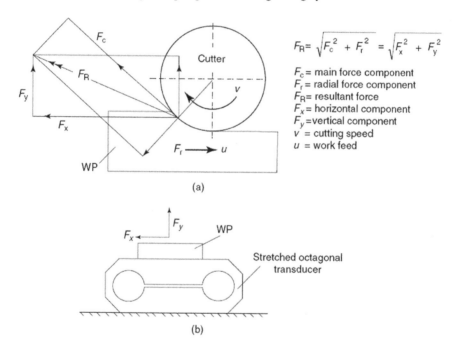

$$F_R = \sqrt{F_c^2 + F_r^2} = \sqrt{F_x^2 + F_y^2}$$

$F_c$ = main force component
$F_r$ = radial force component
$F_R$ = resultant force
$F_x$ = horizontal component
$F_y$ = vertical component
$v$ = cutting speed
$u$ = work feed

(a)

(b)

**FIGURE 11.14** Two-component milling dynamometer: (a) two-dimensional milling and (b) two-component milling dynamometer.

## 11.5 PIEZOELECTRIC (QUARTZ) DYNAMOMETERS

### 11.5.1 PRINCIPLES AND FEATURES

In 1880, the Cuire brothers discovered the piezoelectric effect, in which an electrical charge appears on the surfaces of certain crystals when the crystal is subjected to a mechanical load. Of the numerous piezoelectric materials, quartz is by far the most suitable for force measurement because it is a stable material with constant properties. In its crystalline form, quartz is anisotropic, in that its material properties are

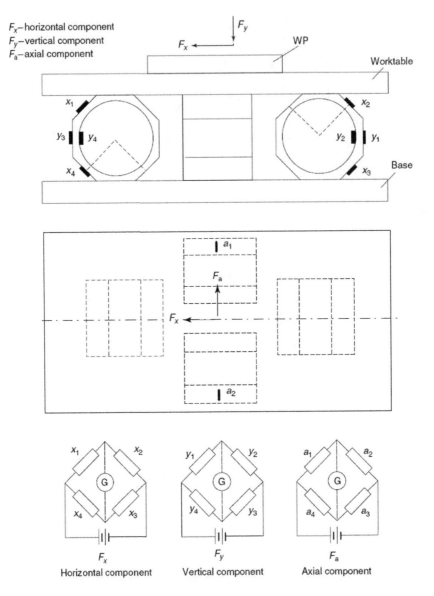

$F_x$–horizontal component
$F_y$–vertical component
$F_a$–axial component

Horizontal component

Vertical component

Axial component

**FIGURE 11.15**  Three-component milling dynamometer.

not identical in all directions. Depending on the position in which they are cut out of the crystal, disks are obtained that are:

1. Sensitive only to pressure (longitudinal effect), as shown in Figure 11.16a, which measure the main force component $F_z$ (brown).
2. Sensitive only to shear in one particular direction (shear effect), as shown in Figure 11.16b, which measures components $F_x$ (blue) and $F_y$ (green), perpendicular to $F_z$, as well as the torque $M_z$ (red). Figure 11.16c illustrates the generalized multicomponents with reference to a Cartesian coordinate system.

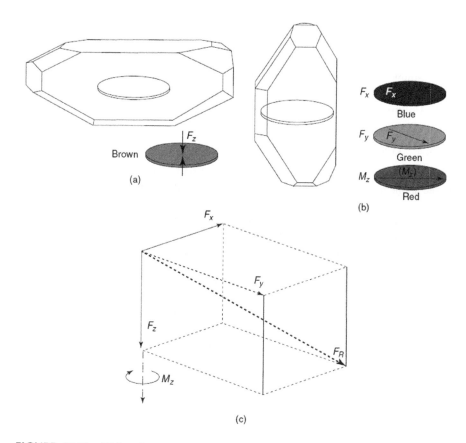

**FIGURE 11.16** Disks of quartz crystals. (a) Pressure-sensitive, (b) shear-sensitive, (c) multicomponents in reference to a Cartesian coordinate system. (From Kistler Winterthur Instrumente AG, CH-8408, Switzerland, Technical Report of Quartz Multicomponent Dynamometers, 2006. With permission.)

The piezoelectric force measuring principle differs fundamentally from previously discussed systems in that it is an active system. When a force acts on a quartz element, a proportional electric charge appears on the loaded surfaces, which means that it is not necessary to measure the actual deformation.

In such a system, the deflection is not more than a few micrometers at full load, whereas with conventional systems, several tenths of a millimeter may be needed. Thus, piezoelectric dynamometers are very stiff systems, and their resonant frequency is high, so that even rapid events can be measured satisfactorily. Moreover, the individual components of the cutting force can be measured directly, such that cross talk between measuring channels is typically less than 1%. Quartz dynamometers require no zero adjustment or balancing of the bridge circuit. It is just a matter of pressing a button, and they are ready for duty. The outstanding features of quartz dynamometers are:

- High rigidity, hence high resonant frequency
- Minimal deflections (few micrometers at full load)

- Wide measuring range
- Linear characteristic (that is, free of hysteresis)
- Lowest cross talk (typically under 1%)
- Simple in operation and without need for bridge balancing
- Compact design
- Unlimited life expectancy

## 11.5.2 TYPICAL PIEZOELECTRIC DYNAMOMETERS

Piezoelectric dynamometers are efficiently used on the majority of machine tools. Three application examples are provided together with the measuring setup.

1. *Two-component piezoelectric drilling dynamometers.* Figure 11.17 shows a two-component drilling dynamometer in which shear-sensitive disks are arranged in a circle with their shear-sensitive axes oriented to respond to the torque $M_z$ (red), whereas pressure-sensitive disks are arranged and oriented to measure the thrust load $F_z$ (brown). A high preload is necessary because the shear forces must be transmitted by friction to measure the torque.

    The two-component dynamometer, shown in Figure 11.17, is suited for operations including drilling, thread cutting, countersinking, reaming, and so on. Torques and forces acting when machining holes from less than 1 mm to over 20 mm in diameter can be measured satisfactorily by this dynamometer. A record of $M_z$ and $F_z$ is illustrated in Figure 11.17, from which it is clearly seen that $F_z$ rises steeply at the beginning (entry of tool chisel),

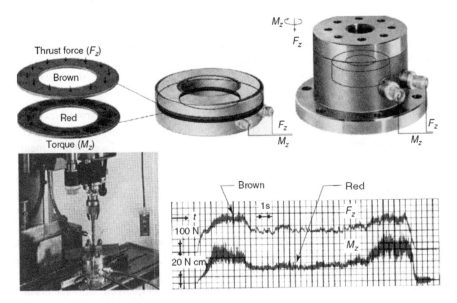

**FIGURE 11.17** Two-component drilling dynamometer. (From Kistler Instrumente AG Winterthur, CH-8408, Switzerland, Technical Report of Quartz Multicomponent Dynamometers, 2006. With permission.)

followed by the gradual rise of the $M_z$ component, as the latter is more affected by the force acting on the two drill lips.

2. *Three-component piezoelectric turning dynamometer.* This model includes several shear-sensitive quartzes, with their shear-sensitive axes oriented to measure $F_x$ (blue ring) and $F_y$ (green ring), respectively. Their shear sensitive-axes are inclined to each other at an angle of 90°, and both are contained in a housing to form a two-component force-measuring element for $F_x$ and $F_y$ (Figure 11.18). Pressure-sensitive quartz disks are contained in a single housing to form a single-component force-measuring element for $F_z$ (brown ring). Another alternative is illustrated in the construction shown in Figure 11.18, where three separate elements for measuring $F_x$, $F_y$, and $F_z$ are sandwiched under high preload between a base plate and a top plate. The dynamometer is mounted on the lathe slide in place of a cross slide. A record of the three components is also shown in the same figure, from which it is clear that $F_x = F_y$, which means that the cut is performed at an approach angle $\chi = 45°$.

3. *Three-component piezoelectric milling or grinding dynamometer.* Whole quartz rings may be employed. Two shear-sensitive quartz pairs, for $F_x$ (blue) and $F_y$ (green), and a pressure-sensitive pair for $F_z$ (brown), can be assembled in a common housing to form a three-component force-measuring element (Figure 11.19). The pressure-sensitive quartzes are arranged in the middle so that they lie in the neutral axis under bending. During milling and grinding, the application point of the force varies a great deal.

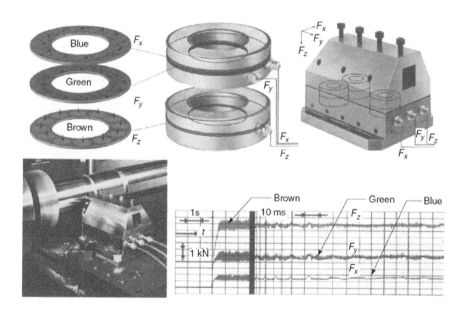

**FIGURE 11.18** Three-component piezo turning dynamometer. (From Kistler Instrumente AG Winterthur, CH-8408, Switzerland, Technical Report of Quartz Multicomponent Dynamometers, 2006. With permission.)

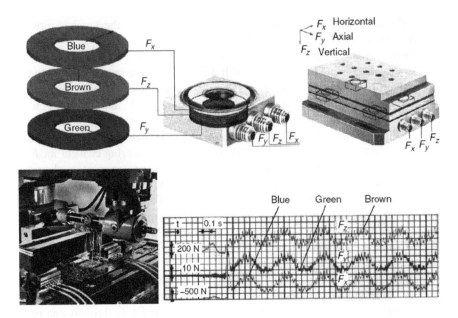

**FIGURE 11.19** Three-component piezo milling dynamometer. (From Kistler Instrumente AG Winterthur, CH-8408, Switzerland, Technical Report of Quartz Multicomponent Dynamometers, 2006. With permission.)

Consequently, dynamometers having four piece three-component force-measuring elements are employed. All the $x$, $y$, and $z$ channels, respectively, are paralleled electrically. This makes the measurement independent of the momentary force application point. For bigger work, two dynamometers, paralleled electrically and mechanically, may be employed together. This system measures correctly independently of the point of force application. An output of milling the three-component dynamometer is shown in Figure 11.19. Milling is performed under the following conditions:

- Up-milling
- Cutter diameter = 63 mm, helix $\beta = 30°$, $n = 90$ rpm, $Z = 12$ teeth
- Feed $u = 53$ m/min
- Depth $t = 3.5$ mm

The severe periodic fluctuation in the measured forces is attributed to an eccentric motion of the cutter shaft. Superimposed are vibrations due to gearing of the machine. It is perfectly clear from the record that the setup shown is far from ideal. The force measure, therefore, sheds light on the machine-tool behavior as well, not just on the actual cutting operation.

## 11.6   REVIEW QUESTIONS

11.6.1   What are the main applications of machine-tool dynamometers?
11.6.2   Explain what is meant by the Rapier parameters of dynamometer design.

11.6.3   State the main requirements of a good dynamometer.

11.6.4   Using a line sketch, show the principles of a Chisholm cantilever dynamometer.

11.6.5   Using a line diagram, describe the principles of a two-channel slotted cantilever turning dynamometer.

11.6.6   Explain what is meant by a strain gauge factor.

11.6.7   State the proper procedure for mounting strain gauges to a dynamometer body.

11.6.8   Give examples of turning and drilling dynamometers that employ strain gauges.

11.6.9   Explain the principles of an octagonal ring dynamometer.

11.6.10  Illustrate how a stretched octagonal ring dynamometer can be used for measuring the cutting and thrust force in plunge surface grinding.

11.6.11  Show a milling dynamometer that is based on octagonal rings and strain gauges.

11.6.12  What are the principal features of a piezoelectric material? Show the principles of piezoelectric dynamometers used for turning, drilling, and milling operations.

11.6.13  1What are the different measures that minimize cross-sensitivity when designing a machine-tool dynamometer?

## REFERENCES

Boothroyde, G 1981, *Fundamentals of metal machining and machine tools*, McGraw-Hill, New York.

Chisholm, AWJ 1955, Progress Report on the Wear of Cutting Tools, H. M. Stationary Office, MERL, Plasticity Report, 106.

Hottinger-Baldwin Meβ-Btechnik 1989, Der Wegzum Meβgröβenaufnehmer, Technical Data, Darmstadt, Germany.

Kistler Instrumente AG Winterthur, CH-8408, Switzerland 2006, Technical Report of Quartz multi-component dynamometers.

Pahlitzsch, G & Spur, G 1959, Einrichtungen zum Messen der Schnittkräfte beim Bohren. *Werkstattstechnik*, 49, Heft9.

Rapier, AC 1959, Cutting Force Dynamometers, NEL Plasticity Report, H. M Stationary Office, 158.

Shaw, MC 1986, *Metal cutting principles*, Clarendon Press, Oxford, UK.

Youssef HA 1971, 'Design of machine tool dynamometers', *Bulletin of the Faculty of Engineering, Alexandria University*, vol. X, pp. 279–303.

# Index

For Product Safety Concerns and Information please contact our EU
representative GPSR@taylorandfrancis.com Taylor & Francis Verlag GmbH,
Kaufingerstraße 24, 80331 München, Germany


Printed and bound by CPI Group (UK) Ltd, Croydon, CR0 4YY

01/05/2025

01858578-0002